Deer of the Southwest

DEER
of the Southwest

*A Complete Guide to the Natural History,
Biology, and Management of Southwestern
Mule Deer and White-Tailed Deer*

Jim Heffelfinger

TEXAS A&M UNIVERSITY PRESS
College Station

Financial assistance for the
publication of this book generously provided by
Dr. Scott Weiss, M.D.
Eldon "Buck" Buckner
Tucson and Flagstaff Chapters of The Mule Deer Foundation
Arizona Chapter of Safari Club International
Their contribution to improving the quality of this book
and continued commitment to deer conservation
in the Southwest is greatly appreciated.

Copyright © 2006 by James R. Heffelfinger
Printed in China
All rights reserved
First edition

The paper used in this book meets the minimum requirements
of the American National Standard for Permanence
of Paper for Printed Library Materials, z39.48–1984.
Binding materials have been chosen for durability.

Library of Congress Cataloging-in-Publication Data

Heffelfinger, Jim, 1964–
Deer of the Southwest : a complete guide to the natural history, biology,
and management of mule deer and white-tailed deer / Jim Heffelfinger.—1st ed.
   p.   cm.
Includes index.
ISBN-13: 978-1-58544-515-8 (flexbound paper : alk. paper)
ISBN-10: 1-58544-515-0 (flexbound paper : alk. paper)
1. Mule deer—Southwest, New.   2. White-tailed deer—Southwest, New.   I. Title.
QL737.U55H43   2006
599.65′20979—dc22
                        2006005035

*Dedicated to the memory of
Dr. Samuel L. Beasom:
mentor, colleague, and friend.
You are sorely missed.*

It is possible to make natural history entertaining and attractive as well as instructive, with no loss in scientific precision, but with great gain in stimulating, strengthening, and confirming the wholesome influence which the study of the natural sciences may exert upon the higher grades of mental culture.

—Elliot Coues (1906)

Contents

List of Figures	xi
List of Tables	xiii
Preface	xv
Acknowledgments	xix

Chapter 1. Southwestern Deer   3
    Taxonomy: What's in a Name? 3
    Classification of Deer 4
    Deer of the Southwest 7
    Differences between Species 15
    Hybridization 18

Chapter 2. Historical Perspective   25
    Origins and Evolution of Deer 25
    Native Americans 32
    Exploration Period 38
    History of Deer Management in the Southwest 42
    History of Wildlife Management in Mexico 53

Chapter 3. Physical Characteristics   56
    Body Weight 56
    Dentition 61
    Digestive Tract 61
    Pelage 63
    Senses 65
    Voice 69
    Scent Glands 70

Chapter 4. Antlers   77
    Antlerogenesis 78
    Factors Affecting Antler Size 80
    Evolution and Function 84
    Abnormalities 88
    Antlered Does 94

Chapter 5. Diet and Water Requirements — 97
  Important Southwestern Deer Foods 97
  Deer Diets 97
  Water Requirements 116

Chapter 6. Density, Home Range, and Movements — 126
  Density 126
  Home Range 129
  Movements 134

Chapter 7. Reproduction — 143
  Rutting Behavior 143
  Timing of Rut 148
  Gestation and Fawning 153
  Reproductive Rate 154

Chapter 8. Mortality — 165
  Mortality Rates 168
  Malnutrition/Starvation 170
  Legal Harvest 171
  Predation 172
  Diseases 180
  Parasites 193
  Unrecovered Deer 200
  Illegal Harvest 201
  Accidents 202

Chapter 9. Deer Management — 205
  Management Authority 205
  Management Data 207
  Deer Management in the Southwest 218
  Antlerless Hunts 237

Epilogue — 241
Appendix — 245
Literature Cited — 249
Index — 276

Figures

1. Distribution of mule and white-tailed deer in North America 6
2. Mule deer 8
3. Distribution of mule deer in the Southwest 9
4. Distribution of white-tailed deer in the Southwest 12
5. Early naturalist Elliot Coues 13
6. Coues whitetail 14
7. White-tailed deer and mule deer 16
8. Whitetail × mule deer hybrid (F_1) buck raised in captivity 19
9. Metatarsal glands of a white-tailed deer, mule deer, and white-tailed deer × mule deer hybrid 20
10. White-tailed deer × mule deer F_1 hybrid shot in southeastern Arizona 22
11. Eurasian *Eumeryx*, an early mammal form 26
12. *Cranioceras, Synthetoceras,* and *Ramoceros,* three early groups of large ruminants 27
13. *Dremotherium* and *Procervulus,* deer-like forms of the Miocene 28
14. *Stephanocemas* and *Dicrocerus,* the first animals to shed their antlers on a regular and recurrent basis 29
15. *Eocoileus,* a direct ancestor of today's mule and white-tailed deer 30
16. Evolution of early deer-like forms 31
17. Deer pictograph found near Caborca, Sonora, Mexico 36
18. Maxillary canine teeth, occurring rarely in deer 62
19. Transition from summer to winter coat 64
20. Tarsal glands, located on inside of hind legs 71
21. Interdigital gland of a Coues white-tailed deer 73
22. Preorbital gland, located directly in front of the eye 74
23. Antler size, increasing with age 82
24. Antler palmation on Coues whitetail buck 89
25. Coues whitetail showing an extra antler growing from top of eye orbit following an injury 92
26. Desert mule deer doe with antlers 94
27. Fairy duster (*Calliandra eriophylla*) following page 108
28. Buckwheat (*Eriogonum wrightii*) following page 108

29. Mountain mahogany (*Cercocarpus* spp.) following page 108
30. Desert ceanothus (*Ceanothus greggii*) following page 108
31. Jojoba (*Simmondsia chinensis*) following page 108
32. Spurge (*Euphorbia melenadenia*) following page 108
33. Skunkbush sumac (*Rhus trilobata*) following page 108
34. Holly-leaf buckthorn (*Rhamnus crocea*) following page 108
35. Cliffrose (*Cowania mexicana*) following page 108
36. Big sagebrush (*Artemisia tridentata*) following page 108
37. Gambel oak (*Quercus gambelii*) following page 108
38. Trembling aspen (*Populus tremuloides*) following page 108
39. Artificial wildlife water developments in desert 120
40. Fibroma on eyelid 185
41. Deer with chronic wasting disease (CWD) 191
42. Nasal bot fly 193
43. Ticks, widespread and recognizable external parasites of deer 195
44. Tapeworm larvae visible on the surface of a deer liver 200
45. Helicopter used to observe and classify a large sample of deer in rugged terrain 212
46. Three generations of Heffelfinger family deer hunters 215
47. Deer habitat management, a crucial issue 243

Tables

1. Taxonomic classification of southwestern deer 4
2. First Arizona game laws, passed in 1887 43
3. History of significant changes in southwestern deer management 44
4. Average buck dressed weights (pounds) reported for mule deer 59
5. Average buck dressed weights (pounds) reported for white-tailed deer 60
6. Seasonal percent composition of mule deer diets 99
7. Most important mule deer foods by season 102
8. Seasonal percent composition of white-tailed deer diets 112
9. Most important white-tailed deer foods by season 113
10. Mule deer densities 130
11. White-tailed deer densities 131
12. Home range estimates for mule deer 132
13. Peak breeding and fawning dates for mule deer 150
14. Peak breeding and fawning dates for white-tailed deer 152
15. Average number of fetuses per doe by age class for mule deer 156
16. Average number of fetuses per doe by age class for white-tailed deer 157
17. Average percentage of females pregnant by age class for mule deer 158
18. Average percentage of females pregnant by age class for white-tailed deer 159
19. Comparison of survey and harvest data used by state wildlife management agencies 219

Preface

Deer have long been the most popular big game animal in the Southwest. They were an important source of protein for Native Americans, and later served as an important staple for the residents of mining camps and military forts, and other early settlers who came west to carve an existence out of the dusty Southwest.

Unfortunately, the deer herds and other wildlife suffered from this heavy use without the benefit of conservation laws and law enforcement to limit the harvest to levels that could be sustained indefinitely. The initial decline of deer abundance in the Southwest reached its low point in the 1890s when years of unregulated market hunting, overgrazing, and a severe drought brought big game populations crashing down. Conservation-minded people at the time lobbied successfully to enact laws to limit overexploitation of big game to levels that would allow recovery. Money generated from those who hunt under this regulated system now funds a large part of the world's most highly successful system of wildlife conservation. Thus, abundant deer populations are important to everyone who enjoys wildlife.

Remarkably, those interested in the deer of the Southwest are left wanting for up-to-date information. In the 1950s, 1960s, and 1970s the Arizona Game and Fish Department (AGFD) produced three small books about deer in Arizona: *The Mule Deer in Arizona Chaparral* (1958), by Wendell Swank; *The Kaibab North Deer Herd* (1964), by John Russo; and *The Arizona Whitetail Deer* (1977), by Ted Knipe. Each of these books covered a single species or limited geographic area and discussed different aspects of deer biology, research, and management. The New Mexico Department of Game and Fish published a book on New Mexico wildlife management in 1967 and a small booklet in 1957 called *The Deer of New Mexico* (Lang 1957). In 1998 Carlos Galindo-Leal and Manuel Weber published a book in Spanish on the Coues whitetail of the Sierra Madres of northern Mexico (*El venado de la Sierra Madre Occidental*). Mule deer management in west Texas was also addressed more recently in an excellent publication by the Texas Parks and Wildlife Department (Cantu and Richardson 1997). However, there is no source for contemporary information on white-tailed and mule deer throughout the Southwest.

Much has been learned about deer in the last several decades, but most of the recent information is buried in technical scientific papers, graduate theses, and annual reports of the state wildlife agencies. The present book consolidates and condenses historical and contemporary information pertaining to the deer of the North American Southwest and presents it in a consistent fashion that is solidly based in research and yet easily understood. The technical information is presented with as little jargon as possible. In the interest of readability, scientific citations were kept to a minimum. Scientific research can be used to anchor nearly every sentence in some areas of the book, but such a practice makes the text nearly unreadable. The purpose of this book is to provide biologists, natural resource managers, students, hunters, landowners, and nature enthusiasts a source for nearly all things related to southwestern deer in a format and style that is easy to digest.

Deer do not recognize the international boundary, so available information from management and research in Mexico has been added where appropriate. The use of *Southwest* in the title refers to the southwestern region of North America, including not only the southwestern United States, but also northern Mexico.

During the preparation of this book, many of the researchers and managers who spent much of their careers working with deer in the Southwest were routed out of retirement and badgered with questions. These questions sometimes yielded information that has not been captured in any previous publications and would otherwise have been lost to the deer world. Their comments, open discussion, and field notes bolstered many portions of this book. The historical deer information from the literature and experiences of past biologists has been intertwined with results from the most recent research on deer not only in the Southwest, but throughout the country.

There has been some confusion regarding how many different kinds of deer live in the Southwest. Mule deer were called "blacktails" by some early pioneers, and rumors persist of a smaller version of the Coues whitetail called a "Mexican fantail." Rocky Mountain and desert mule deer are described as different, but little has been written about how they differ. Biologists have written papers and drawn distribution maps of a different kind of mule deer called a "Bura" deer. Hybrids between white-tailed deer and mule deer are reported by hunters each year. These deer variations and their origin are discussed in detail. Elk are also members of the deer family (Cervidae), but they differ too greatly in biology, behavior, and management to be covered reasonably in this volume. They deserve a similar work devoted to elk alone.

This book was written with special attention to the questions that continually arise regarding deer. An entire chapter is devoted to antlers because, more than any other topic, these structures are the source of considerable admiration—and innumerable questions. Other frequently asked questions include the following: Can deer see color? What do they eat? What is the average weight of a mature Coues buck? Do predators affect deer populations? How large is an average home range?

Are "antlered does" really does? Can deer hear deer-repelling car whistles? How do state wildlife agencies actually manage deer populations? What causes nontypical antlers? When is the rut? These questions and more are answered herein.

Several portions of the book contain the most current information in print. For example, the maps of deer distribution are new revisions of distributions for whitetails and mule deer. All previous maps excluded the tribal lands, leaving large white spaces as if there were no deer in those areas. Tribal biologists were consulted, and these revised maps include those lands to give a complete picture of deer distribution. Likewise, biologists in all western states, Canadian provinces, and Mexican states assisted in revising the distribution of both species in North America. The North American deer distribution map given here represents the most accurate currently in print.

How whitetails and mule deer evolved in North America is not entirely clear, and there are many theories being bandied about. One of these theories (that mule deer themselves are hybrids) has been widely published in the popular literature and has caused considerable discussion among deer enthusiasts. The discussion of deer evolution in this book benefits from very recent and important findings in this area. All North American deer evolved from a common ancestor that entered the continent via the Bering land bridge 5 million to 6 million years ago. What this common ancestor looked like is unknown, but a recent work sheds light on a deer species from that time period. Fossil evidence of *Eocoileus gentryorum* provides us with a glimpse of what this ancestor probably looked like and is illustrated here for the first time.

Although an attempt was made to include all pertinent information on southwestern deer, no single source can be completely comprehensive, or have all the answers. Research findings and management experience will continue to accumulate information about deer that will be continually evaluated. Also, research studies, especially those lasting only a few years, cannot possibly capture the complexities of deer ecology. The most obvious examples of this are studies of deer diets and movements, both of which vary dramatically depending on habitat conditions, deer densities, and other land use practices. Many areas of the Southwest experienced high deer populations in the late 1950s and mid-1980s. The latter period of abundance is easily explained by several consecutive years of above-average precipitation. The reasons for the deer population increases in the 1950s are more obscure. There is still a lot to be learned about what drives deer populations. Periodically, deer populations reach very low levels as well. These violent fluctuations in deer abundance are typical in the Southwest, but the extreme lows cause considerable consternation nonetheless.

Many factors contribute to these deer population fluctuations. The overriding factor is always habitat quality, which in the arid Southwest equates mostly to precipitation. When rainfall is at or above average, pregnant does have adequate nutrition to produce healthy fawns born with adequate hiding cover, and predators have plenty of rabbits and rodents to fill their bellies. Below-average rainfall patterns result in weak fawns born on bare ground among hungry coyotes. The challenge to those who care about deer will be to preserve and protect their habitat throughout these changes in deer abundance to prevent population declines from becoming permanent.

Acknowledgments

This book would not have been possible without the contributions of many people. I would like to express my sincere appreciation to numerous wildlife biologists, researchers, and assistants; behind their polished final reports are many frustrating and difficult hours afield. Bob Miles and George Andrejko (Arizona Game and Fish Department), Ted Noon (University of Arizona Veterinary Diagnostic Lab), Clay McCulloch, John Moreguard (USFWS), Yar Petryszn (University of Arizona), Dan Robinett (National Resource Conservation Service), Beth Williams (University of Wyoming), Tom Thorne (Wyoming Game and Fish Department), Gerald Day, Roger Drummond, Joe and Marisa Cerreta, and John Holcomb made available high-quality photos. Randy Babb produced superb illustrations to help enliven and explain the text. Robert and Dolores Heffelfinger offered constant support and encouragement throughout all phases of this project. Kent and Gerianne Hummel provided computer hardware and technical support that proved crucial to completing this book. Todd Rathner helped me focus on the big picture more than once. Eldon "Buck" Buckner opened some important doors and lent unwavering and aggressive support for this book from the minute he saw it. Darius Semmens, Craig Wissler, Todd Black, and Anne Gondor offered their considerable GIS expertise and time to generate the deer distribution maps.

Scott Weiss, Flagstaff and Tucson Chapters of The Mule Deer Foundation, Buck Buckner, Arizona Chapter of Safari Club International, Pat Juhl, Bob Jacobs, Tony Abbott, and especially Todd Rathner came through with critical financial assistance for publication. Their commitment to the dissemination of information and deer conservation in general is to be applauded.

A huge thanks to Francisco "Paco" Abarca, Ian Alcock, Karen Alexy, Bruce Anderson, Lou Bender, Gene Beaudoin, Vern Bleich, Tim Bone, Clay Brewer, Tom Britt, Dave Brown, George Bubenik, Dave Cagle, Neil Carmony, Rogelio Carrera, Matt Cronin, Gerald Day, Susan Flader, Art Fuller, Valerius Geist, John Goodwin, Barry Hale, Jon Hanna, Wally Haussamen, Annette Heffelfinger, John Heffelfinger, Robert Heffelfinger, Bob Henry, Larry Holland, Harry Jacobson, Phil Jenkins, Paul Krausman, Raymond Lee, William Longhurst, Steven Lukefahr, Keith McCaffery, Clay McCulloch, Ted McKinney, Danny Mead, Karl Miller, Amber Munig, Ted Noon, Richard Ockenfels, John Olsen, Carl Olson, Jefferson Reid, Gene Rhodes, Gerardo Sánchez-Rojas, Joan Scott, Harley Shaw, Norm Smith, Wendell Swank, Pamela Swift, Brian Wakeling, S. David Webb, Darrel Weybright, Gary White, and others for their helpful review of parts of the manuscript and for sharing previously unpublished information. The book improved incrementally with the addition of each of their thoughtful comments. A special thanks goes to Dave Brown and Neil Carmony for instilling in me their indomitable skepticism. Nancy Reis, Cristina Jones, Paul Greer, Pat Devers, Annemarie Grace, and Antonio T. helped round up and copy publications, and their assistance is appreciated.

Improvements in deer distribution maps were made by Carlos Alcalá-Galván, Sergio Alvarez, Scott Bailey, Bob Barsch, Chad Bishop, Randy Botta, Gene Byrne, Dave Cagle, Jorge Cancino, Rogelio Carrera, Joaquin Contreras, Mike Cox, Mark Elkins, John Ellenberger, Steve Flinders, Brad Fulk, Art Fuller, Bill Glasgow, John Goodwin, Ed Gorman, Barry Hale, Joe Hall, Jon Hanna, Ian Hatter, Bob Henry, Bill Jensen, Rolf Johnson, Tom Keegan, William Longhurst, Lee Luedeker, Carl Lutch, Enrique "Indio" Marquez-Muñoz, Ken Mayer, Kathleen McCoy, John McGehee, Philip Merchant, John Millican, Jerry Nelson, Les Rice, Russ Richards, Joe Sacco, Mike Shaw, Dan Smith, Billy Tarrant, Mike Welch, Darrell Weybright, Joe Williams, and Clayton Wolf. Bob Miles lent consistent support for this book from its inception and frequently provided intellectual grease to keep the project moving forward. David Bertelsen, Clay McCulloch, George Montgomery, Pete Petrie, Joan Scott, Kim Stone, Thomas Hulen, and especially Phil Jenkens assisted greatly in sorting through thorny issues related to southwestern plant identification and taxonomy. Jay Villemarette at Skulls Unlimited prepared skulls for photographing.

The summary of the management of deer throughout the southwestern states and Mexico would not have been possible without all the help and information I received from Francisco Abarca, Carlos Alcalá-Galván, Vern Bleich, Randy Botta, Clay Brewer, Rogelio Carrera, Lou Cornicelli, Mike Cox, Steve Cranney, John Ellenberger, Barry Hale, Jim Karpowitz, Scott Lerich, Pat Mathis, Gerry Mulcahy, and Darrel Weybright.

We are all indebted to hunters who support wildlife research and management by purchasing outdoor equipment and licenses, and also by supplying vital information and biological samples through questionnaires and check stations. All who enjoy watching deer have the deer hunter to thank for decades of contributions to habitat acquisition, protection, management, and wildlife law enforcement, making the American system of wildlife conservation the model for which the rest of the world strives.

Most important, thanks to my wife Annette, for enduring my incessant and obnoxious keyboarding at all hours of the night; and for following me everywhere, even when I didn't know where I was going.

Deer of the Southwest

Chapter 1

Southwestern Deer

Taxonomy: What's in a Name?

TAXONOMY is simply the process of naming, describing, and organizing plants and animals into categories based on similarities and differences. These categories indicate evolutionary relationships because similar animals generally have common ancestors. This structured system of classification helps scientists and biologists discuss relationships among animals.

In 1758 Swedish physician and botanist Carl von Linnaeus finalized an approach to naming plants and animals in a classification scheme (*Systema naturae*). Linnaeus' naming scheme consisted of a hierarchy of seven classifications that grow progressively more specific as one gets closer to individual species (Table 1). This system is called *binominal nomenclature* because it results in two names for each species; the first name is the *genus* and the second is the *species*.

With this naming system, each plant and animal in the world has a unique scientific name (*Genus species*) used by scientists in all countries. Scientific names are sometimes Greek but usually Latin. Because Latin is a "dead" language and not subject to change through time, it is the international language of science. The *subspecies* category was not part of the original classification system, but was added later in an attempt to describe variations (sometimes called *races* or *ecotypes*) within the same species. The subspecies name is added to the end of the two-word scientific name (*Genus species subspecies*). In the field of taxonomy, there are some biologists who are "lumpers" and others who are "splitters." Lumpers prefer to

TABLE 1. CLASSIFICATION OF SOUTHWESTERN DEER

Kingdom	Animalia (animal kingdom)
Phylum	Chordata (animals with a backbone)
Class	Mammalia (mammals)
Order	Artiodactyla (even-toed hoofed animals)
Family	Cervidae (the deer family)
Genus	*Odocoileus* (medium-sized North American deer)
Species	*virginianus* (white-tailed deer)
	hemionus (mule deer)

focus on the similarities among animals and group several similar forms into one category. Splitters, in contrast, prefer to separate even slightly different forms into different taxonomic categories.

Taxonomy kept early naturalists busy. The exploration of new lands resulted in the discovery of animals that had previously been undescribed in the literature. Many new categories were established based on only a few specimens. In some cases, a small and barely discernible difference resulted in the naming of a new species.

Some early naturalists took taxonomic splitting to the extreme. Merriam (1918) examined grizzly bear skulls and declared there were eighty-six species of grizzlies in North America, with twenty-seven species in Alaska alone! Linnaeus himself originally described male and female mallards as different species because of their different color patterns. Many of these early "species" were later reduced to subspecies status or dissolved completely.

Classification of Deer

Several similar species are grouped into a single *genus,* and related *genera* (Latin plural) are grouped into a "family." Similar families are then assembled in a single "order" based on similarities and common ancestry.

Deer are members of the class *Mammalia,* which contains all warm-blooded animals that produce milk for their young, usually have fur, and possess seven neck vertebrae (Table 1). Within the class *Mammalia* are nineteen orders; two of these are made up of animals that walk on hooves. These animals are called ungulates because the hooves they walk on are simply highly developed toenails.[1]

Ungulates with an odd number of toes (one or three) on each hoof belong to the order Perissodactyla (horses, rhinos), while the order Artiodactyla contains all

---

1. The word *ungulate* comes from the Latin *unguis,* meaning "claw" or "toenail" (Gotch 1995).

even-toed (two or four) ungulates, such as cattle, deer, goats, antelope, pigs, and camels.[2] Within Artiodactyla, there are nine taxonomic families, four of which occur naturally in North America: Bovidae (sheep, cattle, goats, bison), Antilocapridae (pronghorn antelope), Tayassuidae (collared peccary/javelina), and Cervidae (deer, elk, moose) (Nowak 1999).

The deer family (Cervidae) comprises all animals that shed antlers annually, including moose, elk/red deer, caribou/reindeer, mule deer and white-tailed deer, as well as several Asian, European, and South American species. Cervids, as members of the family are called, walk on the hooves (toenails) of the third and fourth toes, but no longer have the first digit (thumb or big toe). The second and fifth toes have been reduced and assume a nonfunctioning role and are called "dew claws."

Worldwide, there are seventeen genera in the deer family, but only *Alces* (moose), *Rangifer* (caribou), *Cervus* (elk), and *Odocoileus* (deer) are found in the United States (Nowak 1999). The genus *Odocoileus* includes the only two species of medium-sized deer (less than 300 pounds) in the United States: mule deer (*Odocoileus hemionus*) and white-tailed deer (*Odocoileus virginianus*) (Fig. 1).[3] The mule deer was first described in North America in 1817 based on field notes made by Charles LeRaye while he was held captive by the Sioux tribe on the Big Sioux River in South Dakota (Cowan 1961). The scientific name *hemionus* literally means "half-mule" because of the similarity of the ears to those of a mule. The name *virginianus* reflects the fact that the whitetail was first described by Thomas Hariot during a visit to Virginia in 1584.

Most species vary from one portion of their range to another. These variations come about as individuals in localized areas adapt to habitat, forage, or climatic conditions. These slight variations in some populations are often labeled as subspecies or races. Researchers have described up to eleven subspecies of mule deer and thirty of whitetail. Many of these descriptions were based on only a few individuals and have not been evaluated sufficiently to determine if they are valid. The overlap in characteristics among most deer subspecies is so great that no list of differences can be compiled that will allow biologists to differentiate between most of them (Cowan 1961). Most authorities simply keep using these subspecies names because there is no information available to support or reject the designations.

The whole concept of subspecies has been under attack for some time (Wilson and Brown 1953). Subspecies boundaries, when taken literally (as they usually are), frequently create a nonsensical pattern of geographic differentiation. For example, the map of mule deer subspecies indicates that mule deer in central Arizona and

---

2. The name for these animals that walk on an even number of toes comes from the Greek *artios* (even), and *daktulos* (a finger or toe).

3. The genus name *Odocoileus* comes from the hollow pits, called infundibula, in the chewing surface of the premolars and molars; *odous* means "tooth" and *koilos* means "hollow."

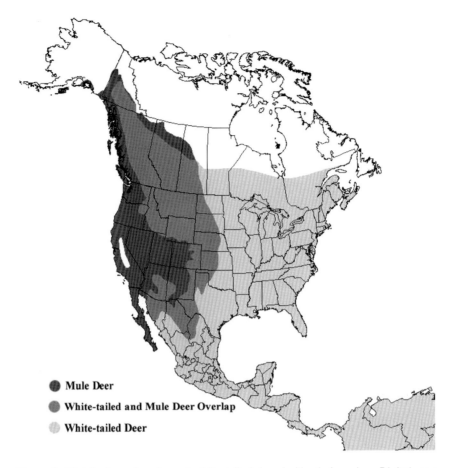

Figure 1. Distribution of mule and white-tailed deer in North America. *Digital map by Darius Semmens*

New Mexico (Rocky Mountain mule deer) are more closely related to those in Yukon, Canada (Rocky Mountain mule deer) than they are to mule deer in southern portions of those states (desert mule deer).

In reality, there are rarely sharp differences between deer at the boundary of two adjacent subspecies (Bowyer and Bleich 1984, Cronin and Bleich 1995). Subtle differences do occur between deer throughout their range, but these differences do not change abruptly as distribution maps indicate. In reality, physical characteristics usually change gradually across the geographic range of the species. Some populations within a species range may differ significantly in size, appearance, behavior, and biology from others of the same species elsewhere. In some cases, recognizing these animals as a different "race" or "ecotype" may be helpful in addressing unique conservation problems facing animals in that area. It can facilitate management actions that are critical for a particular population but that may not be needed elsewhere, such as water availability for "desert" mule deer. However, when the difference between these variations are not well documented, yet receive the official status of a scientific name (subspecies), it can invoke legal repercussions of a magnitude that quickly overshadows the question of taxonomic validity (O'Brien and Mayr 1991, Geist 1992, Cronin 1997).

Molecular markers (genetic analysis) are being employed to identify genetic differences among populations of a species throughout its range, and these will shed light on the validity of the subspecies boundaries that have been in use for years. Genetic analysis has already yielded useful guidance in big game taxonomy (Reat et al. 1999, Stephen et al. 2001), but there is much work ahead for the genus *Odocoileus*. A comprehensive genetic and morphologic evaluation of these subspecies will be the only way to determine which previously described forms really warrant distinction.

Deer of the Southwest

Southwestern deer have been referred to by a variety of names:[4] Sonoran fantails, dwarf deer, rock deer, Mogollon whitetails, Coues deer, desert mulies, blacktails, *venado bura,* and *venado cola blanca.* It is especially common to hear mule deer referred to as black-tailed deer. Early explorers, familiar only with the white-tailed deer, identified this species on the basis of its black-tipped tail. The real black-tailed deer, however, is a type of mule deer native to the Pacific Northwest (*O. hemionus sitkensis* and *O. h. columbianus*).[5] Only two species of deer inhabit the Southwest: the mule deer and white-tailed deer.

Mule Deer (Odocoileus hemionus)

Mule deer occupy a diversity of vegetation associations throughout the Southwest. Mule deer in southern Utah, Nevada, and Colorado, and northern Arizona and New Mexico inhabit pinyon-juniper woodland, ponderosa pine forests, and mixed-conifer forests at elevations ranging up to 11,000 feet above sea level. Deer in this part of the Southwest, and northward through the Rocky Mountains and into western Canada, are usually called Rocky Mountain mule deer (*O. h. hemionus;* Fig. 2). In the summer these deer inhabit higher elevation areas of ponder-

---

4. The term *Southwest* is used in this book to refer to the southwestern region of North America, including not only the southwestern United States, but also northern Mexico.

5. In 1897 early naturalist E. A. Mearns shot a deer in Hidalgo County, New Mexico, that had all the characteristics of a black-tailed deer from the Pacific Northwest. Mearns actually described this as a black-tailed deer (Crook's Blacktail) and wondered why no one had ever seen other blacktails in the Southwest. This specimen created confusion for years; it was later said to be a desert mule deer with some unusual characteristics. However, a reevaluation of this specimen showed that it is not a pure mule deer, but a hybrid between a desert mule deer and a Coues white-tailed deer (Heffelfinger 2000).

Figure 2. Mule deer can be recognized by their black-tipped tails, relatively longer ears, and antlers that fork twice on mature bucks. *Photo by George Andrejko/AGFD*

osa pine, with greatest deer densities in montane meadow areas intermixed with aspen and mixed conifer. Most Rocky Mountain mule deer populations migrate to lower elevations during the winter and subsist on browse (leaves and twigs of brush) when snow covers their summer habitat. Mountain shrubs and pinyon-juniper are important winter habitat for Rocky Mountain mule deer.

The term *desert mule deer* is loosely used for those mule deer found from West Texas, through the southern portions of New Mexico, Arizona, and California, and southward into Sonora, Baja California, Chihuahua, Coahuila, and Durango, Mexico (Fig. 3). Desert mule deer habitat varies from creosote-mesquite–dominated Chihuahuan Desert, with a palo verde–saguaro Sonoran Desert community, to upper desert grasslands, pinyon-juniper, dense chaparral, and oak woodlands. They can be found from 200 feet to 7,300 feet in elevation; however, most desert mule deer are found in semidesert grassland and chaparral vegetation types below 4,500 feet (McCulloch 1972, Krausman 1978a). In Sonoran Desert and desert grassland areas, large mesquite-lined washes are important habitat components for cover and forage. Early taxonomists named several different forms of desert mule deer, but few have been evaluated to see if they are meaningfully different from one another. When most people talk of desert mule deer, they are

Figure 3. Distribution of mule deer in the Southwest. *Digital map by Todd Black and Darius Semmens*

referring to those in southern Arizona, southern New Mexico, West Texas, and northern Mexico (*Odocoileus hemionus eremicus*).[6] Mule deer have expanded their range in the Texas Panhandle, possibly in response to "resting" former cropland in the Conservation Reserve Program (Kamler et al. 2001).

Some authors have designated desert mule deer in southwestern Arizona, southeastern California, and northwestern Sonora as "Bura" or "Burro" deer (Mearns 1897, Brown 1984). A few specimens (including a "type" specimen) collected from Sierra Seri in western Sonora, Mexico, in 1895 differed in coloration and antler configuration, and were designated as a new subspecies; a range was arbitrarily described for that race. These deer are reported to have shorter hair, a black dorsal stripe, lighter forehead markings, and wide antlers. Hoffmeister (1986:541) measured several mule deer specimens from within fifty miles of where the burro deer type specimen was collected and found that skull measurements did not dif-

---

6. Formerly referred to by the scientific name *O. h. crooki*. The representative "type" specimen was shown to be a hybrid. According to taxonomic rules, that finding invalidates the old scientific name and requires the use of the name *O. h. eremicus* (Heffelfinger 2000).

fer from those of other desert mule deer. Most current authorities do not recognize the burro deer as a separate subspecies (Wallmo 1978, Henry and Sowls 1980, Wallmo 1981, Anderson and Wallmo 1984, Hoffmeister 1986).

Other forms of desert mule deer include the "Southern" mule deer (*O. h. fuliginatus*) in southwestern California and northern Baja California, Mexico, and the "Peninsula" mule deer (*O. h. peninsulae*) inhabiting the southern half of the Baja California peninsula. There are no geographic barriers between the northern and southern half of Baja California, so this distinction is suspect. Southern mule deer consistently have a very dark dorsal stripe running down their tail rather than having a white tail with a black tip like most other desert mule deer (Cowan 1933, Bowyer and Bleich 1984). Mule deer in southern Baja California are somewhat smaller than the mule deer on the Sonoran mainland and much smaller than Rocky Mountain mule deer (Cowan 1936, 1961). These deer do not generally have a dark stripe on the dorsal surface of the tail, but rather carry the more typical mule deer tail (S. Alvarez-Cardenas, CIBNOR, personal communication, 2002).

Three island populations of mule deer are found off the coast of Baja California, but have received very little attention. There are mule deer on Tiburón Island, a 466-square-mile island off the coast of Sonora in the Gulf of California. This island has been separated from the mainland since the Pleistocene (Gastil et al. 1983), but is only about one mile from the Sonoran coast. Cowan (1936:238) originally considered these deer to be the same as the desert mule deer on the Sonoran mainland, but Goldman (1939) later differentiated them as *O. h. sheldoni*. The distinction between these two forms of deer was based on a vague mention of color differences and length of the tooth row of three specimens (Goldman 1939). The entire island is owned by the Seri Indians, which makes research on these deer difficult. Most of what is known about these deer comes from a few museum specimens collected many years ago.

Isla de Cedros (Cedros Island), lying forty miles into the Pacific Ocean off the west coast of Baja, has an isolated mule deer population. This 139-square-mile island was separated from the Baja peninsula within the last two million years. These deer are said to be distinct (*O. h. cerrosensis;* Merriam 1898) and are classified as "endangered" on the Red List of Threatened Species compiled by the International Union for Conservation of Nature and Natural Resources (Hilton-Taylor 2000). These mule deer are smaller than any other forms of mule deer and also have a much reduced rump patch, smaller metatarsal glands, and less developed antlers (Cowan 1936:238, Perez-Gil 1981).

The third island population of mule deer has never been studied. Mule deer occupy San Jose Island (75 square miles), just two miles off the eastern Baja coast in the Gulf of California northeast of the city of La Paz (Huey 1964:150). One fawn carcass from San Jose Island observed by the author possessed a very wide black dorsal stripe the entire length of the back. Further investigation may reveal these deer to be different from the peninsula mule deer in southern Baja.

Although most people accept the Rocky Mountain and desert mule deer as different kinds of deer, these subspecies are nothing more than points along a natural gradient of body size that is common in species whose distribution covers a wide latitudinal range. There is no clean break between the northern form and the southern forms; mule deer merely become smaller as one travels southward from Canada to northern Mexico. Also, translocations of mule deer have compromised historic geographic differences in some areas of the Southwest (Etheredge 1949, McCulloch 1968, Cantu and Richardson 1997). For example, more than 800 desert mule deer were translocated from the Trans-Pecos area of West Texas to various locations in the Panhandle of that state between 1949 and 1986 (Clay Brewer, Texas Parks and Wildlife Department, personal communications, 2004).

Mule deer living on islands or in arid, desert environments are smaller in skull and body measurements and in overall body size than those inhabiting the higher elevation forested regions of the Southwest (see chapter 3). These differences reflect the nutritional quality of their habitat, which varies geographically. Through time a population could evolve smaller physical stature through natural selection for smaller body sizes, which are easier to maintain in less productive habitats.

White-Tailed Deer (Odocoileus virginianus)

White-tailed deer are found in scattered populations throughout most of the Southwest from southern Mexico, northward through southeastern and central Arizona and southern and eastern New Mexico, and into Texas (Fig. 4). In most of the Southwest, whitetails occur primarily in partially isolated mountains above 4,000 feet. However, this highly adaptable deer also occupies some lower desert vegetation associations in Mexico, gently rolling sandhills and riparian zones in eastern New Mexico, and riparian areas in northeastern Coahuila, Mexico.

Early naturalist and army surgeon Elliot Coues never actually collected a whitetail in the Southwest. However, Dr. C. B. R. Kennerly shot many along the Mexican border and saved one in 1855, making him the first scientist to procure a specimen of this small white-tailed deer. As was common in the early days, Kennerly named it as a new species (*Cervus mexicanus*). In 1874 another army surgeon, Dr. Joseph Rothrock, collected and saved two specimens from the Santa Rita Mountains in Arizona and stated (correctly) that these were not a new species, but merely a smaller version of the common eastern whitetail (Carmony 1985). He suggested that these deer be referred to as Coues white-tailed deer (*Odocoileus virginianus couesi*) in honor of that pioneering naturalist (Fig. 5).[7]

Coues white-tailed deer occur from central and southeastern Arizona, to southwestern New Mexico, and southward into Mexico, including the states of

---

7. According to descendants of Elliot Coues, the correct pronunciation of Dr. Coues's last name sounds like "cows" (W. Coues, personal communication, 2000), although it is frequently pronounced "cooz."

Figure 4. Distribution of white-tailed deer in the Southwest. *Digital map by Todd Black and Darius Semmens*

Sonora, western Chihuahua, Sinaloa, Nayarit, and Durango. In the United States and in the Sierra Madre Mountains of Mexico, whitetails occupy relatively rough, wooded terrain with steep canyons (Fig. 6). Typical whitetail habitat in the northern part of the range is mixed oak woodland, but they can be found anywhere from ponderosa pine/mixed conifer at 10,000 feet down to the upper limits of semidesert grassland (White 1957, Findley et al. 1975, Knipe 1977, Galindo-Leal and Weber 1998). Although elevations with the highest deer densities vary among different mountain ranges, most Coues whitetails are found between 4,000 and 7,000 feet (Ligon 1927, White 1957, Welch 1960). At lower elevations, there is considerable overlap in habitat use with desert mule deer (Anthony and Smith 1977).

There are small, isolated populations of whitetails in the low Sonoran Desert of southwest Arizona (Ajo, Sauceda, Sikort Chuapo, and Sand Tank Mountains) and extreme western Sonora and Sinaloa, Mexico (Burt 1938; B. Henry, AGFD, personal communication, 1997). The distribution and persistence of these pockets of white-tailed deer in the Sonoran Desert is limited by seasonal drought (Brown and Henry 1981).

Figure 5. Early naturalist Elliot Coues, for whom the diminutive southwestern white-tailed deer is named. *Photo courtesy of Library of Congress*

There is a common misconception that within the range of Coues whitetails several different local types exist, especially an extra-small whitetail (Rock, Sinaloan, Sonoran Fantail, Dwarf) that is said to occur in localized areas of the Southwest (Ligon 1927). Young deer, with small 3×3 racks, are often the cause of such rumors because observers mistake them for unusually small, mature bucks. Another contributing factor is the wide variation in the color of the dorsal (back) side of the tail of Coues whitetails. The dorsal surface of the tail may appear gray/brown (same as the animal's back), reddish, blond, very dark brown, or black. These are not an indication of different types of deer, but instead are color variations found in some individuals.

Coues whitetails in southwestern New Mexico are found primarily in the Animas, San Luis, and Peloncillo Mountains, extending up to the Mogollon Moun-

Figure 6. Coues whitetails are a southwestern version of this widespread species and considered by some to be wiliest of all whitetails. *Photo by Joe and Marisa Cerreta*

tains and Black Range. These whitetails are geographically separated from, and do not interbreed with, the population found in the Sacramento Mountains in south-central New Mexico and in scattered, isolated pockets in the eastern half of New Mexico and West Texas (Findley et al. 1975; J. Nelson, New Mexico Department of Game and Fish, personal communication, 2000; TPWD Files). All white-tailed deer east of the Rio Grande in New Mexico are more similar to those in South Texas (*O. v. texanus*), which are measurably larger in body and antler size (Bailey 1931, Raught 1967).

Although the smaller Coues form and the Texas form of whitetail are separated in the United States, they may interchange across a series of populations south of the border (Findley et al. 1975). Whitetails gradually become larger as one moves north and east out of Chihuahua through West Texas, Coahuila, Nuevo Leon, and then into South Texas (Krausman et al. 1978). Whitetails in northern Coahuila, Mexico, and the Chisos Mountains of West Texas have been described as a different subspecies, the Carmen Mountain white-tailed deer *(O. v. carminis)*. These deer are somewhat isolated from both the Coues and the Texas whitetail at the extreme northern portion of their range in West Texas. Krausman et al. (1978) supported the recognition of Carmen Mountain whitetails as a distinct subspecies, based on skull and body measurements of Coues, Texas, and Carmen Mountain whitetails. However, there is no geographical separation of Carmen Mountain whitetails in the central and southern part of their range. In fact, Coues whitetails from western Chihuahua gradually become larger in size throughout their scattered distribution into Coahuila and then South Texas.

Farther south in Mexico, the Coues whitetail intergrades imperceptibly into populations of whitetails that have been described as several different subspecies. No study has ever been conducted of how these whitetails differ from Coues whitetails, or from each other. These deer probably represent a gradient of characteristics throughout the range of this species.

Differences between Species

The vegetation associations occupied by desert mule deer and white-tailed deer in the Southwest are just the opposite of what occurs in the rest of the Rocky Mountain states. In most areas of the West where both deer species are found, mule deer inhabit the higher mountain areas and whitetails occupy the lower valleys and river systems. This habitat preference is reversed in the Southwest, where whitetails are found in the mountains, generally above 4,000 feet, and the desert mule deer occupy the lower elevation valleys and foothills (Ligon 1927, White 1957, Welch 1960, Anthony and Smith 1977). Because of the interspersion of whitetail and desert mule deer habitat, the Southwest has an extensive zone where the two species overlap and coexist. This results in the animals being in proximity to one another throughout the year, including the breeding season. At times, mixed groups containing whitetails and mule deer are observed, and hybridization is known to occur (Heffelfinger 1999).

Being able to accurately identify both species is especially important in areas where the ranges of whitetails and mule deer overlap. This is important for hunters because hunt permits are generally prescribed separately for each deer species. Mule deer differ from whitetails in several characteristics; however, there is enough variation of these characteristics in each species to periodically present some interesting specimens that cannot be quickly identified. Some of the identifying characteristics, when used singly, can be confusing or yield an incorrect identification. It is important to use all the information available when differentiating these deer species.

Tails

Whitetails have a wide, flattened tail that is broad at the base and narrower at the tip (Fig. 7). The pure white underside contrasts with the darker (brown/gray/reddish/black) dorsal (back) side. The darker tail is edged with white fringe hairs that are an extension of the white underside. White-tailed deer tails are at

Figure 7. White-tailed deer (*left*) have wide tails that are brown on the back side and small, white metatarsal glands. Mule deer (*right*) have a larger white rump patch, smaller rope-like tail, and long brown metatarsal glands. *Photo by Pat O'Brien/AGFD*

least 7½ inches long, which makes them longer than mule deer tails (less than 7½ inches). Additionally, white-tailed deer lack a large, conspicuous white rump patch.

Mule deer tails appear cylindrical, or ropelike, and are usually (but not always) white on the dorsal side with a distinctive black tip surrounded by a large, obvious white rump (Fig. 7). Some mule deer may have a thin black line running down the dorsal surface of the tail.

Antlers

Antlers are the least informative characteristic to use when trying to differentiate between these two deer species. This is because of the extensive variation in antler shape and conformation in both species. However, there are differences, which can be used in combination with other characteristics to help distinguish the species.

Mule deer antlers have small brow tines (also called eye guards), if they have them at all. The main beams sweep out and upward, forking once, and then each fork divides again in mature bucks (Fig. 2). Mature bucks typically have eight total points, or ten if the brow tines are present ("eastern count"). The same buck would be considered a four-point buck with "western count" (number of points on one side excluding the brow tines). Typical whitetail antlers have several antler tines that arise independently off a main beam that sweeps outward and forward from the bases (Fig. 4). The brow tines are nearly always present and usually prominent. Mature whitetail bucks usually have eight total points including the brow tines (three-point bucks, or 3 × 3 in western count).

It is not unusual for whitetails to have forked tines like a mule deer or to see a mule deer with all tines arising from the main beam. The whitetails in the Sierra del Carmen Mountains of northern Mexico seem to exhibit a high degree of forked antlers like mule deer (Leopold 1954).

Facial Markings

A whitetail's forehead is usually about the same coloration as the rest of the face, although it is sometimes slightly darker. The white eye rings and nose markings are prominent. A mule deer has a distinctive black forehead, or "mask," that contrasts sharply with a light gray face. Their lighter facial coloration results in less obvious eye rings and muzzle.

Body Size

Mule deer are much larger than whitetails in most of the Southwest, which aids identification under some circumstances. Mature Coues whitetail males rarely weigh more than 110 pounds field dressed (all internal organs removed). In contrast, even small desert mule deer bucks (two years old) weigh more than 135 pounds field dressed (see chapter 3).

Ears

Coues whitetail ears average about 7 inches in length and are considerably shorter than those of mule deer, which average about 9½ inches. This difference is not entirely because of the difference in body size; mule deer ears are also proportionately longer than those of whitetails. The ears of a whitetail are generally two-thirds the overall length of the head (back of head to nose), while those of a mule deer are three-quarters the length of the head.

Flight Behavior

When alarmed, a whitetail usually throws up its tail, exposing the fluffy white underside, which alerts all other deer in the area of the apparent danger. It then runs directly away from the source of danger, using its speed to put as much distance between itself and the disturbance as it can (Lingle 1989).

The mule deer does not "flag" its tail, and it often bounces away in a motion called "stotting" in which all four hooves land and push off at the same time. Their escape is not as fast as a whitetail's, but it is effective in quickly moving through rugged terrain. Both species may stop and look back at the source of the danger before disappearing over a ridge, but this behavior is much more typical of mule deer.

Metatarsal Glands

The best diagnostic characteristic for differentiating between whitetail and mule deer is the size and location of the metatarsal glands. This gland differs greatly in these two species; however, it is not readily observable on a free-ranging deer. The metatarsal gland of both species is located on the *outside* of the lower portion of the hind leg and is sometimes confused with the tarsal gland on the *inside* of the leg (hocks).

Whitetails have metatarsal glands that are 1 inch or less in length and always encircled with white hair. In whitetails, this gland is at midpoint or below midpoint on the lower shank of the leg (figs. 7 and 9). Coues whitetails have metatarsal glands that are usually only ½-inch long.

Mule deer have much larger metatarsal glands, which measure 3–7 inches in length and are not encircled with white hair. The mule deer metatarsal gland starts at the ankle joint and extends downward toward the hoof. It appears as a large, long tuft of fur (Fig. 7). The metatarsal glands of desert mule deer average about 5 inches and are proportionately shorter than those of their northern counterparts.

Preorbital Glands

The preorbital gland is situated in front of the eye socket and differs considerably between the two species (Fig. 22). The preorbital gland of whitetails is very small, appearing as a small slit with a maximum depth of ⅜ inch. In mule deer, this gland is comparatively large, forming a substantial pocket with a depth averaging ¾ inch. The mule deer preorbital gland commonly has a small ball of yellow, waxy substance, composed of plant material, skin cells, and possibly glandular secretions, inside the skin fold.

Hybridization

Different species of animals, even those closely related, are normally kept from breeding with one another by being geographically isolated or by using different types of habitat. If the animals coexist in the same habitat, then they generally breed at different times or have different courtship and breeding behavior to prevent hybridization (Geist 1981).

In the case of whitetails and mule deer, courtship and breeding behavior is different enough that body language and scent cues given off by a female mule deer during rut are not normally "understood" by a male whitetail and vice versa. This system of species segregation has worked remarkably well throughout their evolutionary coexistence. However, in rare cases this system breaks down and hybrid-

Figure 8. Whitetail × mule deer hybrid (F_1) buck raised in captivity. *Photo by Gerald Day*

ization occurs, resulting in a deer that is half whitetail and half mule deer. This hybridization between the two different deer species is extremely rare but does occur throughout the West where their ranges overlap (Heffelfinger 1999).

Hybrid deer show characteristics that are intermediate between mule deer and whitetails. Body size is indicative of mule deer, but the tail usually is dark chocolate brown or black on the dorsal side and white underneath. The tail of a hybrid is often similar to a whitetail's, but usually is much darker (Fig. 8). Ears are larger than those of whitetail, but smaller than a mule deer's. The preorbital gland is also intermediate between the deep pits found in mule deer and the shallow depression of whitetails.

Most hybrids have whitetail-like antlers, but it is impossible to tell a hybrid by antlers alone. There is simply too much variation in antlers to serve as a reliable indicator of hybridization. The best feature to determine if a deer is a hybrid is the size of the metatarsal gland on the outside of the lower portion of the rear legs (Fig. 9). Unfortunately, this characteristic is nearly impossible to evaluate in the field from a distance. A whitetail × mule deer hybrid has metatarsal glands that are intermediate between the long, brown mule deer glands (over 3 inches) and the small white glands of a whitetail (less than 1 inch). Hybrid metatarsal glands measure between 2 inches and 4 inches, and may or may not be encircled with white hair (Fig. 9).

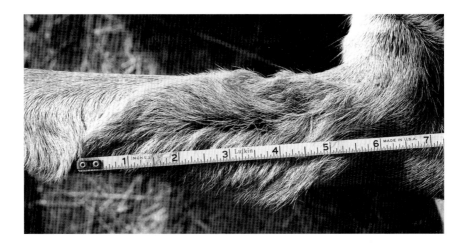

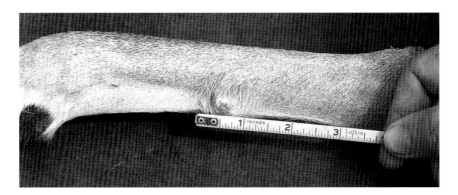

Figure 9. Metatarsal glands of a mule deer (*top*), white-tailed deer (*middle*), and white-tailed deer × mule deer hybrid (*bottom*). *Photos by author (top and middle) and John Holcomb (bottom)*

Two-year-old male mule deer are most frequently mistaken for hybrids. This is because of their smaller antler development and because dichotomous branching (producing the fourth main tine on each side) usually does not occur until the buck is at least three years old. Young mule deer sometimes give the appearance of a very large white-tailed deer, especially if the tail has a dark stripe down the back, as sometimes occurs.

Hybrids have been reported from captive facilities as early as 1898, when a whitetail × mule deer cross was produced at the Cincinnati Zoo (Gray 1972). Occurrences were later reported from the zoo in Minot, North Dakota, deer pens in Alberta, and other locations. The male hybrids are usually sterile, as is the case in mammals; however, female hybrids are fertile when bred back to one of the parent species.

In the 1930s Arizona Game and Fish Department biologists produced hybrids in research pens by mating mule deer bucks to whitetail does and vice versa (Nichol 1938). These matings resulted in nine hybrid fawns, of which only four survived the first few months. The research ended abruptly, and the deer had to be released before any meaningful data could be collected. In the 1970s AGFD researcher Gerald Day (1980) also produced hybrids in captivity. Ten hybrids were born, but only four lived past six months of age.

Hybrids have a much lower survival than purebred whitetail or mule deer. Even captive hybrids, which are protected and well fed, have a high mortality rate. To complicate matters, hybrids inherit predator avoidance strategies from both parents; however, whitetail and mule deer have different techniques for escaping predators. This difference in predator avoidance does not always work well in combination and can result in an ineffective flight response.

Mule deer have developed their bouncing escape (stotting), which allows them to navigate their rugged but open terrain. Research using captive animals in Alberta showed that stotting is so specialized that even a hybrid that is one-eighth whitetail and seven-eighths mule deer does not use this bouncing locomotive style (Lingle 1989:96). The hybrid's escape behavior was reported as chaotic; hybrids approached the threat and jumped around in confusion. Such behavior is not conducive to a long life in the wild.

Whitetail × mule deer hybrids have also been reported in the wild from Alberta, British Columbia, Saskatchewan, Nebraska, Kansas, Colorado, Washington, Wyoming, Texas, and Arizona (Heffelfinger 1999). Southwestern biologists have documented the presence of hybrids in the wild on a few occasions (Fig. 10). The relative scarcity of confirmed hybrids among the thousands of deer examined by field biologists each fall illustrates how rare they are. Retired Arizona wildlife manager John Holcomb documented three hybrids harvested by hunters in the Galiuro Mountains near Willcox during his sixteen-year tenure there (Fig. 9). Longtime AGFD biologist Ted Knipe reported seeing four hybrid bucks in his thirty-four years of field work (Knipe 1977).

Figure 10. White-tailed deer × mule deer F_1 hybrid harvested in southeastern Arizona by Rudy Alvarez in 1997. *Photo by author*

Every year numerous hunters report seeing "hybrid" deer. AGFD researcher Gerald Day, who had produced captive hybrids, investigated over 200 reports of "hybrids" and did not find a single legitimate whitetail × mule deer hybrid.

Recent advances in DNA analysis technology provide better diagnostic tools to identify hybrids than looking at ears and antlers. Genes regulate the production of proteins in the body. By analyzing differences between some proteins, researchers can identify what species a sample of tissue came from. Serum albumin is a protein that has proven particularly useful (McClymont et al. 1982). Analyzing this protein with a process called electrophoresis produces horizontal bands on a gel surface. This protein produces a band in a different location on the gel for whitetails than for mule deer. When a first-generation hybrid is tested, both the whitetail *and* the mule deer bands are present. Mixing whitetail meat with mule deer meat and testing the mixture can produce the same result. There seem to be some exceptions to the unique banding patterns, but this test is at least 95 percent accurate in identifying first-generation crosses.

Cronin et al. (1988) in Montana used both albumin and mitochondrial DNA (mtDNA) analysis to determine the extent of hybridization and found that very little, if any, had occurred in that state. In West Texas managers have reported an increasing trend in the number of hybrids they see on their ranches. In the early 1980s whitetails and mule deer in a five-county area were tested using serum albumin, and on average 5.6 percent of the deer tested were hybrids. Individual ranches ranged from none to 24 percent (Stubblefield et al. 1986).

At the same time, Carr et al. (1986) were busy analyzing the genetics of whitetails and mule deer on a ranch in West Texas using mtDNA. Mitochondrial DNA contains only a few genes that regulate factors that are not generally visible (that is, not tail color or antler shape). An animal inherits this type of DNA only from its mother. Analysis of mtDNA is very useful because a hybrid that has a whitetail father and mule deer mother will have only mule deer mtDNA (the nuclear DNA will be from both mother and father). If the hybrid is a female, it will continue to pass mule deer mtDNA through its daughters and their daughters even if bred by whitetail bucks. After a few generations, the results of these matings would look like white-tailed deer but would have pure mule deer mtDNA from their mother and grandmother.

And that's apparently what has happened: these researchers found that the mule deer on the ranch had mtDNA that was indistinguishable from the whitetails on the ranch. They concluded that hybridization was more common in this area. Even more surprising, analyses showed that the mtDNA of mule deer on this ranch was more closely related to *whitetails* than to blacktails (a type of mule deer) from northern California. This was initially seen as an indication that hybridization must have occurred between mule deer bucks and whitetail does because the whitetail mtDNA the deer carried was inherited from their mothers (Carr et al. 1986).

This conflicts with the conventional theory that most hybridization occurs between a whitetail buck and mule deer doe. This theory is based on the different breeding strategies of the two species. The whitetail buck is accustomed to chasing a whitetail doe relentlessly until she allows him to breed. The mule deer breeding behavior is much more relaxed, with the doe only moving a few steps if she is not ready (see chapter 7). A whitetail doe would run far away from a pursuing mule deer buck, confusing him, but a mule deer doe would not run far from a whitetail buck in pursuit. It seems more likely, then, that a persistent whitetail buck would likely breed with a mule deer doe. An additional consideration is the small size of Coues whitetail does and the problems they would have giving birth to a larger hybrid fawn.

Indeed, more recent analysis of mtDNA sequencing of deer in West Texas indicates that it is more likely that white-tailed bucks bred mule deer does to produce the hybrids they observed (Carr and Hughes 1993). This direction of hybridization (whitetail bucks breeding mule deer does) was later confirmed by analysis of the Y-chromosomes of bucks in the area (Cathey et al. 1998).

A different type of genetic analysis called "microsatellites" is fast becoming the method of choice for diagnosing hybrids. This method analyzes nuclear DNA (half from the mother and half from the father) and allows a sample from a deer of unknown species to be assigned to either "mule deer" or "white-tailed deer." The analysis assigns true hybrids as 50 percent mule deer and 50 percent whitetail.

We know that whitetail × mule deer hybrids occur. However, they are extremely rare in the wild and almost impossible to accurately identify because of the large variation in characteristics in each species. Some whitetails have characteristics (tails, forehead) that look like mule deer, and some mule deer may have whitetail-like features (no antler forks, dark fur on back of the tail). Hybrids cannot be identified with certainty at a distance, and it is highly unlikely that a person will ever see a hybrid while afield. The low number of interspecies matings and the low survival rate of hybrid offspring greatly reduce the chance of encountering one in the wild.

Chapter 2

Historical Perspective

Origins and Evolution of Deer

THE EARLIEST hoofed animals with an even number of toes (artiodactyls) appeared in the fossil record during the Eocene Epoch, 34–56 million years ago. Rabbit-sized ungulate ancestors, such as *Diacodexis* and others in the family Dichobunidae, were distributed throughout North American and Eurasia (Romer 1966). *Diacodexis* possessed a unique ankle bone, called the astragalus, which acts as a double pulley, providing great flexibility in the hind foot. This feature marks this animal unmistakably as the first known artiodactyl; all even-toed ungulates have this bone, which allows remarkable flexibility of motion. These animals possessed long limbs for running and, although they had four toes with hooves on them, they supported most of their weight on the two central toes (hooves) on each foot.

Primitive artiodactyls diversified and increased in abundance throughout the Eocene. The close of the Eocene saw the development of a group of primitive ruminants, which were precursors to the cattle, pronghorn, camel, and deer families. Evolutionary development of these ruminants continued through the Oligocene Epoch (24–34 million years ago) with the appearance of *Leptomeryx* in North America and *Eumeryx* in Eurasia. *Leptomeryx* already possessed many characteristics that are seen in today's deer, and also in bovids: upper incisors absent, incisor-like lower canines, brachyodont molars, and much reduced first lower premolars. The more highly advanced Eurasian *Eumeryx* represents the transition between

Figure 11. The Eurasian *Eumeryx* represents an early mammal form that eventually branched out into the deer and cattle families. *Illustration by Randall Babb*

the primitive artiodactyls and the increasingly deerlike forms that followed (Fig. 11; Stirton 1944).

The end of the Oligocene and the beginning of the Miocene (about 24 million years ago) saw a remarkable flush of diversity in the higher ruminant families (deer, bovids, pronghorns) in North America and Eurasia (Romer 1968). The most interesting North American forms are represented by groups such as Dromomerycidae, Protoceratidae, and Merycodontinae, which consisted of medium to large grazing and browsing animals that had large, bony hornlike structures called ossicones extending upward from the skull over the eye sockets (Scott 1937, Scott and Janis 1987). These were not shed as are antlers and do not appear to have been covered by a keratinous horn sheath as in pronghorn (*Antilocapra americana*). The horns were probably covered simply with a layer of skin as are the cranial knobs in present-day giraffes and okapi (Romer 1966). Dromomerycidae and Protoceratidae included animals such as *Cranioceras* and *Synthetoceras,* respectively (Fig. 12). These are just a few examples of a remarkably diverse assemblage of ruminants that roamed North America at that time. Unfortunately, none of these spectacular animals remained at the close of the Miocene (5 million years ago). The merycodonts were an interesting group that diversified into many species with unique horns, such as *Ramoceros* (Fig. 12). Despite the very antlerlike form of some merycodont ossicones, these animals were not ancestors of deer, but rather primitive types of pronghorn. The pronghorn that we see today is the only surviving remnant of this large and diverse group.

The North American *Blastomeryx* and the Eurasian *Dremotherium* are examples of primitive deer with no antlers, but exaggerated tusklike canines that appear

Figure 12. *Cranioceras, Synthetoceras,* and *Ramoceros* represent three spectacular groups of large ruminants that roamed North America during the Miocene Epoch. *Illustration by Randall Babb*

in the Miocene Epoch (Fig. 13). An Asian form of these sabre-toothed deer, such as *Dremotherium,* is the most probable ancestor of all true deer (those with antlers). These large canine tusks may seem out of place on a deer, but the present-day musk deer (*Moschus*) and Chinese water deer (*Hydropotes*) of Asia are remarkably similar and probably represent direct descendants of these primitive deer.

Despite the abundance and diversity of horned ungulates in North America during the Miocene, none of these forms gave rise to North American deer. There is no record of true deer in North America until the close of the Miocene (Webb 2000). All deer in North America today are descendants of the Eurasian deerlike animals, such as *Dremotherium* or *Procervulus* (Fig. 13). *Procervulus* possessed not only large canine tusks, but also forked antlers that may have been shed at irregular intervals. Thus, *Procervulus* is positioned precisely at the genesis of the deer family—the evolution of deciduous antlers.

Figure 13. The increasingly deer-like forms such as *Dremotherium* and *Procervulus* appear in the Eurasian fossil record during the Miocene. *Dremotherium* had no antlers, but the males had large, saber-like canines not unlike the Chinese water deer living today. The horns/antlers of *Procervulus* may have been shed at irregular intervals. *Illustration by Randall Babb*

The earliest true deer (Cervidae) appeared in Eurasia during the Miocene (Scott and Janis 1987). One of these ancestral deer had small antlers that normally formed a single fork (*Dicrocerus*). Another Miocene deer, called *Stephanocemas*, had tusklike canines and antlers that formed a bowl-shaped palm (Fig. 14; Kurtén 1971). The antlers of these early deer were elevated on long antler bases, much like the present-day muntjac (*Muntiacus*) of Asia.

With the evolutionary development of elaborate antlers, the occurrence of tusklike canines was much reduced in the deer family (Eisenberg 1987). The antlerless water deer, musk deer, and mouse deer (Tragulidae) have prominent canines, while in other antlered deer the canines have been lost entirely or very much reduced (as in elk). The muntjac and tufted deer (*Elaphodus*) of Asia are intermediate, with small antlers and small canines. The reduction of large canines may have occurred because the development of elaborate antlers supplanted the need for these teeth as sexual display organs. Alternatively, the reduction of these enormous canines may have been caused simply by a change to a browse-dominated

Figure 14. *Stephanocemas* and *Dicrocerus* were the first animals to shed their antlers on a regular and recurrent basis. All of today's true (antlered) deer arose from early deer such as these. *Illustration by Randall Babb*

diet and the need to grind food with side-to-side jaw movements (Obergfell 1957, in A. B. Bubenik 1990).

It was an early deer species similar to *Dicrocerus* that crossed over the Bering land bridge to North America about 5–6 million years ago. The earliest fossils of Cervidae in the Americas are represented by *Eocoileus gentryorum* (Fig. 15), the most probable ancestor of *Odocoileus* (whitetails, blacktails, and mule deer) and all South American deer species.

Evidence of deer evolution throughout the late Pliocene/early Pleistocene (600,000 years to 4 million years ago) has mostly been lost as a result of repeated glaciations that scoured the landscape for thousands of years during the Pleistocene. *Odocoileus* remains have never been found in Eurasia, indicating this genus evolved solely in North America from *Eocoileus* or a similar ancestor. A single European deer species, the roe deer (*Capreolus*), is related to the genus *Odocoileus* and probably arose from the same ancestor that migrated to North America in the late Miocene (Fig. 16).

During the early Pleistocene, deer were not as abundant or widely distributed as they are today (Guthrie 1984). The diversity and sheer abundance of other large ungulates at that time led to intense competition for resources. The last glacial retreat, at the close of this period (8,000–11,000 years ago), coincided with mass extinctions of these large animals throughout the world. Most of the large mammals

Figure 15. Sometime prior to 5 million years ago, an early deer ancestor crossed the Bering Land Bridge, and thus true deer were introduced into North America. Fossils of *Eocoileus* indicate it is a direct ancestor of today's mule deer and white-tailed deer. *Illustration by Randall Babb*

that were native to North America died out in a remarkably short time (camels, giant sloths, mastodons, sabre-toothed cats, long-horned bison, native horses). Many theories have been proposed for the cause of these extinctions, but all have serious flaws (Martin and Klein 1984). Deer did not become the widespread, dominant ungulates we know today until the multitude of large grazing mammals disappeared, allowing them to expand their ranges and assume vacated niches.

The archaeological evidence from the Pleistocene that does exist reveals fossil *Odocoileus* that are essentially the same as those of today (Kurtén 1988). Little differentiation can be made between mule deer and whitetails from bone fragments because the most obvious differences are small portions of the skull that are not well preserved after thousands of years. Other cervids, such as the elk and moose, migrated to North America through the Bering Strait more recently (late Pleistocene) and therefore more closely resemble their European counterparts (red deer, European moose).

We know very little about the evolutionary relationships and divergence of mule deer and white-tailed deer (*Odocoileus*). The most widely accepted theory is that a primitive *Odocoileus* split into a western species (blacktails) and an eastern species (whitetails) during the Pliocene. The mule deer then originated as an offshoot of the blacktail line, becoming larger and more elaborate in antlers and body

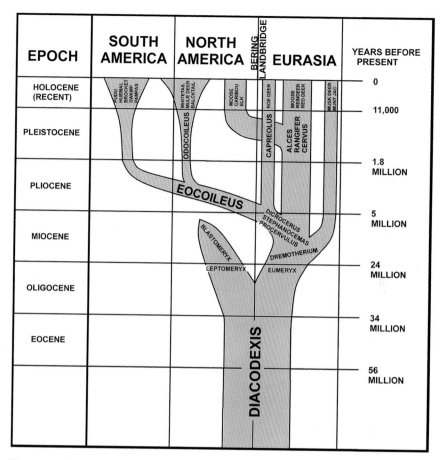

Figure 16. Stylized representation of the evolution of early deer-like forms, showing not only the time they appeared, but also their colonization of various continents. *Illustration by Randall Babb*

markings as they spread out into the fertile habitat left in the wake of receding glaciers (Geist 1981).

Genetic analyses have revealed that mitochondrial DNA (mtDNA) from whitetails and mule deer is very similar (indistinguishable in some cases) and that both are very different from blacktails (Cronin et al. 1988, Cronin 1991a, 1991b). This genetic relationship was unexpected because mule deer and black-tailed deer are the same species (different subspecies). This evidence has given rise to an alternative theory: that mule deer are a relatively "new" species, resulting from the hybridization of female whitetails and male blacktails brought together after the retreat of the glaciers and mass extinctions at the close of the Pleistocene Epoch 10,000 years ago (Geist 1994, 1998).

There are several reasons why this alternative theory is less plausible. First, hybrids between mule deer and whitetails have low rates of survival (Nichol 1938, Day 1980, Lingle 1989), which is inconsistent with the evolutionary success necessary for this mode of speciation to occur (hybrids would need to have a survival advantage rather than a disadvantage). In addition, the animals resulting from this hybridization would then have had to evolve relatively quickly to account for the differences we now see among mule deer, blacktails, and whitetails. There are diagnostic differences between white-tailed deer and mule deer skulls in front of the eye sockets (the lachrymal fossa). Frick (1937:194) illustrates a fossil deer skull found in California from the Pleistocene that clearly has a deep mule deer lachrymal fossa. This shows that deer with mule deer characteristics were already present in the Pleistocene. More recent analyses of Y chromosomes, mtDNA, and proteins representing nuclear DNA clearly show that mule deer and blacktails are very similar genetically and are different from whitetails (Derr 1990:86, Cathey et al. 1998, Cronin 1991b, Carr and Hughes 1993). This, along with morphological differences (Cowan 1936), argues against a Pleistocene hybrid origin of mule deer and supports the traditional scenario of mule deer differentiating from ancestral blacktails.

Native Americans

Big Game Hunters

Sometime before 13,000 years ago, small bands of humans crossed the Bering land bridge and spread across the continent hunting and gathering their food supplies (Morlan 1986). Owing to their nomadic way of life, these earliest Americans left little evidence of their existence. Archaeological evidence indicates that these hunters had reached the Southwest and northern Mexico by 11,500 years ago. These people hunted with distinctive spear points, called Clovis points after a site in Clovis, New Mexico. What we know about this early Clovis culture comes from several big game kill sites that contain spear points and other tools, such as scrapers and knives. Clovis peoples probably obtained much of their daily subsistence from small game and plants, but some of the animals they hunted were "big game" in every sense of the word. Two sites in southern Arizona (Murry Springs and Lehner) revealed mammoth remains with several Clovis points embedded in their skeletons. There is evidence that Clovis hunters were already using the atlatl, a spear-throwing stick that gave extra leverage for throwing their projectiles much harder and farther. These early hunters roamed a landscape with the dire wolf, sabre-toothed cats, horses, lions, camels, ground sloths, bison, and 100-pound beavers, all of which were native to North America before disappearing at the close of the Pleistocene Epoch. Deer were not as common and widespread at this time, but their remains have been found in association with Clovis artifacts (Cordell 1997:81). By about 10,500 years ago, Clovis points disappear from the archaeological record of the American Southwest, indicating that those who used them changed their way of life or moved to another area.

A hunting society, referred to as the Folsom complex, follows the Clovis culture in the eastern edge of the American Southwest (Cordell 1997). Folsom hunters occupied eastern New Mexico into the Great Plains. This culture is named after the Folsom, New Mexico, site that, in 1926, yielded the first distinctively fluted spear points that characterize it (Lister and Lister 1983). These people also gathered plant material to supplement their diets, but were too nomadic to develop elaborate seed or vegetable processing procedures. There is little evidence that deer were hunted extensively in these early times (Frison 1986). The Folsom people were primarily hunters who seemed to concentrate on a species of large bison (*Bison antiquus*) as their staple, much as the Plains Indians did in later periods. One archaeological site contained many bison skeletons with most missing the tail bones, indicating the hunters had skinned the animals and removed the tails from the site with the hides (McGregor 1965). Around 4,000–5,000 years ago, this large form of ancestral bison underwent a rapid evolutionary transformation to the smaller bison (*Bison bison*) we know today (Gingerich 1993). At that time, these formerly nomadic hunters began to focus their hunting activities on the smaller bison and on medium-sized game. Deer began to assume greater importance as a source of food and materials, a trend that persisted to recent times. From this time forward there appeared several hunter-gatherer cultures that occupied what is called the Archaic Period.

The Archaic Period

The Archaic period in Arizona lasted from about 7,500 to 2,000 years ago and is characterized by cultures that varied in their subsistence patterns (whether primarily hunters or gatherers) depending on the wild plant and animal resources available to them.

Bison were not available to those people in the western portion of the Southwest, so they had to rely on varying combinations of small and medium-sized game and plant resources. The San Dieguito–Pinto culture (southern parts of California and Nevada, and southwestern Arizona) and the Oshara culture (southern parts of Utah and Colorado, and northern Arizona and New Mexico) took advantage of a wide variety of animals, such as deer, antelope, bighorn sheep, cottontails, jackrabbits, lizards, rodents, and freshwater mollusks (Cordell 1997:93).

In the more arid portions of the Southwest, big game was less available, and Archaic inhabitants relied to a greater degree on small game and plants. This is evidenced by the extensive use of grinding stones (metates) and smaller projectile points. In contrast, hunting cultures emphasized the production of larger spear points and made less used of seed-grinding stones.

One example of a culture less reliant on hunting is the Cochise culture, which occupied southern Arizona and southwestern New Mexico. The Cochise culture was primarily a sparsely distributed gatherer society, changing locations to take advantage of shifting resource availability. Although they hunted small to medium-sized game (rabbits and deer), they left abundant grinding stones, scrapers, and choppers in areas they inhabited, revealing their emphasis on collecting and grinding seeds and other plant materials (Cordell 1997:110).

About 3,000–3,500 years ago, the Cochise culture obtained corn from Mexico (Cordell 1997:129); with the advent of agriculture, corn and human beings began to domesticate each other. By now, some cultures had also developed more elaborate building structures and were making pottery (another import from Mexico). At this point human populations became more sedentary and left behind more clues about their lifestyles and culture. Archaeologists generally recognize four major post-Archaic cultural traditions in the Southwest: the Mogollon, Anasazi, Hohokam, and Patayan (Cordell 1997:219). These cultural traditions later diversified into more than forty-six recognizable tribes at the time of European contact (Cordell 1997:9). An additional outside influence was added when an Athabascan-speaking culture arrived in the Southwest from the Pacific Northwest just prior to 1500 CE.

Mogollon Culture

The early stages of the Mogollon culture were indistinguishable from the late Cochise culture. These people were both farmers and hunters, and occupied the rugged country in eastern Arizona and southwestern New Mexico, extending down into northern Mexico. The agricultural practices among the Mogollon became more elaborate, and corn was supplemented with beans and squash (300–500 CE). These agricultural developments induced the people to settle down in small villages near the fields they worked. Hunting was done primarily with the atlatl until about 1000 CE, when the bow and arrow came into common use. This improvement in weapons undoubtedly increased their deer-hunting effectiveness. Agriculture remained important, but diets were supplemented extensively with wild fruits, nuts, rabbits, turkey, and deer. Deer provided not only important protein and fat, but also raw materials for clothing, tools, jewelry, and ritual objects.

Evidence from the Grasshopper Pueblo near the Mogollon Rim provides information about how these people hunted and used game animals such as deer in the 1200s. One bone from a deer sternum was unearthed that had a 1-inch stone arrowhead embedded in it (Olsen 1990). The bone material had grown around and encased the projectile point, indicating that this was not a fatal wound. A shoulder blade showed evidence of having been penetrated by an arrow. Many antlers were found in the pueblo that had been shed naturally. Although antlers were used extensively for tools, the fourteenth-century hunters probably talked about and admired antlers as much as we do today.

Both mule deer and whitetail bones were found at the Grasshopper site, but a majority of the 9,525 cervid bones were reported to be mule deer (Olsen 1990).

The remains of at least 997 individual deer were catalogued, making up more than 25 percent of all animal remains found. Deer supplied many useful products, including sinew for bow strings, hide, and material for rings and awls. Antlers were commonly fashioned into "billets" and "flakers," which were used for manufacturing stone arrowheads. Although no leather goods survived the ages, leg bones of deer recovered from the site had cuts and nicks made when the skin was removed. Other cuts around the joints provide evidence that the animals may have been butchered and quartered away from camp and packed back to the Grasshopper Pueblo. Human burials sometimes included ritual artifacts, such as hair pins and wands, and in one case a whole skull from a mule deer buck.

In the later stages of occupation of this site each deer was processed more completely, which probably indicates a growing scarcity of deer. This is supported by tree-ring data indicating that northern Arizona experienced an extended drought between 1335 and 1355.

Anasazi Culture

The Anasazi culture occupied the canyon country from southeastern Nevada to the Rio Grande in New Mexico and from central Arizona/New Mexico northward into southern Utah and Colorado. This culture is thought to have derived from the Oshara culture of the Archaic period. The Navajo term *Anasazi* has been loosely translated as "enemy ancestors" (Reid and Whittlesey 1997). The early Anasazi people lived in pithouses and were known for their exquisite basket making, earning them the name Basketmakers. Later they developed pottery, and the appearance of small, basal-notched projectile points indicates they may have been using the bow and arrow for hunting game (Cordell 1997:193). During this period the Anasazi hunted rabbits and deer extensively, and they kept turkeys in a semidomesticated state, primarily for the feathers (Morris 1980). Deer were valuable for a wide variety of uses beyond meat. Bones and antlers were fashioned into scrapers, awls, and spatulas. Leather from deer hides was tanned with deer brains and used to make clothing, pouches, rope, and sandals.

The mastery of agriculture and the development of improved hunting weapons allowed the Anaszi to swell in numbers and form large population centers living in elaborate masonry structures and cliff dwellings. With the advent of these more permanent structures, they became what we now call Pueblo peoples. The population reached its peak between 1100 and 1250 CE and then declined rapidly; the large population centers were abandoned completely sometime between 1250 and 1450. The reasons for this abandonment are complex; many theories have been proposed, but none adequately explains this phenomenon. Drought appears to explain some cases of abandonment (Lang and Harris 1979), but this relationship was not universal (Cordell 1997). Some Anasazi populations moved south and east into the mountains (Lister and Lister 1983). With this shift, the Anasazi and Mogollon cultures intermingled in what is sometimes referred to as the Sinagua culture. The Hopis of northeast Arizona are thought to be the present-day descendants of the Anasazi/Mogollon cultures (Reid and Whittlesey 1997).

Hohokam Culture

The Hohokam culture of southern Arizona is indistinguishable from the early Mogollon and is either derived from the Cochise culture of the Archaic period or possibly the result of a migration from Mexico (Reid and Whittlesly 1997). *Hohokam* is a Pima word meaning "those who have vanished." The Hohokam began to differentiate from the Mogollon before the year 1 CE (Lister and Lister 1983). They differed in their close reliance on agriculture and the development of an elaborate irrigation system in the Gila and Salt River valleys. They also began to excel in pottery-making. Although their culture was based largely on agriculture, they also hunted small and medium-sized game in the nearby mountains (Fig. 17). With an increase in agriculture, small game attracted to the fields became a more important part of their diet. Deer and rabbits made up a majority of the meat used by the Hohokam (Szuter 1989).

An excellent archaeological record of the Hohokam comes from Ventana Cave near Sells, Arizona, where Hohokam artifacts were found in the substrate of the cave floor. Below the Hohokam artifacts was evidence of occupation by the earlier Cochise culture, complete with deer remains at least 5,500 years old (Haury 1950:138). Underlying this layer were remains of extinct animals found with pro-

Figure 17. Pictograph found near Caborca, Sonora, Mexico. The importance of deer to early inhabitants is shown by their prevalence as subjects of pictographs throughout the Southwest. *Photo by Yar Petryszn*

jectile points. The present-day Pima and Tohono O'Odham are believed by some to be descendants of the Hohokam (Reid and Whittlesey 1997).

Patayan Culture

The Patayan remains the most poorly documented of the four southwestern cultural traditions. Originally called the Yuman culture, the Patayan occupied sites from central Arizona west to the Colorado River valley. With sparse resources in this arid region, these people were undoubtedly very nomadic, which accounts for the absence of evidence for permanent structures (Cordell 1997:212).

Spier (1933) reported that the tribes in western Arizona were primarily hunters of small game, especially jackrabbits. Deer were present in low numbers in the mountains to the south and west, but other food items were more easily obtained. When deer were hunted, they were ambushed at springs or tracked by an experienced tribal member. Small game was cooked by the men, while the women cooked deer and mountain sheep (Spier 1933).

Athabascan-speaking Tribes

Sometime before 1500 CE, the Southwest experienced its second major cultural influx from the Northwest. Newcomers arrived in the Southwest speaking a language related to the Athabascan tribes of northwestern Canada. These newcomers were the Navajo and Apache, which were not recognized as different from each other at the time of Spanish contact.

The Navajo were influenced by the Pueblo people they encountered in northern Arizona, particularly in regard to agriculture. Their agricultural ties were responsible for their name, which is derived from the Tewa word *Navahuu*, meaning "the arroyo with the cultivated fields" (Dutton 1975). With the acquisition of sheep and goats, the Navajo began to abandon their nomadic lifestyle to live closer to the fields they tended.

Despite their pastoral reputation, the Navajo were accomplished hunters. Deer hunts were highly ritualized, with special observances beginning days before the hunt. Several different Navajo hunting styles have been described; all are variations of still hunts, stalking, or driving deer.

There were many rules and taboos regarding the handling and preparation of meat. When a deer was killed, it was turned with its head toward camp before it was field dressed. Deer were usually butchered in the field, and only the meat was packed back to camp, where it was either eaten fresh or dried into jerky. Fresh deer meat from a kill by a mountain lion or wolf was considered a gift from the gods and eaten.

Once distinct from the Navajo, the Apache cultures diversified into many tribes in the 1600s, each relying primarily on hunting and gathering, supplemented by raiding the villages of the agricultural tribes for supplies (Bahti 1973). With the acquisition of horses from the Spanish the raiding intensified, which contributed to the movement and fortification of the early Pueblos. This strife with the Pueblo peoples resulted in the name *apachu,* which means "enemy" in the Zuñi tongue (Dutton 1975).

The Apaches were also skilled hunters, with the men spending most of their time hunting. It has been estimated that 35–40 percent of their diet was meat (Goodman 1987). The hunters quartered and butchered the deer in the field and packed the meat back on horseback. Deer skins provided the raw material for arrow quivers, wrist guards, knife sheaths, pouches, moccasins, bow cases, and sometimes elaborate buckskin clothing (Ferg and Kessel 1987). Hides were processed with tools such as scrapers made from the lower leg bones of deer.

It is clear that deer were a very important part of life to the diverse and widely scattered groups of Native Americans occupying the region from the mid-Archaic period (about 5000 BCE) to recent times. Deer continued to be of great importance to Native Americans as the early Spanish explorers and American miners, military, and cattlemen began to filter into the Southwest.

Exploration Period

The beginning of the Spanish exploration period in the Southwest is marked by the entrance in 1539 of Fray Marcos de Niza and an ex-slave from Morocco named Estevanico Dorantes. They traveled northward through southeastern Arizona, reaching west-central New Mexico, and returned after Estevanico was killed outside one of the Zuñi towns (Riley 1997:5). The next year, enticed by tales of great wealth in this new land, Francisco Vásquez de Coronado and his entourage of more than 1,000 men began their famous exploration of the Southwest. Upon reaching the Zuñi Indian pueblos in northwestern New Mexico, they found no riches and so smaller subgroups were sent west to the Grand Canyon and northeast onto the Great Plains. In the spring of 1542 they returned to Mexico without riches, but with the experience of traveling the Southwest for two years (Cordell 1997:429).

Not until 1582 did the Spanish again make a foray into the American Southwest. In that year, Don Antonio de Espejo and fourteen soldiers explored much of Arizona and New Mexico and established several valuable mines in northwestern Arizona (McGregor 1965:34). Some Spanish settlement in the Southwest continued throughout the 1600s and 1700s, with most occurring in the Rio Grande Valley of central New Mexico and the Santa Cruz Valley of southern Arizona.

A Jesuit priest, Father Eusebio Francisco Kino, roamed much of present-day Arizona around 1700, preaching to native people. Father Kino's copious diaries left us much ethnographic information, but decidedly little on deer abundance or distribution (McGregor 1965:36).

Information on wildlife and their distribution was not recorded consistently until the 1820s, when trappers began to trap beavers from southwestern river systems. The most famous of these men was James Ohio Pattie, who made three trips along the Gila, San Pedro, and San Francisco Rivers. Although there are some serious discrepancies in Pattie's accounts (Batman 1984), his observations provide clues to the abundance and distribution of wildlife at that time.

Some settlement was occurring in the upper Rio Grande area of northern New Mexico, but unresolved and relentless conflicts with the Apaches stifled the settlement of much of the Southwest until the mid-1800s. It wasn't until the late 1840s that this region saw any significant occupation. At that time military forces were dispatched to the Southwest to secure a travel route to California. Among the first American military expeditions through the Southwest was that of Stephen Kearny and William H. Emory who, in 1846, were charged with the mission of producing the first detailed map of the land between the Rio Grande and the Pacific Ocean (Davis 2001:27). Emory recorded wildlife information and distribution in detail for the first time. His notes on deer distribution 150 years ago are not unlike the diary of a present-day deer hunter, finding abundant deer sign, but amazed at how shy the deer were.

With the discovery of gold in California in 1849, the stage was set for a flood of settlers passing through southern New Mexico and Arizona en route westward. Those travelers who kept diaries of the trip recorded deer as abundant and as important additions to their rations in some cases. In other cases, parties were eating their horses to survive because of the lack of wild game (Mearns 1907:182, Longhurst et al. 1952:10, Hancock 1981).

In that same year a boundary survey was commissioned to establish the official United States–Mexico boundary (Davis 2001:59). This and successive boundary surveys provided the first systematically collected wildlife information from the Southwest. Boundary surveys continued—some haphazardly—under various commanding officers until 1854 when the Gadsden Purchase was signed. This purchase by the United States of the land south of the Gila River established the present international border.

With this purchase there was an urgent need to accurately survey the new international border. A survey commission was established and placed under the direction of William Emory, accompanied by an army doctor and naturalist, Dr. C. B. R. Kennerly. The survey was completed in little more than a year, with Kennerly carefully documenting the status of the habitat, as well as the distribution and abundance of various forms of wildlife. Dr. Kennerly was one of the first to differentiate between whitetails and mule deer. He noted particularly that white-tailed deer were "observed in the valleys of all the streams passed by us in our journey from El Paso to Nogales, as well as in the various mountain ranges, particularly the San Luis and Sierra Madre. In the valley of the Santa Cruz River and the adjacent country we found them in such numbers as to influence the belief that a few skilful [sic] hunters might have supplied our entire party with fresh meat" (Baird 1859:50).

During this time of the boundary surveys, others were traversing the Southwest in search of the best transportation routes for railroads and wagon roads. Expeditions by Capt. Lorenzo Sitgreaves, Lt. Amiel W. Whipple, and others through northern Arizona and New Mexico recorded deer as "plenty" and "abundant" around the San Francisco Peaks, Mount Sitgreaves, and Bill William's Mountain (Davis 2001:91).

Although there is some contradiction regarding the abundance of deer in the Southwest at the time of exploration and settlement, most records speak of a relative abundance of deer. Although "abundant," venison in the stew pot was by no means assured; indeed, it appears the same effort was required to harvest a deer 100 years ago as it is today. Reports indicate deer became less abundant as one traveled southwest into the drier desert areas. Deer populations throughout the Southwest undoubtedly fluctuated substantially throughout the several decades of exploration, just as they do today.

With better maps and the establishment of travel routes, immigrants began to trickle into the Southwest after the mid-1850s. Most were miners, ranchers, or men associated with one of several military posts. In 1858 a young man named Phocian Way came to make his fortune in the mines of the Santa Rita Mountains in southern Arizona. Way hunted at every opportunity and shot his first deer on July 25, 1859, later recording in his diary: "I neglected to state that it is customary in this country to set up with a man the night after he has killed his first deer, and make him treat the party present to a gallon of whiskey. I was fortunate enough to escape this ordeal as most of our company were in Tubac and there was no whiskey to be had without travelling 20 miles for it" (Way 1960:359–60).

In 1864 Fort Whipple, near Prescott, Arizona, was established and a twenty-one-year-old army surgeon named Elliot Coues was among one of the first groups of military personnel to be stationed there. In Coues's day it was quite common for frontier doctors to be interested in natural history. Coues was an exceptionally acute observer of wildlife and already began to see the effects of an ever growing human population on the natural resources of that new state: "Both naturalists and hunters distinguish two species of deer in Arizona, called the Black-tailed and the White-tailed. Of these the former is by far the most abundant and characteristic; although, judging from accounts formerly given of it, it has considerably decreased in numbers owing to the persecution to which it is subjected so constantly from both the native tribes and the white settlers. It is . . . also called the Mule Deer, from the length of its ears. . . . This deer forms no small share of the food and clothing both of the Indians and white settlers" (Davis 2001:172).

Venison was important to the early settlers and provided a much needed supplement to their scarce rations. This is illustrated by Phocian Way, writing in 1859: "We have been out of fresh meat for several days and we cannot purchase any without losing so much time and going a long distance, and then we would probably have to pay an exorbitant price. We had better kill our own meat if we can"

(Way 1960:288). With the increasing pressures of settlement and overexploitation, deer became less abundant. The scarcity of deer was more pronounced near the major areas of human occupation because of the lack of harvest restrictions, accelerated by the commercial sale of venison to nearby towns, mining camps, and military posts.

In 1892 Edgar A. Mearns began his second tour of duty in Arizona as a medical officer/naturalist assigned to yet another United States–Mexico International Boundary Commission. Mearns collected more than 30,000 specimens along the way, but published only a fraction of this information in a 500-page book titled *Mammals of the Mexican Boundary of the United States*. In his book he stated, "The whitetail's range is mainly south of the Gila, where it is still abundant though its numbers are decreasing" (Mearns 1907:182). Mearns specifically noted that this decline in deer populations was concurrent with the rising human population in Arizona.

Unregulated hunting was not the only negative impact brought about by the increasing human population. The arid ranges of the Southwest were stocked with cattle and sheep in great numbers during the late 1870s (Cooper 1960). A period of intense overgrazing ensued, causing unprecedented range destruction and landscape-scale vegetation changes (Allred 1996). The abuse continued until a two-year drought, starting in 1891, caused massive die-offs of cattle on the overstocked ranges (Bahre 1991). These changes were described in 1901 by C. H. Bayless, who had owned a large ranch north of Tucson, Arizona, since 1886:

> The present unproductive conditions are due entirely to over-stocking. The laws of nature have not changed. Under similar conditions vegetation would flourish on our ranges today as it did fifteen years ago. We are still receiving our average amount of rainfall and sunshine necessary to plant growth. Droughts are not more frequent now than in the past, but mother earth has been stripped of all grass covering. The very roots have been trampled out by hungry herds constantly wandering to and fro in search of enough food. The bare surface of the ground affords no resistance to the rain that falls upon it and the precious water rushes away in destructive volumes, bearing with it the lighter and richer particles of soil. That the sand and rocks left behind are able to support even the scantiest growth of plant life is a remarkable tribute to our marvellous climate. Vegetation does not thrive as it once did, not because of drought, but because the seed is gone, the roots are gone, and the soil is gone. (Bahre 1991:112)

It is not hard to envision what impact this widespread habitat destruction had on the deer populations that rely on the forbs and browse.

Deer populations at the turn of the twentieth century probably reached their lowest point in historic times. Those species that proved themselves the most "useful" to early settlers were in danger of being lost from the landscape due to rampant overuse with complete disregard for the future (Mearns 1907:29–38). Luckily this abuse of the nation's natural resources did not go unnoticed. Some enlightened individuals understood that the present rate of harvest would not last long, and they rallied against the overexploitation of the big game herds. The first step in saving the large game animals from extermination was to stop the uncontrolled harvest and begin managing the exploited animals for future generations.

History of Deer Management in the Southwest

The uncontrolled use of wildlife resources came to an end very slowly. It took several decades to stop the momentum and turn things around. As early as 1852 California closed the deer season for six months of the year. This was necessary in California at this early date because many of the people who flooded into that state to find fortune in the gold mines soon found that they could make more money as market hunters than as miners (Longhurst et al. 1952:13). Others states (Nevada, 1861; Colorado, 1867; Utah, 1876) followed suit with closed deer seasons to protect deer during the spring and summer months. The slaughter of bison on the Plains brought to light the issue of wasting game meat. In 1874 Colorado established a $25 fine for unlawfully killing or wasting the meat from bison, deer, elk, mountain sheep, or antelope (Barrows and Holmes 1990).

The 1880s saw several fledgling attempts to begin conserving deer as a renewable natural resource. During this decade New Mexico restricted the deer season to six months (1880), and California initiated buck-only hunting (1883). The beginning of wildlife management was marked by the establishment of fish commissions in Arizona (1881) and California (1878). At that time the commissions were only concerned with the promotion of fishing opportunities, but the management infrastructure would be carried through to the present. Fish commissions eventually gained authority over the game animals, marking the beginning of what can be considered the era of management. The first game and fish regulations in Arizona occupied less than one page of the Arizona Revised Statutes (Table 2). Except for their brevity and lack of complexity, the 1887 regulations resemble today's, with restricted seasons (October 2 to January 30), restricted methods of take, and total protection of certain species. Although most states had not yet introduced bag limits and buck-only hunting, these early restrictions formed the nucleus around which current hunting regulations developed.

In 1893 California reduced its deer season drastically, to six weeks, and Arizona banned the transport of game out of the territory by "common carrier" (Table 3). This latter law was an attempt to halt the mass shipments of native wildlife to markets in larger population centers. New Mexico banned the killing of deer "for hides" (1895), prohibited the sale and export of game (1897), outlawed the killing of does (1899), and finally established a bag limit of one deer (1899).

Arizona was already expanding its conservation infrastructure by 1897, with the addition of twelve assistants to the three fish and game commissioners; these assis-

TABLE 2. FIRST ARIZONA GAME LAWS, PASSED IN 1887 (IN THEIR ENTIRETY)

Title XVI   1887
Preservation of Game and Fish

Section 993.	It shall not be lawful for any person or persons to take, kill or destroy any elk, deer, antelope, mountain sheep, mountain goat, or ibex in the territory of Arizona at any time between the first day of February and the first day of October in each year.
Section 994.	It shall not be lawful for any person to buy, sell or have in his or her possession any of the game animals enumerated in the preceding section within the time the taking or killing thereof is prohibited, except such as are tamed or kept for show or curiosity.
Section 995.	It shall be unlawful for any person to take or catch any fish in any stream, lake or river, pond or pool in the territory of Arizona with any seine or net.
Section 996.	It shall be unlawful for any person in the territory of Arizona to shoot or kill any partridge, wild turkey, goose, brant, swan, curlew, plover, snipe, rail, or ducks of any kind between the first day of March and the first day of September of each year, except on his own premises. And it shall be unlawful for any person or persons at any time to trap or net any quail, partridge, wild turkey, grouse, or prairie chicken in this territory, except on his own premises.
Section 997.	It shall be unlawful for any person or persons in the territory of Arizona to buy, sell, or have in his or her possession any of the game birds named in the preceding section within the time of killing or shooting thereof is prohibited, except such thereof as may be tamed or kept for show or curiosity.
Section 998.	It shall be unlawful for any person or persons at any time to take, kill or destroy any fish with giant powder or any other explosive substance.
Section 999.	It shall be unlawful for any person or persons at any time within five years after the passage of this act, to shoot or kill any grouse or prairie chickens in the territory of Arizona.
Section 1000.	Any person violating any of the provisions of this title is guilty of a misdemeanour, and shall be fined in a sum not exceeding one hundred dollars, nor less than fifteen dollars, and the costs of prosecution, and in case such fine is not paid the person or persons so convicted shall be imprisoned in the county jail until said fine is paid; provided, such imprisonment shall not exceed one day for each dollar of such fine.

tants acted as volunteer game wardens. In that year, the commission enacted the first buck-only law in that state, allowing the killing of male deer only from August 1 to December 15. Game management was evolving slowly to incorporate what was known about the biology of the species: that there are surplus males that can be removed without affecting the reproductive potential of the population. An important addition to the Arizona game laws that year was the prohibition of selling game meat, which curtailed the practice of selling the meat to local markets in cities, mining camps, and military installations.

TABLE 3. HISTORICAL TIMELINE OF SIGNIFICANT CHANGES IN SOUTHWESTERN DEER MANAGEMENT

Year	Event
1852	California legislature closes deer season for 6 months per year in 12 counties
1861	Nevada sets first deer season for July 1–December 31
1867	Colorado closes deer season January 15–August 15
1874	Colorado establishes $25 fine for unlawfully killing deer or "wasting" deer meat
1876	Utah establishes 6-month season for big game
1878	California Board of Fish Commissioners assumes authority over game animals
1880	New Mexico outlaws killing of deer between May 1 and September 1
1881	Arizona establishes a three-member "Fish Commission"
1882	Colorado Game and Fish Protective Association formed
1883	California bans killing of does (a "buck-only" law)
	Texas establishes a 5-month closed season for deer, but also exempts many counties from game protection laws
1887	Arizona Fish Commission receives authority to manage game (first game laws)
1888	Colorado State Fish and Game Commission formed (citizen group)
1891	Colorado sets first bag limit of 5 deer and outlaws transportation of game meat by common carrier
1893	Arizona bans transport of game out of the territory
	California reduces deer season to 6 weeks per year
1894	Mexico passes legislation restricting take of wild game
1895	New Mexico restricts hunting to 3 months and bans killing of deer "for hides"
	California legislature appoints county wardens for $50/month
1897	New Mexico bans market hunting and prohibits sale and export of game meat; governor is given authority to appoint wardens
	Arizona institutes a "buck-only" law and bans sale of game meat; 3 commissioners now have 12 assistants
1899	New Mexico establishes bag limit of 1 buck and bans killing of does
1901	California allows hunting for 2 months, establishes annual bag limit of 3 bucks, and prohibits sale and export of deer meat and hides
	Nevada closes deer season for 2 years
1903	Arizona deer season reduced to 2½ months, with bag limit of 1 buck per day and 3 per season
	New Mexico establishes Territorial Department of Game and Fish with a territorial game warden. Only 1 buck may be killed and it must be in November or December
	Colorado requires a hunting license for the first time
	Texas sets bag limit on deer at 6 and outlaws selling of game meat and skins
	Nevada restricts deer season to 2 months or less, depending on county, and restricts doe harvest
1905	Arizona requires hunting license for nonresidents
	New Mexico season is moved to September 15–October 31
	California reduces annual bag limit to 2 bucks
1907	California requires hunting license and uses that money for law enforcement
	Texas adopts bag limit of 3 bucks per year and reduces deer season to 2 months (November–December)
1908	Utah closes season for 5 years

TABLE 3. (*CONTINUED*)

Year	Event
1909	New Mexico creates "Game Protection Fund" supported by big-game license fees ($1 residents, $25 nonresidents)
	California Fish and Game established with 3 commissioners
	Nevada requires hunting licenses
	Texas adds game responsibilities to existing Fish Commission and sells hunting licenses to pay for enforcement of game laws
1912	Arizona and New Mexico become states
1913	Arizona names first state game warden
	Arizona resident hunting licenses are required and a "Game Protective Fund" is established
	Arizona institutes bag limit of 2 bucks per year
	New Mexico establishes first chapter of Southwestern Game Protective Association
	Nevada establishes first state game reserve
	Utah reopens deer hunting season but limits harvest to bucks with antlers 5 inches or more
1915	Texas outlaws use of deer calls for "calling or attracting the attention of any deer"
1916	Arizona Game Code initiative passed limiting open season to 2 months and reducing bag limit from 2 bucks to only 1
	New Mexico Game Protective Association is formed with Aldo Leopold touring southern New Mexico
	Mexico establishes Departmento de Caza y Pesca (Department of Game and Fish)
1917	Utah establishes its first game refuges
	Mexican Constitution declares all wildlife to be property of the "nation"
1919	California bans harvest of spike bucks
	Texas hires 6 game wardens to patrol entire state
1921	New Mexico establishes a 3-member commission
1923	Arizona Game Protective Association is formed
	Colorado establishes its first antler point restriction by making legal only bucks with 2 or more points on each antler
1925	New Mexico Game and Fish Commission is given authority to hire state game warden
	Nevada restricts harvest of bucks to those with 2 or more points
	Texas outlaws hunting with dogs and artificial light
1928	Arizona voters repeal State Game Code of 1916 to prepare for a more flexible system whereby Game and Fish Commission has authority to set game laws annually
1929	Arizona Game and Fish Commission acquires authority from legislature to regulate hunting and fishing
	Nevada requires deer tags for the first time
	Texas restricts hunting of mule deer ("black tails") west of Pecos River to the last 2 weeks in November
1931	New Mexico Game and Fish Commission acquires authority from legislature to regulate hunting and fishing
	New Mexico begins hunting does to control some populations
1934	Mexico outlaws sale of meat and hides of white-tailed deer
	Utah issues first doe permits to control some deer populaitons

(*continued*)

TABLE 3. (*CONTINUED*)

Year	Event
1937	United States passes Federal Aid in Wildlife Restoration Act (Pittman-Robertson)
	Colorado legislature passes bill establishing 6-member commission
1938	Colorado establishes first deer and elk check stations
1939	Colorado establishes first doe hunt to control overpopulation of deer and approves archery as a legal method of take
1940	Mexico passes first federal game law (Ley Federal de Caza)
1941	Arizona reaches maximum number of game refuges (81)
	Nevada establishes county big-game committees
1946	Arizona requires tags for hunting big game
1949	Arizona authorizes "any-deer" hunts in some areas to control increasing deer populations
1951	Utah institutes "either-sex" hunting to control increasing deer populations
1952	Colorado uses Game Management Units to manage deer
	Mexican federal game law revised by Mexican Game Law of 1952
1953	New Mexico authorizes take of does on wider scale to control increasing deer populations
1954	Colorado holds first statewide archery season
1955	Arizona begins special bow hunts
1958	Arizona deer management zones I, II, and III are dissolved in favor of 36 GMUs with new numbering system (still in use), beginning of new "wildlife manager" system
1959	Colorado outlaws use of artificial light to hunt game
1963	Colorado legalizes muzzle-loading weapons for deer
1969	Arizona discontinues "any-deer" hunts except in specific management areas
1971	Arizona North Zone/South Zone deer management format abandoned in favor of statewide permit system for all deer hunts in 55 GMUs
1974	Arizona legalizes any buck with antlers rather than "fork-antlered" bucks only
	Utah discontinues widespread doe hunts
1975	New Mexico restricts harvest with a 16-day season in 3 hunt periods and discontinues widespread doe hunts
1976	Nevada implements quota system for all deer tags in state
1978	Arizona establishes first designated muzzle-loader hunt
1983	New Mexico discontinues doe hunts in almost all areas
	Texas passes Wildlife Conservation Act, shifting wildlife conservation authority in all counties to Texas Parks and Wildlife Department
1986	Arizona issues "left-over" tags that can be used to harvest a second deer
1988	Texas deer season in Panhandle and Trans-Pecos areas is increased from 9 to 16 days
1992	Arizona initiates "juniors-only" deer hunts
1994	Utah Divison of Wildlife Commission reduces number of deer permits to 97,000 (only 53% of 180,000 permits issued in 1992)
1999	Colorado begins permitting all deer hunts statewide
2000	Mexico revises structure of wildlife administration with new general law of wildlife (Ley General de Vida Silvestre del 2000)
2004	New Mexico commission adopts "3 point or better" antler point restrictions statewide

Americans were starting to realize that the only way to save wildlife was to build a constituency that would fight against unregulated destruction in favor of conservation through regulated use. These early laws were a good start, but without effective enforcement they were nothing more than words. Without adequate money and more personnel, these new game laws could not be enforced. With the drought of the 1890s in full swing and unregulated harvest continuing because of a weak law enforcement presence, deer populations continued to decline.

At the turn of the twentieth century, California instituted some changes, establishing an annual bag limit of three bucks to be taken in a two-month season and prohibiting the sale and export of deer meat and hides (1901). Arizona followed suit two years later and instituted a bag limit for the first time—only one buck per day and no more than three per season (September 15 through December 1). During this year (1903) the New Mexico Territorial Department of Game and Fish was established with a territorial game warden. This department became the New Mexico Department of Game and Fish.

In 1905 the game laws in Arizona underwent a major overhaul. Section 1 of chapter 25 of the Arizona Revised Statutes that year stated that all previous game and fish laws "are hereby expressly repealed" and replaced with a new set. Section 2 clarified an extremely important point, which is the cornerstone of modern wildlife conservation policy: "All game or fish now or hereafter within this Territory not held by private ownership legally acquired and which for the purpose of this Act shall include all the quadrupeds, birds, and fish mentioned in this Act, are hereby declared to be the property of the Territory, and no right, title, interest or property therein can be acquired or transferred, or possession thereof had or maintained except as herein expressly provided." As property of all citizens of the territory (and later, the state), wildlife could not be owned by anyone because it was owned by everyone.

New Mexico wildlife also belonged to the citizens of the state, but New Mexico law provided for an unusual exception. Based on political pressure, the new state legislature passed the Class A Park and Lake Licensing Act in 1912. This act allowed private ranches that were licensed under the act to actually own the fish and wildlife on their land. This system has been changed somewhat, but still allows landowners to own native wildlife as if it were livestock.

It was becoming clear that wildlife laws in and of themselves were not going to be effective unless there was a way to raise money and hire personnel to enforce them. The solution to this came in the form of the hunting license. Territories and states began to require the purchase of hunting licenses; income from license sales was then used to run the developing system of game management. Starting in 1905, anyone who was not a resident of Arizona Territory was required to purchase a nonresident hunting license, the proceeds from which could be used by the Game and Fish Commission to carry out its duties. California (1907), Nevada

(1909), and New Mexico (1909) started to require resident hunting licenses and established a "Game Protective Fund" to deposit and disburse these funds.

In 1913 the newly appointed Arizona state game warden was responsible for collecting and recording all money received from hunting licenses in the state. This money went into the state's Game Protection Fund, which then paid the warden's salary ($1,800 per year) and related expense. It also covered the costs of transport, propagation, and release of game and fish in the state, and all printing, postage, and office supplies. For the first time, residents of the state who wanted to hunt were required to purchase a hunting license. A resident hunting license sold for 50 cents, a nonresident was charged $25, and "aliens" paid $100 (Sec. 670). Duplicate licenses would set you back an additional 10 cents. Section 661 of the game codes clearly stated that if a posse had to be formed to catch a chronic game violator, the violator was liable for the total cost of the posse. This was the beginning of the highly successful system of wildlife management that was developing throughout the country, whereby money is generated for wildlife by the regulated use of a few species.

The public gradually became aware of the plight of game animals, and there was a slow but steady increase in concern over the future of the state's big game species. Concerned citizens began to gather in local groups called "game protective associations." The first such group was established in 1913 in Silver City, New Mexico, out of concern that politics was getting in the way of proper wildlife conservation. Local game protective associations were established in Arizona starting in 1916 and were consolidated into the Arizona Game Protective Association in 1923.

An increasingly interested public was growing discontented over the inflexibility of regulations that were at that time governed by the legislature. As was seen in state after state, wildlife management cannot operate properly in the political arena. These groups held regular meetings and began to lobby hard for the separation of wildlife management and politics, and they made some remarkable headway in early wildlife conservation in the Southwest. They pushed through the passage of laws that prohibited the legislature from diverting wildlife funds, established the presence of game commissions to make decisions, and transferred game law authority from the governor to the commission.

It was during this time period that the experience on the Kaibab Plateau in northern Arizona underscored the necessity of biologically based and flexible deer management programs. No discussion of the history of wildlife management in general, or southwestern deer management specifically, is complete without an examination of the lessons learned on the Kaibab Plateau.[8] The Kaibab became the center for deer management knowledge for early biologists all over the country. "The lesson of the Kaibab Plateau" is still taught, mostly incorrectly, to every university student majoring in wildlife management throughout the country.

---

8. The term *Kaibab Plateau* is understood to include not only the plateau proper, but also the adjacent lower elevation intermediate and winter range.

The story of the Kaibab starts in 1906, when after a visit to the area, Theodore Roosevelt created the Grand Canyon National Game Preserve. This preserve contained more than 1 million acres, encompassing the entire Kaibab Plateau and surrounding lands. Immediately after the establishment of the game preserve, an intensive predator control program was implemented; between 1905 and 1931, a total of 781 lions, 4,849 coyotes, 30 wolves, and 554 bobcats were reported to have been removed (Mann and Locke 1931).

The deer population increased dramatically and peaked in 1924 before crashing to a level that was probably well below the original population. The habitat was severely overbrowsed, and the deer population collapsed under its own weight. This predator-prey correlation has been used as an example of what happens when you remove predators. The interrelationships of ecosystems are rarely this simple, however, and other changes were taking place during those years on the Kaibab.

By presidential proclamation, no deer hunting was allowed from 1906 to 1924. Although removing only bucks will not keep a deer population from growing, the establishment of the game preserve also stopped the harvest of female deer. Fires were not suppressed as effectively as they are today, and this natural disturbance altered habitat in such a way that it favored species such as deer that thrive on disturbed habitats (brushy cover and browse with abundant weeds).

During this period, the amount of livestock grazing on the plateau was also greatly reduced. Mann and Locke (1931) reported 20,000 cattle and 200,000 sheep grazing the Kaibab Plateau and surrounding desert country in 1887–89. At the turn of the century, there were also several dairies and large numbers of horses on the same range. By 1924 grazing pressure had been reduced to about 4,000 cattle and 3,500 sheep. The earlier numbers were only rough estimates and doubtless exaggerations, but there is no doubt the intensity of grazing was reduced substantially during this period. The reduction of grazing pressure left more forage available for deer, further aiding the build-up of the deer population.

By 1918 the vegetation was showing signs of damage from overbrowsing by the burgeoning deer population. Inspections of winter range for several years after this returned reports of a rapid deterioration of the range. Conditions worsened until Henry C. Wallace, secretary of agriculture, assembled a special investigative committee of experienced game biologists from around the country to report on the condition of the deer herd and its habitat. This committee entered the Grand Canyon National Game Preserve on August 16, 1924, and reported that "the conditions of forage throughout the Preserve can only be characterized as deplorable, in fact they were the worst that any member of the Committee had ever seen. When not only the leaves but the annual growth of trees, bushes, shrubs, and grass are so closely cropped that seeding is impossible the condition of the range moves swiftly towards utter destruction" (Cutting et al. 1924:20). The final report issued by this committee contained several recommendations aimed not only at preserving the Kaibab deer herd in a healthy condition, but also at halting and restoring the range damage from overbrowsing. The report stated the committee's belief that "as an immediate remedy for the present situation no reduction of less than 50% of the existing deer herd would be effective. We, therefore, recommend that one half of the existing herd be removed and that this removal be accomplished as quickly as possible" (Cutting et al. 1924:23).

Curiously, the committee's first recommendation for accomplishing this was to trap deer and move them to other deer habitat elsewhere (Russo 1964:51). This recommendation is understandable when placed in context, since most other mule deer populations were at low levels at that time. A trapping program was started in 1925, but no deer were trapped. In 1926 eighty-five trapped deer were shipped off the Kaibab, followed by twenty-four the next year. Obviously, this method never contributed to a reduction in the deer population.

The committee also recommended opening the area to hunters as the only efficient and effective way to solve the problem. The report also suggested, as a last resort, that government officials directly shoot deer to reduce the population, though it acknowledged that "the official killing of game when not absolutely required by the failure of other methods is contrary to all existing principles and theories of American Game Conservation and would not recommend the adoption of such a plan except as a last resort." (Cutting et al. 1924:31). A policy of hiring paid professional hunters instantly converts wild deer from a valuable natural resource to a public liability.

Almost immediately after the committee's final report in October 1924, the Forest Service announced it would open a hunt on November 15, 1924. Arizona Gov. George Hunt heard about the Forest Service's plans to open a hunt and was vehemently opposed to federal control of the deer herd. The hotly contested issue of jurisdiction is clearly illustrated in an undated status report written by state game warden G. M. Willard: "The long prevailing idea that the deer were the property of the federal government as long as they remained within the reserved area, but the property of the state whenever they stepped over an invisible border line is, to put it mildly, a legal absurdity. Title cannot be made to depend upon which way the deer jumps—whether to one side or the other of an invisible line" (U.S. Forest Service files). To maintain control of the situation, Governor Hunt sent the Coconino County sheriff to arrest anyone leaving the Kaibab Forest with a deer because the state of Arizona had not opened a deer hunt on the Kaibab. In five days, about 200 hunters shot one deer each before the arrest of three hunters on November 20 by the sheriff put an end to the hunting. On November 28 an agreement was reached between game warden Willard and R. H. Rutledge, Forest Service ranger, allowing hunters to harvest up to three deer each from December 1, 1924, to January 5, 1925.

Some concerns were expressed over this hunting season and the effect it might have on the famous deer drive that was planned for December 1924 (Carmony 2002). This drive was part of the plan to relieve the range of deer. The plan was to sweep the plateau with a line made up of men on foot and horseback, and physically move thousands of deer off the Kaibab Plateau, across the Colorado River, and up the south rim of the Grand Canyon. A group of 125 enthusiastic drivers pushed through the deer habitat in a large drive line; the deer were driven from the men, but not in any consistent direction. The drive was abandoned when it was evident there were many more deer behind than in front of the drivers (Russo 1964, Carmony 2002).

An early snow in the fall of 1924, following a dry summer, caused massive winter kill as weakened deer were forced onto poor winter range earlier than usual. Another cooperative agreement to hunt the second year was reached, but the meager number of deer removed was inadequate to protect the deer herd from itself.

In 1925 a program of fawn rearing was instituted in yet another attempt to remove deer from the Kaibab (Russo 1964:51). Fawns were captured and raised for transplanting to other locations. Many fawns died in captivity, and only about 100 were transplanted per year before the program was discontinued in 1930.

Annual agreements between state and federal authorities allowed hunts to continue from 1925 to 1927, but the harvest was woefully inadequate to reduce the deer population sufficiently, and the deer herd incurred serious and obvious losses each winter. The road from the nearest sizeable city (Flagstaff, Arizona) to the Kaibab was 170 very primitive miles and took several days to complete. This remoteness, and the fact that the Colorado River had to be crossed by ferry, resulted in much less interest in deer hunting on the Kaibab than there is today. The intense predator control program continued throughout this multiyear deer die-off and the concurrent effort to drastically reduce the number of deer on the range. Wildlife officials were spending time and money organizing and administering fawn-rearing projects, deer "drives," deer trapping, legal hunting, and government shooting, and yet thought nothing of the ongoing program to remove as many deer predators as possible. During the years of the worst episodes of massive deer starvation (1924–30), 107 lions, 1,849 coyotes, 19 wolves, and 434 bobcats were removed from the Kaibab Plateau, including the Grand Canyon National Park (Mann and Locke 1931). This continued removal of predators seems perplexing in hindsight, but at the time predator control was being driven primarily by the livestock interests on the plateau (Young 2002:193).

In 1928 a lawsuit reached the U.S. Supreme Court to settle once and for all the issue of who had jurisdiction over the Kaibab deer herd. The Supreme Court ruled that the Forest Service had the right to protect the range, which it was responsible for managing (Trefethen 1967). That winter, the Forest Service authorized government shooters to remove deer, but the public outcry was so intense that this practice was discontinued the next year.

It was becoming increasingly obvious that wildlife management controlled by the politicians was not working. The 1924 committee report on the condition of the Kaibab herd and habitat (Cutting et al. 1924:28) foresaw this problem and prophetically stated, "The scientific management of game by a special authority such as a commission should have the power to change regulations from year to year in order to meet the changes in the local situation. The special authority charged with Game Management should have sufficiently elastic powers to control not only the open seasons, the shooting areas, the refuges, and the bag limits, but also the absolute numbers which may be killed in any given locality each season. Upon such principle of Game Management does the future of much of our wild life depend."

This recommendation became a reality, but only through a few legal maneuvers. Since the Arizona State Game Code of 1916 had been passed by initiative (not by the legislature), it could not be repealed except by a vote of the people. In 1928 the Arizona Game Protective Association was successful in getting the legislature to place a referendum measure on the ballot that would repeal the existing rigid game code so it could be replaced with a more flexible one. The referendum passed. In the spring of 1929 the Arizona legislature passed a new state game code, which established a three-person commission that had the authority and flexibility to set bag limits and season lengths annually.

The various estimates of Kaibab deer abundance throughout this period of population increase and crash are poorly supported and represent mere guesses, sometimes made by individuals who spent less than two weeks there. After scrutinizing the methods by which these population estimates were derived (or lack thereof), a careful reader has no choice but to discard these oft-repeated numbers for a more general discussion of relative abundance.

Because many factors were involved, it is difficult to assess what affected the increase in deer numbers preceding the die-off. Given what we know about other ungulate population irruptions, it seems likely that habitat conditions (increased precipitation, reduction of livestock grazing, fire) played the primary role in the initial build-up of the deer population. By most accounts it appears that the Kaibab herd had been reduced enough by 1934 to allow a recovery of the browse. With the help of several wet years in the late 1930s (McCulloch and Smith 1987: fig. 6), deer body condition improved, and browse plants showed vigorous annual growth (Mann 1941). Lessons learned on the Kaibab—both biological and political—set the stage for the further refinement of deer management throughout the West.

Concurrent with the Kaibab situation was an almost identical, though less famous, deer population irruption in the Black Canyon Game Refuge on the Gila National Forest in western New Mexico. Black Canyon and the Kaibab illustrated that a large, inviolate refuge was not the answer. A larger number of smaller refuges, it was thought, would act as areas of propagation where deer populations could build up and "overflow" into the surrounding areas that were hunted (Flader 1974:63). Game and fish commissions of the day established statewide systems of refuges based on that premise. In Arizona, the commission started with twenty-three new refuges, totaling more than 2 million acres; the number of refuges peaked at eighty-one in 1941. Eventually, wildlife officials realized that these small and widely scattered game refuges did little for game populations on a large scale, and they were abandoned.

In 1937 wildlife management in America reached a turning point with the passage of the Federal Aid in Wildlife Restoration Act. An excise tax on guns and ammunition had been collected since 1932, but that money was funneled into the general funds and not earmarked for any specific purpose. When Congress was discussing discontinuing this tax in 1937, conservationists requested it be retained and used for wildlife restoration. Sen. Key Pittman and Rep. A. Willis Robertson sponsored a bill to collect an 11 percent excise tax on all guns and ammunition. This proved very popular, and the Federal Aid in Wildlife Restoration Act (also known as the Pittman-Robertson Act) passed with very little opposition. Later additions to this act added pistols and archery tackle to the list of taxable items.

Today, this tax generates over $100 million each year, which is collected by the U.S. Fish and Wildlife Service and distributed back to the states based on the number of licensed hunters and the size of the state. Since 1937, this tax has raised over $4 billion for wildlife conservation, allowing state wildlife agencies to hire professionally trained wildlife biologists to conduct annual abundance surveys and long-term research to properly manage the wildlife of the state. With this money state agencies rapidly obtained research knowledge and management experience, and began running down the road to modern wildlife management.

At this point in the history of southwestern deer management each state began to evolve along a slightly separate path. The different trajectory that is so obvious today among state wildlife agencies was shaped by differing local politics and cultural values, diverse habitat and deer population responses, and sometimes for no other reason than that state agencies operated independently.

History of Wildlife Management in Mexico

Unfortunately, establishing an effective wildlife administration infrastructure in Mexico has been a much more difficult task than in the United States. Mexico passed legislation as early as 1894 restricting the take of wild game (Federal Forest Law). However, like the United States at that time, it lacked financial resources to enforce those laws (Leopold 1959). The establishment of the Departmento de Caza y Pesca (Department of Game and Fish) in 1916 mirrored similar developments north of the border at that time and provided the infrastructure (but still not the funds) to administer wildlife conservation.

The Federal Aid in Wildlife Restoration Act of 1937 was starting to generate funding for research and management in the United States, but these benefits did not extend south of the border. In 1940 the first game law was passed (Ley de Caza), establishing a legal foundation for wildlife administration in Mexico. This law reiterated the declaration in the 1917 Mexican constitution that all wildlife was to be the property of the nation. Establishing wildlife as property of the citizens at the federal level differs from the situation in the United States, where wildlife is held in public trust at the state level. This game law was revised by the Federal Game Law of 1951 and stood for nearly fifty years as the main authority for the conservation of wildlife in Mexico. Several proposals have been advanced to improve the wildlife administration in Mexico (Alessio-Robles 1959, Beltran 1966); however, none has so far been adopted.

One problem hampering the administration of wildlife conservation in Mexico has been the frequent shifting of responsible agencies and reorganization of departments. Depending on the wishes of various past presidents, wildlife has been administered by several different agencies (R. Carrera, personal communication, 2003). For thirty years after the Game Law of 1951, SARH (basically covering agriculture and livestock) administered wildlife, but from 1982 to 1988 the responsibility shifted to an agency called SEDUE (an odd mix of urban development and ecology), and it then shifted to SEDESOL (social development) from 1988 to 1994. Those responsible for Mexican wildlife conservation were always buried in a small, underfunded department within a larger agency, whose responsibilities centered on what many thought were more pressing issues (urban development, agriculture, livestock, social development). In 1994 significant changes took place in Mexico's wildlife administration. For the first time, wildlife administration resided in a federal agency responsible for the environment, natural resources, and fish, referred to as SEMARNAP (Secretaría de Medio Ambiente Recursos Naturales y Pesca).

Although SEMARNAP maintained an office in the capital city of each Mexican state, only the General Wildlife Office in Mexico City was authorized to issue hunting permits for game species. This is still the case today, and it represents an administrative challenge for hunters residing outside Mexico City. In addition, strict gun control in the country, tied with the high fees hunters must (by law) pay to outfitters every time they hunt, is a big barrier to sustaining a solid customer base, necessary to support long-term wildlife management programs.

Another serious impediment to a robust system of Mexican wildlife conservation is the paucity of structured wildlife programs at the university level. Students interested in wildlife do not have the educational options available to enable them to build a strong foundation of knowledge in wildlife conservation. Some dedicated students are able to persevere and become biologists, but unlike the situation in the United States, environmental agencies in Mexico do not have a large pool of young, university-educated wildlife biologists to staff their agencies.

The Federal Game Law of 1951 was long obsolete in many respects, but the Annual Hunting Regulations (Calendario Cinegético Anual) helped fill the gaps (C. Alcalá-Galván, personal communication, 2003). In 2000 more changes occurred affecting wildlife management in Mexico. First, another small reorganization moved fisheries ("Pesca") out of SEMARNAP and into another agency; thus the name changed to SEMARNAT (Secretariat of Environment and Natural Resources). Second, a new general wildlife law was passed (Ley General de Vida Silvestre del 2000). This new law was intended to be all-encompassing, to include the management of all living organisms (plant and animal), not just game animals. It also moved toward decentralizing wildlife to the state and even the private landowner level. While some decentralization may be beneficial, private ownership of wildlife is contrary to the structure of successful conservation paradigms.

Although there is a firm legal basis for wildlife administration in Mexico, the government infrastructure has historically lacked stability and continuity. In addition, there continues to be no effective state or federal wildlife law enforcement in Mexico, rendering the wildlife laws rather pointless. Many people in Mexico are concerned about the future of wildlife conservation in their country and would like to see the establishment of an effective and well-funded administrative system. Such changes, however, are restrained by deep ties to economic and social issues.

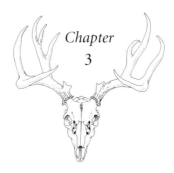

Chapter 3

Physical Characteristics

Body Weight

DEER WEIGHTS have been reported in a variety of ways. Terms like *field dressed, eviscerated, hog dressed, gutted,* and *live weight* make it difficult to meaningfully evaluate what people mean when they talk about weights. *Live weight,* as the term implies, is what the animal weighs as a live animal, or "on the hoof." *Hog dressed* refers to the removal of all internal organs, as is usually done by deer hunters in the field. Most people call this *field dressed,* but field dressing technically means that only the stomach, intestines, liver, and reproductive organs are removed; the heart and lungs are kept intact, protected by the diaphragm. Field dressing was more popular in the past as a way of keeping the inside of the carcass clean until it reached camp. Since the term *field dressing* is so often used incorrectly (heart and lungs are rarely retained), the term *dressed weight* is preferable to indicate a fully eviscerated (organs removed) carcass.

Estimates of deer live weight based on dressed weight depend on many things, like how much food was in the large rumen (stomach) and intestines. Researchers have proposed various conversion factors, but in general dressing the animal reduces the overall live weight by one-fourth to one-third. Live weights then can be estimated from dressed weights simply by dividing the dressed weight by 0.7, which accounts for the organs that were removed (Anderson et al. 1974). About half of the deer's live weight is muscle tissue (Hakonson and Whicker 1971), but not all this muscle tissue is reasonably retrievable in the butchering process. As a rule, about half of the dressed weight can be converted to boneless venison in the freezer.

Body weights fluctuate throughout the year and between years in response to changes in the deer's nutrition and level of activity (Day 1964). Although bucks are usually about 20 percent heavier than does, they may lose a substantial amount of weight during the rut, as they eat less then and spend their time actively seeking females. In dry years, or in populations that are nutritionally stressed, deer weigh less than they do when forage is more plentiful. Body weights can also vary greatly among animals of the same age, simply because of genetic factors.

Trends in deer weights can tell a lot about how the deer population is doing in relation to the carrying capacity of the habitat (how many deer the habitat can support). If there is an overabundance of deer and they dramatically reduce the amount of food available, the nutritional stress will be obvious when one looks at the average body weights in that population. This is just one of several ways a deer manager can tell if there are too many deer in the population, without ever knowing how many deer are actually there. Because deer continue to gain weight throughout their lives, it is important to look at trends in body weight by individual age category. If the harvest is made up of 80 percent yearlings one year and only 20 percent yearlings the next, the average weight of harvested deer will increase, but that will have nothing to do with deer abundance or the quality of the habitat. To monitor the health of deer populations, it is especially useful to measure trends in the body weights of yearling bucks (1.5 years old). These deer have grown from fawns to nearly adult size in the preceding year, making their weight a useful index to the quality of the deer habitat during that time.

Mule Deer

Even within the Southwest, Rocky Mountain mule deer in southern Colorado/Utah and northern Arizona/New Mexico weigh more and have a slightly larger skeletal frame than their desert cousins to the south (Lang 1957, Hoffmeister 1986). The best data set of deer weights in this area of the Southwest comes from the extensive harvest information collected on the Kaibab Plateau in northern Arizona. This check station has been in operation since the 1920s, and the weights of nearly 6,000 bucks were summarized by McCulloch and Smith (1987). Mule deer bucks in this region have an average dressed weight of about 102 pounds for yearlings, 142 pounds for 2.5-year-olds, 166 pounds for 3.5-year-olds, 185 pounds for 4.5- to 5.5-year-olds, and 194 for bucks over 6 years old (Table 4). Bucks estimated at 6.5 years old varied from 142 to 265 pounds (McCulloch and Smith 1991). They also found that body weights of bucks and does increased until the age of 6.5 years and then stabilized. Yearling does from the same period averaged 83 pounds, 2.5-year-old does averaged 91 pounds, and those in the 3.5- to 5.5-year category averaged

97 pounds. Weights of Rocky Mountain mule deer in northern New Mexico were reported to be lower (Lang 1957), but these weights did not come from a large, consistent data set as did those from the Kaibab Plateau.

Desert mule deer from southern California, Arizona, New Mexico, West Texas, and northern Mexico weigh less than their cold-adapted cousins. Several good sources provide information on dressed weights of desert mule deer. The best sample of weights comes from the Sacramento Mountains of southern New Mexico, where Howard and Eicher (1984) weighed 1,095 mule deer from three check stations in the late 1970s. They reported that yearlings weighed 74 pounds, 2.5-year-olds weighed 97 pounds, bucks 3.5 to 5.5 years of age weighed 125 pounds, and bucks 6 years old or older weighed 148 pounds. Desert mule deer weighed throughout the 1980s and 1990s in southern Arizona were consistently 10–20 pounds heavier than those in southern New Mexico (Table 4). Desert mule deer exceeding 200 pounds (dressed weight) are harvested by hunters every year, but this is near the upper range of body size under normal habitat conditions. Does, of course, weigh somewhat less, with yearlings at 67 pounds, 2- to 3-year-olds at 74 pounds, and does 4 years old and older averaging 79 pounds (Wallmo 1961).

Mule deer in Baja, Mexico, and on nearby islands are consistently smaller than those on the Mexican mainland (Leopold 1959:504, Cowan 1961:355). Ramón Pérez-Gil Salcido (1981) is the only researcher to have focused on any of the three island populations of mule deer in Mexico. He conducted an overall evaluation of the basic biology and ecology of mule deer on Cedros Island off the west coast of Baja. During the course of his field work, he estimated the height of deer in the field, measured skulls and skins, and showed that this form of mule deer was indeed smaller than those in southern California and the Baja peninsula. No reliable weight or size data exist for the other two island populations of desert mule deer (on Tiburón and San Jose Islands), but if they are effectively isolated from the mainland, we would expect the deer to be smaller, as is the case with most insular populations of large mammals.

White-Tailed Deer

Some extraordinarily large whitetails weighing 350–400 pounds have been documented in the eastern and midwestern United States, but the Coues whitetail is much smaller. The Coues whitetail in the Southwest stands about 32 inches tall at the shoulder, which makes it one of the smallest forms of white-tailed deer. Whitetails from central and southern Mexico may be smaller than Coues, but few studies have evaluated body size in those populations.

The Arizona Game and Fish Department (2000a) weighed more than 650 whitetails in southeastern Arizona during the 1980s, and this data set provides the best age-specific body weights for this deer. Dressed weights of Coues whitetail bucks average 53 pounds for yearlings, 67 pounds for 2.5-year-olds, and 85 pounds for bucks 3.5 to 5.5 years old. Bucks over 6 years old average 93 pounds,

TABLE 4. AVERAGE BUCK DRESSED WEIGHTS (POUNDS) REPORTED FOR MULE DEER IN THE SOUTHWEST

Age (years)				Sample		
1.5	2.5	3.5–5.5	6+	size (n)	Years	Source
69.7 (n=5)	112.2 (n=5)	145.9 (n=5)	132 (n=4)	19	1972–81	Big Bend NP, W. Texas (Krausman et al. 1984)
76.0	113.6	133.0	140.3	...	1957–60	Black Gap Area, W. Texas (Wallmo 1961)
99.9 (n=15)	129.4 (n=19)	154.8 (n=176)	174.4 (n=144)	354	2002–2003	Texas Trans-Pecos (Bone 2004)
100	124.3	167.6	192.3	33	2001–2002	Texas Panhandle (Bone 2003)
93	110	[135–186]	[193–200]	...	...	N. New Mexico (Lang 1957)
64	79	104	[125–145]	...	...	S. New Mexico (Lang 1957)
72.3 (n=167) [51–101]	←——	116.0 (n=330) [66–193]	——→	497	1954	Sacramento Mt., S. New Mexico (Anderson 1964)
66.1 (n=134) [42–93]	←——	95.4 (n=191) [56–174]	——→	325	1954	Guadalupe Mt., S. New Mexico (Anderson 1964)
73.7 (n=399)	96.7 (n=551)	125	148	1,095	1975–78	Sacramento Mts, S. New Mexico (Howard and Eicher 1984)
102 (n=3317) [60–153]	142 (n=1290) [98–227]	173 (n=1107) [102–248]	194 (n=184) [129–249]	5,898	1971–85	Kaibab, Arizona (McCulloch and Smith 1987)
88 (n=98)	118 (n=24)	136 (n=33)	153 (n=3)	158	1981–89	SE Arizona (AGFD 2000a)
85 (n=77) [70–105]	117 (n=57) [89–140]	136 (n=44) [107–199]	174 (n=6) [166–178]	184	1992–97	SE Arizona, Altar Valley (AGFD 2000a)
84 (n=16) [55–120]	106 (n=35) [52–130]	148 (n=64) [101–210]	154 (n=16) [121–169]	131	1989–94	SW Arizona, GMUs 39–45B (Bob Henry, Pers. comm.)
80.7 (n=72)	98.8 (n=87)	←—— 111.3 (n=58)	——→	217	1984–86	SW California, Camp Pendleton (Pious 1989)
...	←——	[112–200]	——→	...	...	Mexico (Leopold 1959)

Note: Sample size (n=) and range [in brackets] are given when available. In some cases, weights were originally recorded according to broader age categories than those indicated by a single column in this table. In such cases, arrows indicate the span of years to which a particular value applies.

with very few over 100 pounds (Table 5). Adult does generally average about 50 pounds dressed (Day 1964).

Dressed weights of 881 Coues deer taken on the Fort Huachuca Military Reservation from 1958 to 1963 averaged 78 pounds for adult (more than one year old) bucks and 55 pounds for adult does (Pratt 1966). A few bucks harvested during

TABLE 5. AVERAGE BUCK DRESSED WEIGHTS (POUNDS) REPORTED FOR WHITE-TAILED DEER IN THE SOUTHWEST

Age (years)				Sample		
1.5	2.5	3.5–5.5	6+	size (n)	Years	Source
...	←——[75–80]——→			...	...	Western New Mexico (Lang 1957)
...	←——[85–90]——→			...	...	New Mexico (Raught 1967)
40 (n=2)	←————64————→ ←———(n=7)———→			9	1957–63	Chiricahua Mts., Arizona (Day 1964)
56 (n=69) [47–69]	66 (n=84) [51–83]	82 (n=111) [56–112]	94 (n=35) [75–114]	299	1972–74	Arizona (Knipe 1977)
...	←————78————→ ←——(n=881)——→			881	1958–63	Ft. Huachuca, SE Arizona (Pratt 1966)
53 (n=281)	67 (n=192)	85 (n=154)	93 (n=26)	653	1981–89	SE Arizona (AGFD 2000a)
50 (n=17) [44–59]	67 (n=29) [52–85]	79 (n=40) [66–90]	84 (n=1)	87	1992–97	Altar Valley, Arizona (AGFD 2000a)
...	87.9 (n=2) [85–91]	108.5 (n=3) [90–137]	...	...	1972–74	Big Bend NP, Texas (Carmen Mts.) (Krausman and Ables 1981)
...	←———[64–100]———→			...	...	Mexico (Leopold 1959)

Note: Sample size (n=) and range [in brackets] are given when available. In some cases, weights were originally recorded according to broader age categories than those indicated by a single column in this table. In such cases, arrows indicate the span of years to which a particular value applies.

this period were very large for Coues whitetails. One particular buck, shot by Diane Watton in 1964, weighed 145 pounds dressed and was suspected by Jerome Pratt (Fort Huachuca biologist) to be a hybrid between a whitetail and mule deer. Known hybrids reportedly were released from captivity in this area in 1963 (Pratt 1966). From the photo of this buck (Knipe 1977), it does appear to have a metatarsal gland consistent with hybrids produced in captivity.

Knipe (1977) reported a twenty-year sample of dressed weights from whitetails in southern Arizona. These animals averaged 80 pounds, with bucks exceeding 100 pounds (dressed) making up only 8 percent of the annual harvest.

Whitetail body size increases at the eastern edge of Coues whitetail distribution. White-tailed deer in eastern New Mexico are remnants of a larger form that was formerly more widely distributed. In West Texas and Coahuila, Mexico, the Carmen Mountains white-tailed deer show an intergradation to the larger whitetail found in Texas (Krausman and Ables 1981).

Dentition

Deer have no upper incisors, and the eight lower incisor-like teeth are pressed against a hard upper pad, or palate, to pinch and tear off plant parts. Upper canines are absent except in rare cases. Only six of the eight lower "incisors" are true incisors; the outside pair of lower front teeth are actually canines. Through evolution, these lower canines have moved forward in the jaw to look and function like incisors. The lower jaw has three premolars and three molars on each side that match up with their counterparts above to grind and regrind food. Fawns are born with all eight lower incisor-like teeth, all three premolars, and one molar on each side of the jaw. All incisor-like teeth and premolars are replaced with adult teeth before the age of two years, which provides a method of ageing deer according to their tooth eruption and replacement patterns (Severinghaus 1949, Russ 1993). There are unique characteristics that allow individual deer to be aged accurately up to 2.5 years old, but by the time they reach 3.5 years, age is assigned by relative wear on various parts of the tooth row, and this becomes much less reliable. Because tooth replacement and wear has been shown to be inaccurate for placing mature deer in single-year age classes, many agencies group deer into broader age categories, for example: 0.5 (fawn), 1.5 (yearling), 2.5, 3–5, 6–8, and 8+ years (Heffelfinger 1997).

On rare occasions deer can grow upper canine teeth. These canines growing out of the maxillary bones of the skull are evolutionary holdovers from a time when deer ancestors had well-pronounced fangs (see chapter 2). Maxillary canines have been found in white-tailed and mule deer throughout their geographic ranges, including the Southwest (Watkins and Urness 1972, Krausman 1978b, Pederson 1983). When present, these canines are not large, but are generally small, peg-like teeth just breaking the gum line (Fig. 18). Many may be missed because some canines are too small to break through the gums and are not visible by a look in the mouth. In populations where these teeth have been documented, from 0.05 percent to 18 percent of the deer had upper canines (Ryel 1963). It has been surmised that the incidence of upper canines in whitetails increases as one travels south. If this is true, they may be more common in Coues whitetails than we yet know.

Digestive Tract

Deer are herbivores, which means they obtain the nutrients necessary for survival by eating plants. Plant material contains a large amount of cellulose and lignin that makes up the plant cell walls and is not digestible with the normal enzymes and acids that are used in other digestion systems. To process this fibrous plant material, deer and other ruminants (cattle, sheep, goats, pronghorn) have a four-chambered stomach consisting of a rumen, reticulum, omasum, and abomasum.

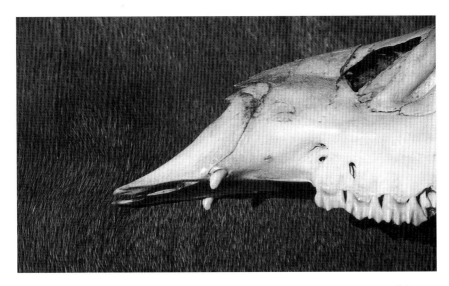

Figure 18. Maxillary canine teeth occur rarely in deer and may be so small they are obscured by gum tissue. *Photo by author*

When food is swallowed, it enters the rumen, which holds about 75 percent of the total amount of food in the digestive tract and is lined with small fleshy projections called papillae (Moen 1973:141). These papillae dramatically increase the surface area for absorption of nutrients. This first chamber and its associated muscles contract in a synchronized manner that produces a kneading motion that continually mixes the contents of the rumen to aid digestion.

The rumen contains numerous bacteria and protozoans that break down the cellulose, and produce vitamins and volatile fatty acids that are important nutritionally to deer. Every ounce of rumen contents contains billions of bacteria and hundreds of thousands of protozoans. The rumen serves as a large fermentation tank maintained at the right temperature and pH for bacterial growth (Moen 1973). Most of the actual breakdown of food particles occurs in the rumen before they are passed to the reticulum.

The reticulum is a softball-sized organ with a honeycombed lining that is partially separated from the rumen by a fold of skin. Food particles are passed back and forth between these two chambers, where the regurgitation or "cud chewing" originates. A complex series of muscles brings a mouthful or "bolus" of nearly liquid stomach contents back up the esophagus to the mouth. The liquid is then squeezed out and swallowed, leaving the food particles in the mouth to be re-chewed with more saliva and then swallowed again. Deer can chew their cud standing, but prefer to be lying down. Cud chewing helps to further break down food particles, which allows the bacteria and protozoans better access to the cellulose.

After food passes the reticulum, it moves to the omasum, an organ with many internal folds that increase the surface area. Here 60–70 percent of the water is removed from the food (Moen 1973). The drier digested matter is then passed to the abomasum, where the secretion of acids and other gastric juices kills the microorganisms (bacteria and protozoa) and breaks down food particles further. The abomasum functions similarly to the stomach of nonruminants. From the abomasum, the food enters the large and small intestines, which are over 65 feet long in an adult mule deer (Short 1981). Most of the absorption of nutrients occurs in the small intestine, while the large intestine removes the remaining water and forms the pellets characteristic of deer feces. The entire digestion cycle takes about twenty-six to thirty hours.

This feeding strategy provides a measure of protection for ruminants because it allows a deer to quickly fill its rumen and return to the safety of cover. Once secluded, the deer can rechew the food it collected and digest its meal.

In rare cases, a "stomach stone" is found in the rumen. These stones (also called bezoars, mad stones, enteroliths, or beazle stones) are generally smooth, oval lumps that develop from a foreign object in the stomach, which accumulates layer upon layer of mineral deposits (primarily calcium), much like the development of a pearl. The most common term for these, *bezoar,* comes from a Persian word meaning "against poison." In ancient times, these stones were believed to possess magical, curative powers (Steiner 1982). A bezoar was found in the stomach of a 4.5-year-old whitetail buck shot in 1992 in the Winchester Mountains near Willcox, Arizona. This stone measured 2.5 inches in diameter and was composed of layers of calcium phosphate and calcium oxalate surrounding a small piece of red granite that was apparently ingested by mistake.

Pelage

Deer molt and replace their fur (*pelage*) twice per year, resulting in a summer coat and winter coat. Like many natural annual cycles, the timing of shedding and replacing the hair coat is controlled by hormones that are regulated primarily by photoperiod (day length). There is, however, a strong genetically based component that interacts with photoperiod (Jacobson and Lukefahr 1998). Other factors, such as temperature, diet, and physical condition of the deer, also play a role in some cases.

The winter coat is grown in August and September each year as the summer coat is shed (Fig. 19). This winter coat is brownish gray in mule deer and steel-gray in whitetails, with long guard hairs and short insulating underfur. The short underfur hairs are about half the length of the guard hairs and are twice as numerous (Bubenik 1996). The guard hairs are hollow, so that they trap air and serve as an effective layer of insulation. Research indicates that photoperiod regulates the

Figure 19. In autumn, the long, reddish summer hairs are lost in patches as the thick, brown winter coat fills in from below. *Photo by author*

molting of the guard hairs, but the growth of underfur responds to a lowering of temperatures in the fall (Bubenik 1994). Small muscles at the base of the hair shaft, called *arrector pili*, allow the deer to erect each hair, thereby increasing the amount of trapped air to serve as an insulating layer.

The summer coat is reddish brown, and its growth begins during April–June as the winter coat is shed. The summer pelage consists of shorter guard hairs that are not hollow. This coat does not have the woolly layer of underfur insulation, so it allows heat to be lost easily. The summer coat of Coues whitetails is not as red as other forms of whitetail in the eastern United States, but it still looks more reddish brown than the steel-gray winter coat. As a rule, deer in arid environments are pale in comparison to their northern and eastern relatives, and the difference between summer and winter coats is less striking. As with all wildlife coloration, there is considerable variation within a population. Although some deer have been reported with unusually woolly coats and horse-like manes, these are less common than color abnormalities.

Albino deer are the most striking of the color anomalies reported. Albinism is a recessive genetic condition, which means that if both parents possess this rare genetic trait, some of the offspring will be albinos. Purposely breeding deer known to carry this trait will increase its occurrence. Albino deer lack melanin in their cells so that all hair, hooves, and skin are without color pigments. The nonpigmented skin with blood vessels running through it looks pink to the observer. These deer have white hair, pink eyes, pink noses, cream or pink hooves, and even white antler velvet. Albinos cannot see well because of the lack of pigment in their eyes, which would help block sunlight. Both albino whitetails and mule deer have been reported in southwestern deer herds, but they are exceedingly rare (Allen 1893, Gallizioli 1956, Knipe 1977).

Piebald, or "pinto," deer are partially or mostly white, but are not albinos. The piebald condition is also a genetic abnormality in which some of the cells can produce melanin. These deer can have pigmented hooves, eyes, and skin, and lack pigment only in some areas of their pelt. Besides varying degrees of white fur, piebald deer may have other genetic abnormalities, such as short legs, bowed nose, arching spine, short jawbones, and oddly formed internal organs (Davidson and Nettles 1997:99). The most severely deformed die soon after birth, but those with minor physical problems may live to adulthood (Mierau and Schmidt 1981). Like albinos, these unusually colored deer would be seen easily in open country, but are rarely reported, which suggests their scarcity.

On the opposite end of the spectrum are deer that produce too much melanin pigment, resulting in very dark or black fur. This coloration is even rarer than "white deer," and it would seem such an individual would be at a great disadvantage during the hot summer months. Melanistic deer generally have some white hairs, usually under the tail and on the tarsal and metatarsal glands.

Senses

Deer hunters are well aware of the deer's keen senses. Deer, like many prey animals, have been under constant pressure throughout evolution to develop and fine-tune their ability to detect danger and escape from it. These selective pressures have removed animals with below-average senses from the gene pool, leaving only the sharpest individuals to produce the next generation. Learning more about the senses of deer helps build an appreciation for the keenness that has resulted from eons of natural selection.

Sight

Differences between predators and prey are often illustrated with a simple rhyme: "Eyes in front, the animal hunts—eyes on the side, the animal hides." True to form, a deer's eyes are set on the side of the head, allowing them to monitor almost a complete circle (310°). With a slight turn of the head, the deer can watch for danger in all directions. Because of the location of the eyes, deer have weaker binocular vision and thus reduced depth perception. Depth perception is very important to predators, which need to accurately target a moving food item, but less important for those animals that feed on plants.

Because detecting movement of approaching predators is important, a deer's vision is extremely sensitive to movement. Deer sometimes unknowingly approach within feet of humans who remain perfectly still, yet they may sneak off to safety after detecting a hunter walking on a ridge line a mile away. Remaining motionless is the key to escaping detection. When a deer spots an object that it cannot identify, it generally freezes. Anyone who has spent a lot of time in deer country has been caught in a stare-down with a deer. With their remarkable patience, deer usually win these contests, remaining motionless for as long as it takes to get the person to move. If humans stand motionless and have the wind in their favor, the deer may go back to feeding, satisfied they are not in danger. Deer have in their repertoire a hoof stomp that they use in such situations, in part to elicit a response (movement) from a strange object. A deer may bob its head back and forth for a better view or move cautiously to a downwind position to use its sense of smell to identify what the eyes cannot.

Deer have excellent night vision. A specialized membrane, called the *tapetum lucidum,* behind the retina reflects light back through the retina a second time. This increases the amount of light the eye can use and increases the deer's ability to see in the dark. This reflective membrane is what causes the nighttime eye-shine when a deer is seen with a spotlight or vehicle headlights.

With round pupils, humans can focus only on one spot at a time because the light is reflected to a single small location on the retina at the back of the eye. In bright daytime light, the deer's pupil is shaped as a horizontal slot, rather than a small hole as in humans. This pupil shape and other differences in eye structure may allow the deer to focus on the entire horizon at once, rather than only one spot at a time.

One of the most common questions asked about deer is whether they are color blind. It was initially thought that mule deer and white-tailed deer were color blind (Cowan 1961, Severinghaus and Cheatum 1961), but research indicated that some members of the deer family could see some colors. Scott (1981) showed that elk consistently differentiated blaze orange buckets containing food, but he did not control for related factors, such as brightness and hue.

Research on color perception of whitetails has been contradictory. The retina of the eye contains two types of photo receptors, rods and cones. The rods enable the animal to see in low light, and the cones are necessary for daylight and color vision. Witzel et al. (1978) analyzed the retina of whitetails and reported a high density of cones in the retina, indicating that deer have the physical structures necessary to see colors. Staknis and Simmons (1990), using an electron microscope, failed to locate any cones in the retina of whitetails; however, a more recent study showed that white-tailed deer possess two classes of photo pigments on the cones of their retinas, indicating that they can see some color (Jacobs et al. 1994). Pigments in these cones allow deer to discriminate between short wavelengths (blue-green) and middle wavelengths (yellow or orange) (Smith et al. 1989).

A deer's sensitivity to blue is similar to that of humans, although humans have a yellow filter in their eyes to filter out ultraviolet light; deer lack this filter. This has sparked concern that deer can see ultraviolet light reflected by certain "brighteners" in laundry detergent. Some outdoor companies market products to neutralize these brighteners in camouflage clothing. These companies assert that the detergent brighteners make the camouflaged hunter appear (from the deer's perspective) to glow in relation to the surrounding foliage and rocks. Although possible, it is not clear if this is true. Jacobs et al. (1994) tested two whitetails and detected no substantial response to ultraviolet light.

Information gathered thus far indicates that white-tailed deer can see several colors ranging through the spectrum from borderline ultraviolet, blue, and green, to yellow or yellow-orange. A deer's eyes are not sensitive to the color orange. Although they can see some colors, a hunter dressed in orange would not appear as obvious to deer as to other humans.

Hearing

The external ears of deer move independently to precisely locate the origin of any interesting sound. Deer can direct one ear cup forward while the other remains vigilant to noises behind them. Any strange noise will immediately draw the attention of the ears, even if the deer remains at ease.

Windy days mask noises and movements, reducing the deer's main defense strategies. Deer movements may be reduced on windy days, in part because of their inability to hear. Deer that are active on windy days appear very nervous and cautious.

Deer hear about the same range of frequencies as humans do. Ken Risenhoover (personal communication, 1998) reported a potential hearing range of 1–8 KHz in whitetails; this is not surprising since deer vocalizations range between 1 and 9 KHz. Deer in his study did respond to higher and lower frequencies (0.5–16 KHz), but only at extremely loud levels (high decibels). Researchers from the University of Arizona found very similar results in desert mule deer in the 4-KHz range they tested (DeYoung et al. 1993).

High-frequency whistles mounted on vehicles to reduce collisions with deer have received some attention in recent years, but research conducted at the University of Georgia indicates that these whistles do not produce a measurable sound when mounted on vehicles (Schildwachter et al. 1989). If hand-held, these devices could be made to produce a high frequency (16–20 KHz) sound by blowing into them, but this is above the hearing range of deer. Previous researchers were unable to elicit a response from deer standing only a few yards away.

Smell

The sense of smell is so highly developed in some animals that humans cannot even begin to appreciate the information gathered by this one sense. Much of the communication that occurs in the deer's world occurs via pheromones or scents. Females do not always recognize their own fawns by sight alone, but can immediately distinguish them by scent.

The long nasal passages of deer are lined with a tissue (*epithelium*) that contains mucus-producing cells. These cells maintain a moist environment in the nasal cavity to aid in the collection of scent. When a deer inhales, it draws in airborne molecules that land on an area of moist nasal lining, and through a chemical reaction, messages are sent to the brain for identification of the scent. Because of the importance of moisture, increased humidity improves a deer's ability to smell.

The ability to analyze airborne scent provides deer with their greatest defense mechanism. Hunters and predators probably see only a fraction of the deer that smell them. Usually people only hear the warning blows, or snorts, of a deer that has caught their scent. Deer receive and identify a complex array of scents each time they inhale. When a foreign or potentially dangerous scent is detected, deer may sneak off or hold in thick cover in hopes of remaining unseen. Strange and curious objects or noises are usually investigated with a downwind approach.

Deer use their sense of smell extensively in their search for food. Deer can smell unseen mushrooms, fungus, and puffballs, and can locate oak leaves and acorns under several inches of snow (Knipe 1977). Dixon (1934) reported that deer could determine good acorns from wormy or hollow ones by smell.

Taste

Taste is not an important defensive sense, but deer use it extensively to select which plants to eat (Fowler 1983). Deer perceive flavor through a combination of taste and smell. Research has illustrated that domestic ungulates react differently to various tastes, a point probably obvious to anyone who has offered a sugar cube to a horse. Taste was considered to be the most important sense used by domestic sheep to select which plants to eat, according to one study (Krueger et al. 1974).

Nichol (1938) reported whitetails and mule deer tasting different forages before eating them. Deer pick up any new food item, move it around with their lips until it becomes moist, and then drop it on the ground. They may repeat this procedure several times before either eating it or abandoning it. Taste may also help deer limit their intake of toxic plant species. Many plants that contain toxic compounds are bitter tasting to most animals.

Vomolfaction

A lesser known sense called vomolfaction is a mixture of taste and smell. This involves a specialized organ in the roof of the deer's mouth called the *vomeronasal*. A buck approaches a spot where a doe has urinated, takes a sample of the doe's urine in his mouth, and performs a lip-curl, or *flehman*. During this lip-curl, the upper lip is pulled upward and the urine is drawn into the vomeronasal organ, where it comes into contact with special receptor cells. Nerves then send messages to the brain that can decipher chemical codes present in the urine. This contact with receptor cells is not unlike the contact of scent molecules on the mucus lining of the nasal passages. Since bucks lip-curl most often during rut, the vomeronasal organ is thought to play a role in determining how close the doe is to estrus. Recent work, however, suggests it may have a role in synchronizing the reproductive physiology of the buck (K. V. Miller, personal communication, 1999). Whitney et al. (1992) felt that readiness of the doe is most likely tested through normal scent detection in the nasal passages.

Voice

Many people do not realize that deer use vocalizations to communicate. Although they do not vocalize frequently, deer have an assemblage of calls that convey different meanings. Very little research has been conducted on mule deer calls, but work with whitetails has identified twelve different types of vocalizations in five categories (Atkeson et al. 1988).

Mother-Fawn

Does and their fawns use four types of calls: *maternal grunt, mew, bleat,* and *nursing whine*. The *maternal grunt* is a low, short grunt that can only be heard within 50 yards of the doe making the sound. It is used by the doe when approaching a bedded fawn. In response to this call, the fawn leaves its bed and approaches the doe.

The *mew* is a high-pitched but low-volume call given by the fawn to attract the attention of the doe. It is audible within only a few yards and sometimes is given in response to the doe's maternal grunt.

The *bleat* is similar to the mew but conveys a more urgent desire for the doe's attention. The bleat can be heard as far as 100 yards from the fawn and intensifies as the need for attention increases. When fawns are disturbed, they may bleat, which usually draws the attention of the doe.

The fawn emits a *nursing whine* when suckling or searching for the nipple. Richardson et al. (1983) speculated that the nursing whine may identify the fawn, reaffirm the maternal bond, transmit pleasure, and solicit more care or continued attention.

Alarm

The *snort* of whitetails and mule deer probably serves as an alarm and may be the most recognized of all deer sounds. Snorts are produced, singly or in a series, when the deer blows air forcefully out of its nasal passages.

Deer also use a *foot stomp* as a form of audible communication, though this is not a vocalization. The foot stomp is frequently used in association with the snort. Often the hoof stomp signals the deer's departure. Like the snort, it serves to warn other deer of nearby danger. Additionally, it may elicit a movement from an object the deer cannot identify or let a predator know it has been spotted and further stalking will be useless.

When in distress or being traumatized, deer *bawl* loudly. This vocalization is only given when the deer is grasped or is badly injured. The duration of the vocalization may be very prolonged depending on the intensity of the trauma. The pitch of this call tends to lower in older deer.

Mating

The *tending grunt* is used by the male when courting a doe in estrus. It is a moderately low sound of longer duration than other grunts. Another sound associated with mating is the *flehman sniff* (see above, under "Vomolfaction"). Like the snort, this sound is not produced by the vocal cords. As the name implies, the flehman sniff is produced by sniffing forcefully through the nostrils. This sound probably does not convey a message, but rather is a by-product of the buck performing a lip-curl (flehman).

Aggressive

The *low grunt* is used by both sexes of mule deer and white-tailed deer at all times of the year. It represents a moderate warning to other deer and is frequently accompanied by aggressive body language (ears back, head-high posture). It is usually a single, brief grunt of low intensity given by a dominant animal. Elevating the aggression, deer combine a grunt and snort into the g*runt-snort*. As the name implies, this is a low grunt followed by from one to four quick snorts through the nostrils. Does use this call infrequently. It is most often used by bucks during the breeding season.

If an antagonistic encounter continues to escalate, male whitetails and mule deer produce a *grunt-snort-wheeze* or a *snort-wheeze*. This call is the most intense aggressive call produced by adult bucks during rut. The grunt and snort(s) are followed by a long-drawn-out (3–10 seconds) wheeze through pinched nostrils. Any buck not backing down from a grunt-snort-wheeze is in for a serious fight.

Contact

When whitetail does have been separated into different pens, they have been heard to emit a grunt of moderate volume and duration. It has been termed a *contact call* because it helps deer maintain contact when they are out of sight of one another.

Scent Glands

Two types of specialized skin glands that are found over the surface of a deer's body play a role in scent communication. *Sebaceous glands* are usually associated with hair follicles and produce a waxy and/or oily substance that helps lubricate and waterproof the skin and hair. These secretions are delivered to the surface of the skin through hair follicles or small pores. The waxy/oily material itself may not produce scent, but it can act as a vehicle to carry scent produced by bacterial action or resulting from compounds produced elsewhere in the body. The second type, *apocrine sudoriferous glands,* can produce a scent that is socially significant (pheromone).

Certain areas of a deer's body have a high concentration of these specialized skin glands and serve as centers for scent communication. These areas of dense glandular tissue are commonly referred to as "glands." Researchers have identified at least eight glands on the body of white-tailed and mule deer.

Tarsal

Tarsal glands are familiar to many hunters because of the role they play during rut. These glands are located on the inside of the deer's leg at the hock, and appear as whitish tufts of hair (Fig. 20). Skin glands underneath these tufts, or tarsal brushes, produce an oily substance. During rut bucks perform a "rub-urination,"

Figure 20. Tarsal glands are located on the *inside* of the hind legs, are similar in mule deer and whitetails, and play an important role in scent communication in deer. *Photo by Joe and Marisa Cerreta*

in which they place their hocks together and urinate on these glands. It is not clear where most of the scent originates, but it may be that urine, interacting with bacteria, produces most of the noticeable odor associated with this gland. Well-developed *arrector pili* muscles at the base of these tarsal hairs allow deer to flare this gland to release more scent or provide a subtle visual signal.

This gland and its odors appear to help identify individual animals and relay important information on breeding status and dominance rank. There is no difference in the structure of the glandular tissue between the sexes, but Quay and Müller-Schwarze (1970) reported that the thickness of this gland increased with age. During rut, these glands on bucks turn dark brown to almost black from being stained with soil and urine. Mature bucks may have a black crust from the hocks to the hooves during the peak of rut. Regardless of the appearance of these glands, it is not necessary to remove them while skinning a deer, and doing so may only spread the scent to the meat.

Metatarsal

Metatarsal glands are found on the outside of the lower legs between the hocks and the hooves. Mule deer metatarsal glands appear as a long tuft of brown fur on the upper half of the lower leg. Below this long tuft of fur is an equally long ridge of hard black material (keratin) less than ¼ inch wide and 3–6 inches long. Surrounding this black ridge is an area of concentrated sebaceous and sudoriferous glands.

Metatarsal glands of whitetails look very different from those of mule deer (Fig. 9). The black keratinous material is less than 1 inch long, encircled with white hair, and located halfway between the hock and the hoof or slightly below midpoint. These glands do not emit a strong odor, but touching the hard, black glandular material will reveal a waxy, oily substance with a noticeable odor.

The purpose of these glands has not yet been deciphered, and there apparently is no difference in this respect between bucks and does. Quay (1959) pointed out that this gland is pressed into the ground each time the deer beds down, which may serve to mark bed sites with individual odors. This gland also is in an optimal position to brush scent onto vegetation as the deer walks from place to place. Observations of captive deer of both species by Nichol (1938:36) in Arizona failed to "indicate a use or function of the metatarsal glands." Quay and Müller-Schwarze (1970) theorized that in black-tailed deer the odor serves as an alarm signal to other nearby deer. Volkman (1981) concluded that scents from the metatarsal gland did not cause alarm behavior in whitetails; however, more recent research indicates that female mule deer become alarmed or alert in the presence of this scent (Müller-Schwarze et al. 1984).

Interdigital

As the name implies, the interdigital gland lies between the two digits of each foot. The gland is a hollow sac between the hooves, with some yellowish, waxy

Figure 21. Interdigital gland of a Coues white-tailed deer. *Photo by author*

substance present in the gland and on the surrounding hair (Fig. 21). The odor becomes obvious when one spreads the toes apart to investigate this gland.

The purpose of this gland seems to be to allow deer to track one another and themselves, because each time a deer puts its foot down, it leaves a scent mark. In addition, research has shown that secretions from this gland act as an antibiotic to reduce infections to the feet (Wood et al. 1995).

Laboratory analysis of the secretions from white-tailed deer interdigital glands identified a minimum of forty-six different compounds; eleven of those occurred in higher concentrations in mature bucks (Gassett et al. 1996). Higher levels of male hormones have been shown to increase sebaceous gland activity. If higher levels of male hormones associated with dominance is what caused the differences found in sebaceous secretions, then the interdigital gland may serve to inform other deer of the presence of a mature buck in the area.

Preorbital

The preorbital gland is located directly in front of the eye and differs substantially between white-tailed and mule deer. The mule deer's preorbital gland appears as a deep pit (about ¾ inch) in front of the eye, while the whitetail's is a very small depression (about ¼-inch deep) (Fig. 22). The sac or pit of mule deer generally contains a yellow, waxy secretion mixed with plant material, eye secretions, skin cells, and other foreign matter.

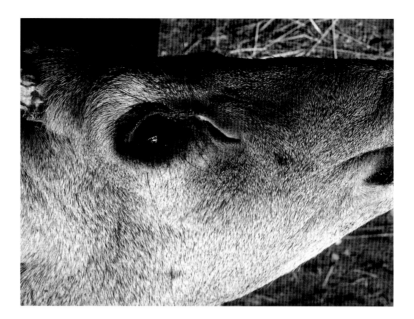

Figure 22. The preorbital gland is located directly in front of the eye and is much larger in mule deer (*top*) than white-tailed deer (*bottom*). *Photos by author*

Muscles associated with this gland can flare the gland open during dominance displays in bucks or when does are nursing. Whitetails rub this gland on overhanging branches above scrapes, presumably to deposit the scent produced by the gland secretions. What information is conveyed by this scent is not known, but Volkman (1981) demonstrated that deer reacted to it. Quay and Müller-Schwarze (1970, 1971) reported that this gland showed no prominent differences between mule deer bucks and does of any age, indicating it may not be of primary importance for mule and black-tailed deer during rut.

Nasal

The nasal glands are small structures located just inside the nostrils. The glandular cells located in these cavities produce a white, fatty substance that does not carry scent well. The purpose of this gland is yet to be determined, but it is probably not important in scent communication because of its location within the nose (Atkeson et al. 1988).

Forehead

The forehead skin of whitetails has a much higher concentration of apocrine sudoriferous glands than the rest of the body. Behavioral observations have shown that black-tailed and white-tailed deer rub this area on small trees and deposit a scent that attracts the attention of other deer (see chapter 7). Male black-tailed deer rub more than does, especially during rut (Müller-Schwarze 1972). The glands under the forehead skin are relatively inactive during the spring and summer, but become very productive during rut in mature whitetail bucks, and to a lesser extent in does (Atkeson and Marchington 1982). That this glandular area is most active in mature, dominant males suggests its importance in advertising and maintaining dominance during the breeding season.

After years of close behavioral observations, Geist (1981) reported that mule deer do not rub saplings in the same way or for the same reason as white-tailed deer. Mule deer appear to use the rubbing of small trees as a noise-producing mechanism during rut rather than for the primary purpose of depositing scent. Histological examination of the forehead skin of mule deer supports these observations. The forehead skin of mule deer has only a slightly increased density of glands when compared to the rest of the body (Quay and Müller-Schwarze 1971). In addition, an increase in glandular activity in mature bucks was not detected in mule deer.

Preputial

This gland was discovered relatively recently inside the end of the whitetail buck's penis sheath. It consists of a dense patch of glandular tissue associated with the follicles of a few hairs that are visible at the end of the sheath. Odend'hal et al. (1992) reported that the glandular tissue in this region does not become more active during rut, indicating that it is not a source of scent used during the breeding season. The purpose of this gland is still unclear; however, because of its location, it might add scent to the urine to convey individual identity and possibly dominance rank. It has not yet been investigated in mule deer.

Caudal

Areas of glandular tissue located at the base of the tail have been described for black-tailed deer, but not for mule deer or whitetails. Quay and Müller-Schwarze (1971) found an area of enlarged skin glands equally developed in fawns and adults, with no difference between bucks and does. This specialized tissue occurs predominantly on the dorsal (back) surface of the tail toward the tip. Although deer are often observed investigating other deer by sniffing in this area, what these glands specifically communicate is unknown.

Chapter 4

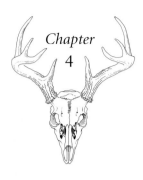

Antlers

HUMANS have been intrigued by antlers since the beginning of time. Although valuable to early man as tools, they undoubtedly elicited the same admiration and curiosity they do today. What factors affect antler growth? How did antlers evolve? Do genetic factors or injuries play a more important role in the occurrence of nontypical antlers? The questions are infinite, but not all can be definitively answered.

Many people erroneously use the word *horns* when describing antlers. The term *horns* is actually reserved for cranial appendages that consist of a bony core sheathed in keratin, a material much like fingernails. True horns on animals such as bighorn sheep, cattle, and bison are never shed and continue to grow throughout the life of the animal.

Antlers are different from horns in origin, form, and function. The branching appendages that define members of the deer family are called antlers and are regenerated each year. Except for caribou, females do not normally grow antlers. The term *antler* is derived from the old French *antoiller*, a derivative of the late Latin *antoculorum*, meaning "before the eyes," because this latter term was originally used to describe the brow tines on European deer (Lydekker 1898:5).

Antlerogenesis

Antlerogenesis is a complicated term for an even more complicated process—the annual physiological production of antlers. Male fawns are born with almost imperceptible bumps on the frontal bone of their skull. At about the age of four months, male fawns experience a rise in testosterone that causes the growth of small platforms, called *pedicles,* on the skull. These pedicles usually appear as small, fur-covered knobs during the fawn's first autumn. At this point, the ends of the pedicles are not polished and therefore are not shed; they remain dormant until the buck's second fall. These buck fawns have small fuzzy knobs and are sometimes referred to as "nub bucks." In areas with exceptional nutrition, some of these buck fawns may produce small spikes that are then polished and shed.

The pedicles provide the structural base for the future development of antlers. If a buck fawn is castrated before his fourth month, he will never grow pedicles, and thus remain antlerless the rest of his life because he lacks sufficient levels of testosterone. Pedicle and subsequent antler growth can be induced, however, with an injection of testosterone.

When a buck fawn reaches the age of nine to ten months, antler growth is initiated from the pedicles, and the youngster thus starts his first real set of antlers. Although testosterone is necessary for the development of pedicles, antler growth does not require large amounts of this hormone. Even moderate levels of this hormone actually inhibit the growth of antlers. Testosterone levels fluctuate throughout the year and are at a very low level in the spring when antler growth starts (G. Bubenik 1990a:296). Testosterone levels appear to increase slightly at the very beginning of antler growth, only to subside again during the growth period. A large number of other hormones, including IGF-1 and derivatives of testosterone, increase and decrease during various phases of the antler-growing period. This complicated orchestration of hormones is not yet completely understood, but serves to promote and sustain antlerogenesis.

In adult deer, antler growth begins within a few weeks after the loss of the last set of antlers. In the Southwest, antler growth is under way by May in Coues whitetails and mule deer (Villa 1954, Hanson 1955, Truett 1971, Hoffmeister 1986, Ockenfels et al. 1991, Weber et al. 1995). The scab that forms over the wound left by the shed antler heals and becomes covered with fine hairs. This finely haired skin forms the beginnings of the velvet that will nourish and protect the growing antlers.

As antler growth begins, the underlying pedicle gives rise to new antler material, which at this point is firm tissue composed primarily (about 80 percent) of protein. This protein matrix of growing antler tissue is cartilage-like and inundated with blood vessels. Most of these nutrient-transporting blood vessels rise up through the velvet lining, but inside the antler are blood vessels that rise through the pedicles. The visible grooves on the base and beam of hardened antlers are impressions left by the blood vessels in the velvet.

The antlers are also richly endowed with a dense network of nerves. These nerves make the velvet antlers sensitive to the touch and thereby protect the soft growing antlers against damage. The nerves also make the buck aware of how his antlers are shaped, which will be useful when sparring against competing males. The antler nerves and the central nervous system likely serve an important, though not well understood, role in determining the species-specific shape of the antlers as well as reacting to injuries. Antlers that are damaged during growth one year may produce an abnormality at that location in subsequent years. Because the antler itself is lost and regrown, G. Bubenik (1990b:340) speculates that the central nervous system has the ability to "remember" and re-create that injury in subsequent years.

Antlers are the fastest growing tissue in the animal kingdom, growing at a rate of over ½ inch per day in some cases. As the antlers reach full size in August, a substantial increase in testosterone and other hormonal changes occur, starting the mineralization (hardening) of the antlers. This process of mineralization progressively replaces the cartilage-dominated antlers with actual bone (Banks 1974).

Because dietary intake cannot supply enough nutrients and minerals to support the rapid rate of growth and mineralization, the buck's body actually mobilizes calcium and phosphorus from the entire skeletal system, and the blood carries these minerals to the antlers. Hillman et al. (1973) documented a dramatic reduction in skeletal bone density during antler growth. This loss of bone material occurred mostly from the ribs. The reduction in bone density is identical to the osteoporosis that occurs in older humans. This has led some researchers to use antlerogenesis as a model to attempt to find a cure for this degenerative bone disease. In deer, however, this osteoporosis is temporary; as soon as the antlers are completely mineralized, the skeletal bone density returns to normal levels (Banks 1966).

In September, antlers of both species become mineralized, and the continually increasing testosterone levels cause the drying and loss of velvet in the latter part of the month (Clark 1953, Swank 1958, Cantu and Richardson 1997). The antlers become fully hardened by late September, and the velvet is stripped off from late September through mid-October, with the peak for Coues whitetail occurring around October 9 (Villa 1954, Ockenfels et al. 1991, Weber et al. 1995). When the tissue has dried somewhat, the buck rubs off the velvet on a small tree or bush. The stripping of velvet occurs rapidly and once started is usually completed in twenty-four to forty-eight hours (Hoffmeister 1986, Ockenfels et al. 1991). When freshly stripped, the antlers are very white. The brown pigment in tree bark, called tannin, along with some residual blood, stains the antlers with the familiar brown color seen in the fall.

Testosterone levels peak during the breeding season and then decline after rut. This postrut decrease in testosterone triggers osteoclasts that erode the base of the antler at the pedicle. With this attachment degraded, the antlers are shed (cast). Mule deer in the northern portions of the Southwest may shed their antlers as early as February (Swank 1958, Bowyer 1986b), but most do so in March (Hanson 1955). Desert mule deer shed their antlers between mid-March and early April (Truett 1971). Coues whitetails shed their antlers from late April through May, with a peak around May 7 (Villa 1954, Ezcurra and Gallina 1981, Ockenfels et al. 1991, Weber et al. 1995). Like other processes in nature there is individual variation, with some bucks shedding antlers very early or very late. Younger bucks shed their antlers later because of the less dramatic decrease in testosterone levels and because their smaller, lighter antlers are more likely to stay in place until the connection is completely eroded. The pedicle wound from the shed antler bleeds very little and soon scabs over. After only a few weeks another cycle of antler growth begins.

The timing of the antler cycle has an underlying genetic basis (Jacobson and Lukefahr 1998), but the annual cycle is driven primarily by changes in the amount of daylight, called photoperiod. Deer, like other animals, have annual biological clocks that are regulated by the number of hours of daylight each day. As the days become shorter in the fall, the eyes transmit this information to the pineal gland in the brain. The pineal gland orchestrates the production and release of hormones that regulate many body processes, including the antler cycle and hair molting.

Researchers have placed bucks indoors under controlled lighting and reproduced two annual cycles of increasing and then decreasing day lengths in a twelve-month period to determine the effect on the antler cycle. Bucks under these artificial lighting conditions grew and shed two complete sets of antlers in one year (Goss 1983). Further experiments duplicating different day-length cycles resulted in bucks growing three and even four sets of antlers in one twelve-month period. Deer could also be induced to grow antlers throughout the winter and shed them in the spring when most deer are just starting their annual development.

Bucks tend to keep the same general antler shape year after year. In other words, bucks with wide racks early in life usually grow a wide set of antlers each year, while the racks of narrow- and tall-antlered bucks also look similar each year. Because of this similarity, it is possible to match two or more consecutive years of shed antlers from the same buck. It is not uncommon to see the same palmation or the same abnormal points year after year in the same buck. Sometimes localized areas, especially relatively isolated populations, will consistently produce a unique antler characteristic. There are areas where forked brow tines or extremely wide racks are consistently produced. It is one of the wonders of nature how a particular set of antlers can develop the same odd "kicker" point on the right side, in the same location, each time the antler regenerates.

Factors Affecting Antler Size

Age

Antler size increases with age until the buck reaches his prime and then declines. Antler development in whitetail and mule deer usually peaks when the bucks reach five to seven years of age, but individual bucks may peak at a younger or older age.

After the peak, a buck's antlers generally lose tine length and have fewer points, but continue to increase in thickness. No matter how good the nutrition and genetics of the area, the population will not contain large bucks unless they are allowed to reach the older age classes.

As a buck matures past the first year, his hormonal system creates a physiological environment for larger antler growth. In addition, once a buck's body is nearly full grown, more energy can be put into antler development. It is very rare for a buck older than a yearling to have only spike antlers (Fig. 23). During times of severe nutritional stress, a few 2.5-year-olds may grow only spikes, but this is an exception to the norm.

Nutrition

Good nutrition is an important prerequisite to growing large antlers. Recognizing a relationship between antler size and diet quality, William Twiti in 1327 remarked that "the head grows according to pasture, good or otherwise." In the Middle Ages, enormous red deer (*Cervus elaphus*) racks decorated castles throughout Europe and were traded among royalty as objects of great value. During the 1930s and 1940s a German chemist named Franz Vogt ran a series of experiments in an attempt to grow red deer antlers that would rival the monstrous medieval heads (Geist 1986). Vogt knew that antler characteristics were somehow passed on to male offspring by their fathers, but chose to ignore this factor and concentrate solely on age and nutrition. Vogt analyzed the chemical composition of antler material and then fed the deer large quantities of protein and identified minerals and watched them grow through the peak antler growth years. His experimental deer grew to great proportions. When the fighting of World War II ended his experiments prematurely, 35 of his 36 stags scored in the top 100 heads ever measured.

Antlers are secondary sexual characteristics that require extra energy to produce each year. Nutrients consumed by bucks go first to body development and maintenance, then to antler growth. This is one reason yearling bucks do not produce large antlers: their nutrient intake is being used almost entirely for body development in their first year. If there is a limited amount of forage available, there may not be enough "extra" nutrients to optimize antler development that year. In years of below-average rainfall or a widespread mast crop failure, bucks will sport smaller racks than in years of abundant forage quality and quantity. Research has shown that dietary restrictions result in a decrease in antler size. Ullery (1983) showed that restricted intake of protein, energy, and calcium by 1.5-year-old whitetail bucks reduced antler volume, beam diameter, main beam length, and number of points. This is the reason a higher proportion of the yearling buck antlers will be spikes in years of poor nutrition (Swank 1958:18). On the Kaibab Plateau the average number of antler points is directly related to the amount of winter rainfall and also the amount of precipitation over the previous two to three years (McCulloch and Smith 1991).

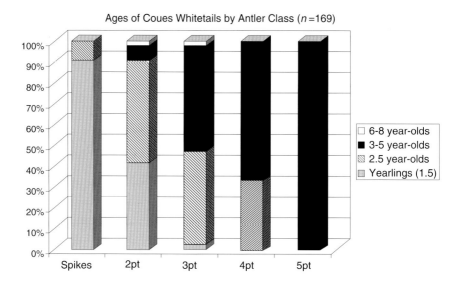

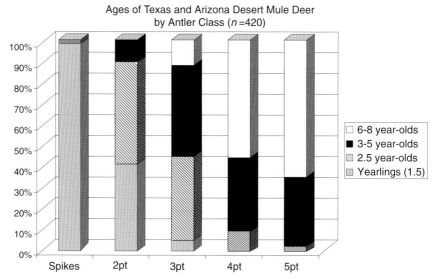

Figure 23. Antler size increases with age in both white-tailed deer (*top*) and mule deer (*bottom*); spikes are nearly always yearlings, and most bucks with four or five points on a side (excluding browtines) are at least three years old. *Data from Russ 1993, AGFD 2000a*

Genetics

Each buck has a different genetic potential for antler growth. Captive bucks of the same age, fed the same diet, show antler conformations that are very different from one another. Some bucks will be superior to others at the same age, and some will never have large antlers, just as some humans never reach six feet tall regardless of diet or age. Every deer population has some individuals whose genetic potential for antler size is far above or below average for that population.

No one has located a single "antler gene" and no one ever will, because genetics is not that simple. Many genes act together to determine the shape, size, mass, and density of a buck's antlers, as well as his ability to efficiently process his nutrient intake and survive to a ripe old age. However, research has illuminated many interesting relationships between genetics and antler growth. For example, the length of the antlers and main beam diameter were found to be significantly related to the level of overall genetic diversity (heterozygosity) in a population of white-tailed deer (Scribner et al. 1989).

In 1973 Texas Parks and Wildlife Department began to conduct a series of progressively more complex experiments on the effects of genetics and nutrition on antler growth in white-tailed deer at the Kerr Wildlife Management Area (WMA) (Harmel et al. 1989). In the first genetic study, eight bucks that were spikes as yearlings and one buck that was a large 10-point (eastern count) at 3.5 years old were bred to does, and the antler development of ten generations of offspring was recorded throughout the study. Researchers concluded that antler size and body weight were genetically based and influenced by environmental factors like nutrition. Data collected from this captive population indicated a buck's future antler size could be predicted by looking at his first set of antlers.

Research continued at Kerr WMA, and follow-up studies confirmed that bucks carrying spike antlers when they were yearlings did not grow antlers as big as those that were forked-antlered yearlings (Ott et al. 1998). Yearling spikes also went on to produce more spike-antlered offspring in their lifetime than bucks with forked antlers as yearlings (Harmel et al. 1989). A more intensive analysis looking specifically at the inheritance of antler characteristics was later conducted on the Kerr WMA data, and most antler characters were considered to have moderate to high levels of heritability (Williams et al. 1994).

A similar heritability analysis was conducted on a data set of antler measurements and pedigrees of captive white-tailed deer housed at Mississippi State University. Analysis of data from this herd from 1977 to 1993 indicated that the occurrence of spike antlers in yearling bucks was related mostly to environmental factors rather than genetics (Lukefahr and Jacobson 1998). These environmental factors may include birth date (early or late in the year), nutritional status, health status, does' maternal ability, milk quantity, and more. Lukefahr and Jacobson (1998) also found that as bucks matured (2+ years), their genetic makeup played a more prominent role than environment in the size of their antlers.

These Mississippi and Texas (Kerr WMA) data have yielded apparently contradictory results, which has led to decades of controversy. Some of the differences in conclusions derived from these research efforts may simply be related to the use of different statistical methods to analyze the data (Waldron 1998). The Mississippi analysis (Lukefahr and Jacobson 1998) intuitively makes sense; one would expect antler characteristics of yearlings to be more dependent on environmental factors because their bodies are still growing and *extra* nutrients available for antler growth may be limited. Further support comes from evidence that yearling bucks born on a later date the previous summer may not grow antlers as large as those with earlier birth dates (Jacobson 1995, Lukefahr 1997). Late-born buck fawns have less time to obtain the nutrients necessary for body development and robust antler growth their first year.

The heritability estimates of several antler traits at Kerr WMA (Williams et al. 1994) differed substantially from those of Lukefahr and Jacobson (1997). Williams et al. (1994) did not account for birth date of yearling bucks, maternal influences, or the fact that many of the sires in the Kerr WMA herd were related to one exceptional sire buck (Lukefahr 1997, Waldron 1998). Environmental factors such as maternal influence on yearling characteristics in domestic animals are well documented (Waldron et al. 1993).

Still, the results of several studies of captive deer at Kerr WMA provide compelling evidence that the antler characteristics of yearling bucks are useful in predicting the size of their antlers as adults (predictability) and those of their future offspring (heritability). Ott et al. (1998) compared antlers of 4.5-year-old bucks and showed statistically that bucks having three or more antler points as yearlings greatly outperformed bucks that were spikes as yearlings. In another study throughout the 1990s, captive-born buck fawns were placed on a low-protein diet to simulate drought conditions, and the following year the yearling bucks with the best antler growth were used as sires to produce more fawns for the next year of research. In this manner, researchers tried to change the average antler size of the captive herd by selecting the largest antlered yearlings as sires each year. In this case, yearling antler size was a good indication of genetic potential for antler development because size-related antler characteristics (gross Boone and Crockett score, number of points, inside spread, main beam length, basal circumference, and total antler weight) all increased significantly as the study progressed (Frels et al. 2002).

To further complicate the relationship between genetics and antler development, consider that does carry half the genetic material for antler size and shape. Females cannot express these genes themselves, but can certainly pass them on to their male offspring. Some captive whitetail does consistently produce male offspring with superior antlers, even when bred to several different bucks (H. A. Jacobson, personal communication, 2000).

Evolution and Function

The origin and primary function of antlers was no doubt debated long before Aristotle in the fourth century BCE wrote about the effects of castration on antler development. Since Aristotle's time, much research has been conducted, and many interesting theories have been offered to explain why antlers evolved. The most commonly cited theories on how and why antlers evolved fall into five categories.

Heat Radiation

Stonehouse (1968) theorized that antlers served to dissipate excess heat from the deer during the hot summer months. The highly vascularized velvet that covers growing antlers is warm to the touch and undoubtedly radiates heat. However, if this were the primary reason for developing antlers, one would expect deer species in warmer climates to have larger antlers, which is not the case. Also, why would only males have developed this method of releasing excess heat? That deer strip off their velvet before the weather gets cold was used to strengthen the argument. However, Anthony Bubenik (1990) pointed out that deer evolved in a relatively warm climate and there would be no reason to lose their velvet seasonally. In addition, European roe deer (*Capreolus capreolus*) and Père David's deer (*Elaphurus davidianus*) grow antlers during the winter and have hardened antlers during the warm summer months.

Protection against Predators

Protection against attacks by predators has been offered as a driving force for the development of antlers. Members of the deer family do use their antlers to defend against predators; however, Geist (1966) aptly points out that females with young would be in greater need of weaponry to deter predators. In addition, males lose their antlers before the late winter/spring period, when they are in the poorest condition and most vulnerable to predator attacks.

Scent Dispersal

Anthony Bubenik (1990) proposed a theory of antler evolution that centers on their function as scent-dispersal structures. According to this theory, velvet-covered "antlers" in ancestral deer evolved as scent posts, situated high off the ground to allow the air currents to disperse the scent widely. The velvet of growing antlers contains a high density of sebaceous (oil) glands that contain pheromones (scents) in other mammals (Bubenik 1993). The elaborate antler designs that developed in many deer species are explained as simply a way to increase the scent-bearing surface area.

Deer sometimes apply scent to their growing, and then hardened, antlers. After velvet is stripped, deer have been observed rubbing secretions from other body glands (such as interdigital) onto their hardened antlers. Sometimes scent is rubbed from the body glands (such as preorbital and forehead area) onto vegetation and then rubbed on the antlers. Another, related behavior that might result in antlers being used as scent posts is the habit of some species of urinating directly onto the polished antlers or into a mud wallow and then digging with the antlers.

As with all other theories, this explanation has some weak points. Antler velvet is lost just before rut, when it would seem scent communication with the opposite sex would be at its peak. Scent communication is undoubtedly important to males during the period of velvet growth to establish their annual dominance hierarchy in bachelor groups; however, scents produced by sebaceous glands in the velvet would not assist them in recognizing one another later in polished antlers. Also, there is no evidence yet that sebaceous glands in the velvet contain any pheromones. Several scent glands on a deer's body produce an odor that is obvious to humans. It would seem that if scent dissipation were the primary driving force for the evolution of antlers, velvet would produce a more detectable odor than other sources of scent from the body.

Weapons for Fighting Other Males

The most popular and obvious explanation for the origin and function of antlers is their use as weapons to battle rival males for access to breeding females (Geist 1966, Clutton-Brock 1982, Goss 1983). It seems logical that throughout the process of natural selection, individuals with small bumps or bony abnormalities on the top of their skull might have an advantage during breeding season when battling other males, and might thereby pass on more genetic material than smooth-skulled individuals. After thousands of years the growth of more elaborate weapons would develop in a species.

That antlers grow in the summer, become polished and hard just before rut, and drop off after the rut seems to support the notion that they function as weapons for use during the breeding season. This theory also explains the presence of antlers in males and not females. It has been suggested that the annual dropping and regrowth of antlers is an adaptation to replace the weapons. Not only are the weapons increased in size each year until the deer reaches its prime, but any tines that are broken while fighting will be replaced with growth of new antlers before the next rut.

There is ample evidence that antlers are used by males extensively for fighting other males during the rut. Geist (1981) reported a high number of injuries to mule deer bucks thought to be caused by the antlers of other bucks. Despite their use as injury-inflicting offensive weapons during fights, they serve in a more important role as structures to engage other bucks in wrestling and shoving contests to establish dominance. The ritualized antler-to-antler shoving matches are an important component of deer breeding behavior.

Even though antlers are used extensively for fighting, this is not necessarily the reason they originally evolved. Fossil evidence indicates that there was very little breakage of primitive antlers in ancestral deer species. A. Bubenik (1983) interpreted this as evidence that antlers did not evolve primarily as weapons.

If antlers evolved solely based on their value as weapons, it is hard to explain the great diversity of antler shapes. Surely if fighting with other males were the only function of antlers, one weapon design would have emerged as the most efficient compromise between an offensive and defensive weapon. A single set of long spikes would be the most effective offensive weapon for inflicting wounds to an opponent. Conversely, large elaborate antlers, such as those of a moose, may be the best design for defending against attacks of other males. In addition, the presence of antlers in female reindeer cannot be explained by this theory. Nonetheless, that all antler designs are terminated with sharp points supports the idea that at least part of the evolution of antlers was based on their use as weapons.

Display and Intimidation

Antlers are luxury appendages that only grow large if other nutritional needs are satisfied. It is logical to assume, then, that in deer of similar age, large-antlered individuals are superior to small-antlered competitors in obtaining or utilizing resources.

One part of this theory states that large antlers on a buck help a female to select a visibly superior individual to mate with and perpetuate her genes. If only the largest antlered males are selected by females to breed, the size of the antlers would naturally evolve larger and larger through time. However, studies of red deer (*Cervus elaphus*) failed to provide evidence that females select the largest antlered males (Clutton-Brock 1982). Females of several deer species mate selectively with older, larger males, but the effect of antler size on their mate selection is complicated by other important factors, such as body size, strength, aggressiveness, and experience of the male. Anthony Bubenik (1983:440) used a dummy head (shoulder mount) of various deer species (moose, caribou, red deer, and wapiti) to test the reaction to different antler sizes. All else being equal, females showed more interest in larger antlers. Rival males paid little attention to small antlers on the dummy head, but became submissive in the presence of antlers larger than their own.

This leads to the second part of this theory—that large antlers can serve to intimidate male rivals before or during the rut. This is closely related to, and not incompatible with, the theory that antlers evolved primarily as weapons. Indeed, antlers would not serve as effective means of intimidation if they were not used as weapons periodically.

Fights for dominance are costly in terms of energy lost and the potential for life-threatening injuries. If a dominance hierarchy, or pecking order, can be established and maintained without excessive fighting, males will increase their survival. Very aggressive individuals prone to violent fighting would not survive to sire as many offspring. If the males with the largest antlers can establish dominance and simultaneously reduce their actual fighting by intimidating rivals, they will make the greatest contribution of their genes to future generations. This would increase the number of offspring sired by the largest antlered males and thereby promote the development of antlers through time.

Rutting behavior of many deer species supports the notion that antlers are used to display to both females and other males (A. Bubenik 1983). There are numerous reports of many different species of deer conspicuously displaying their antlers to females and rival males (Geist 1966). Because antlers are regrown each year, it appears logical that they could serve as visual clues to the maturity, genetic superiority, and condition of the male that particular year.

The largest antlers of all time belonged to the extinct Irish elk (*Megaloceros*), with an antler spread of 10–12 feet. This giant deer was neither exclusively Irish, nor an elk as we know them in North America. It is sometimes used as an example of a deer species whose antlers became so large they caused its extinction. This theory is nonsensical from an evolutionary standpoint, but tempting when one considers 87 pounds of antlers riding on a skull that weighed less than 6 pounds. Because such large antlers with backward-pointing tines would have made unwieldy and inefficient weapons, Gould (1974) invoked this display/intimidation theory as an explanation of the useful function of these enormous structures.

The great diversity in antler shapes and sizes provides strong evidence that more than one factor was involved in shaping antlers throughout the evolution of the deer family. If antlers served only one purpose in all deer species throughout their evolution, it seems one shape would have prevailed as the most efficient. It is most probable that a combination of some of these theories, and possibly others yet to be proposed, contributed to the evolution of these remarkable structures.

Abnormalities

One of the things that makes antlers interesting is the infinite variety of shapes, colors, textures, and unique characteristics. Discussion of abnormal points, such as "kickers," "stickers," "cheaters," drop tines, forks, double brows, and triple beams, consumes many hours of discussion among deer enthusiasts. It is the oddities and abnormalities, termed *nontypicals,* that capture our interest most often. A nontypical set of antlers is technically anything that differs from the normal species-specific antler pattern (figs. 2 and 6).

Because of the interest in nontypical antlers, much research has been directed at determining the causes of antler abnormalities. Many factors cause or affect the expression of antler abnormalities. These contributing factors include genetics, age, nutrition, injury, hormones, and disease.

Genetics

Many odd points and abnormalities are the result of the animal's genotype, or genetic blueprint. Antler characteristics are inherited from the buck's parents. A nontypical buck will frequently produce a disproportionate number of offspring with nontypical points. Remember that for each fawn born, the doe contributed

Figure 24. Antler palmation, as on this Coues whitetail buck harvested by Jeff Derrick, is usually attributed to genetic factors. *Photo by author*

half of the genetic material. Because of this, does that had nontypical fathers may consistently produce buck fawns that grow up to be nontypicals even though the fawns had different fathers. Sometimes a buck fawn grows up with antlers that look amazingly similar to those of his mother's father and not at all similar to those of his own father.

Genetically programmed antler abnormalities can be seen year after year in an individual. A buck may have a small bump on the outside of a rear tine at 2.5 years old, then a 2-inch kicker point (small abnormal point that grows outward) in the same spot the next year, and a 4-inch kicker on both rear tines at 4.5 years old. Palmated antlers, which are "webbed" like those of a moose, are a good example of a characteristic that is usually genetic in whitetails (Fig. 24). Injury to the growing antler can sometimes cause abnormalities for several years, which can be confused with genetic effects (see below, "Injuries").

Robinette and Jones (1959) in Utah observed fourteen mule deer bucks in the 1950s that were missing one or both antlers. They believed this was a genetic abnormality because all the bucks came from one localized area. Ryel (1963) documented four adult whitetail bucks that lacked one antler entirely and sixteen buck fawns with no pedicles by five to six months of age. This was assumed to be a genetic relationship because most of the abnormal fawns did not come from the areas with the poorest nutrition.

Age

Bucks do not usually start expressing abnormal points before at least three years of age. Younger bucks are still channeling most of their nutrient intake to body development, so they have less "extra" antler material to contribute to building numerous nontypical points. Once males obtain a nearly mature body structure at two years old, they then experience more robust antler growth and with that comes nontypical points. In addition, antlers can show epigenetic effects, where some genes do not activate at all until the animal is older and the proper hormonal environment exists.

Nutrition

Good nutrition means the buck has more energy to pass on to antler growth, and more total antler growth allows the buck to physically express his genetic traits. Regardless of the cause of the abnormality, good nutrition is necessary to provide the antler material to express it.

Matschke and Roughton (1977) reported that eleven out of seventy (15.7 percent) yearling bucks examined failed to grow antlers at 1.5 years of age. These bucks possessed the bony pedicles beneath the forehead skin but lacked antler development entirely. Many of these young bucks also lacked functional testes, which is abnormal for this age. The authors hypothesized that malnutrition was responsible for the absence of antlers and for the delay in puberty.

In Scotland, some red deer stags fail to develop antlers until the age of two or three years. This lack of antlers is thought to be caused by the poor level of nutrition. These antlerless stags, called "hummels," are not physically able to cope with the rigors of antler development.

Injury

Physical injury or trauma to the velvet antlers or a major skeletal structure can result in antler abnormalities. Nicks and cuts in the velvet antlers can produce points and oddities. This has led some to suggest that the awe-inspiring double-drop tine might be caused by bucks trying to slip through a fence and getting the underside of their mainbeams caught (and nicked) on a fence wire. As interesting as this sounds, most people agree that drop tines are mostly of a genetic origin. Bucks in velvet do a very good job of protecting their growing antlers because of their sensitivity.

Injuries to the buck's body can also cause antler abnormalities. Injury to a large skeletal structure, such as a broken leg bone, often causes a misshapen antler the next year. If a front leg is injured, either side of the rack may be affected; however, if the rear leg is injured, it is nearly always the antler on the opposite side that is malformed. Some have written this off to coincidence, although a significant amount of evidence from several species of deer shows that this contralateral (opposite-side) effect is real. Studies in Texas by Marburger et al. (1972) showed that amputation of a rear leg stunted the antlers on the opposite side in all six experimental animals. This phenomenon has also been reported in mule deer, whitetail, moose, caribou, elk, and muntjacs.

The cause of this contralateral effect is not known, but many theories have been set forth. Some have related it to the fact that the right side of the brain controls the left side of the body, and vice versa. Others have postulated that while still in velvet, the buck frequently turns to lick the injured leg, thereby repeatedly injuring the opposite side of the rack on nearby brush. This seems unlikely because of the care bucks take to protect their growing antlers; besides, the opposite antler is usually underdeveloped and not just deformed from repeated trauma.

Yet another explanation is that the uneven rack counterbalances the injured leg and allows the animal to move around more comfortably. This would be difficult to prove, but it is interesting to note that after amputation of a rear leg, this contralateral effect occurs annually with each set of antlers for the rest of the buck's life.

Injury to the pedicle (base) itself nearly always causes abnormalities. Extensive trauma to the pedicle before growth begins or soon afterward is the source of many large, freakish racks (G. Bubenik 1990b). Having two mainbeams on one side is usually the result of early damage to the pedicle. If one pedicle is injured severely, that side or both sides will be malformed during the next antler cycle.

In addition, nerves may "remember" an injury to the growing antler and reproduce nontypical antlers for several years (G. Bubenik 1990b). This "trophic memory" only occurs if the injury is substantial and happens in the early stages of antler growth when there is a high density of nerve connections in the growing antler tissue. Experimental damage to growing antlers of a buck that was tranquilized at the time does not produce the same trophic response.

"Acorn points" are a common antler oddity. This is a swelling in the middle of the hardened antler tine. Acorn points are caused when the buck bumps the growing tip of the tine on something hard. The tip is injured and, in the process of repairing itself, deposits more growing antler material at that location and then grows longer. If an antler tine breaks completely while still in the velvet phase, it may stay attached to the rack by velvet and re-fuse with it, leaving a hanging tine that usually has a large, rounded tip.

Researchers working with sika, roe, fallow, and red deer have also succeeded in making antlers grow out of abnormal places on the buck's skull (Jaczewski 1990). By grafting cells from a buck's pedicle to another place on the frontal bone (forehead), researchers have been able to produce deer with a third antler growing between the eyes. In this case, all three antlers undergo the normal sequence of growth, velvet loss, and antler shedding. One experiment involved grafting pedicle cells onto the leg of a European roe buck. Incredibly, a small antler grew on the leg, lost its velvet, and was later shed normally.

Figure 25. Coues whitetail harvested by Scott Pfeiffer in southeastern Arizona in 2002, showing an extra antler growing from the top of the eye orbit following an injury to that part of the skull. *Photo by author*

Trauma to the skull itself can induce antler growth at the site of the injury. Although not fully understood, there are good examples of injury-induced antlers (Fig. 25). Even does are not immune to this effect. In 1894, Blasius reported that a female European roe deer grew a 4-inch antler after a piece of window glass was accidentally driven into her scalp. Researchers have also observed this response in other cases of trauma to the frontal bone of female deer (Jaczewski 1990).

Hormones

The proper production of a variety of hormones is necessary to produce antlers. Abnormal or improperly timed fluctuations of hormones can cause irregular antler cycles or abnormal antler conditions.

Young fawns that are castrated before three to four months of age do not develop pedicles because they never produce the necessary testosterone levels required for this first step in antler development. Late injections of testosterone will initiate the production of a bony pedicle. A buck that is castrated while in velvet will never strip the velvet from his antlers because he lacks the rapid rise in testosterone that occurs in preparation for the rut. Without the subsequent decrease in testosterone after rut, a buck castrated in velvet does not drop his velvet antlers. The buck then continues to grow more antler material in the next antler cycle, never shedding the antlers, until he finally carries a grotesque mass of often stunted velvet antlers. These bucks are sometimes referred to as "cactus bucks" (Mearns 1907:196, Baber 1987).

Antler growth is still possible in these cases because high levels of testosterone are not needed. Low levels of male hormones (androgens) are produced by other structures besides the epididymis of the testicles (such as the adrenal cortex). These low levels are enough to grow antler material, but not enough to stimulate the closure of blood vessels to the antler, which completes velvet drying and antler shedding. These latter two processes require a sharp increase and then decrease in the testosterone level.

Bucks castrated while in the hardened antler stage will drop their antlers within a few weeks because of the sharply falling testosterone level. Bucks castrated after they drop their hardened antlers will grow new antlers the next year, but these will never be polished or shed because of the lack of sufficient hormone levels. In the 1840s naturalist John J. Audubon (1989) described two castrated whitetail bucks he observed: "Their horns continued to grow for several years; the antlers were of enormous length and very irregularly branched, but the velvet was retained on them. . . . they had become very large and when first seen at a distance we supposed them to be elks."

Tumors can affect the hormonal environment and thereby cause antler abnormalities, such as antlered does or cactus bucks. Viruses such as epizootic hemorrhagic disease (EHD) can also cause hemorrhaging in the testicles or other effects that disrupt the proper production and circulation of male hormones that regulate antler growth.

Disease

Besides disease-caused hormonal imbalances, various parasites and viruses can contribute to the occurrence of antler abnormalities by directly affecting growing antlers. Some viruses, such as EHD, can also result in incomplete hardening of the antler tips because of damage to the velvet's blood vessels before antler growth is complete. This disease commonly strikes deer in late July and August, that is, during the last stages of antler growth.

Ticks are common on velvet antlers; blood-sucking parasites can cause malformed antlers if they disturb the velvet and disrupt the flow of nutrients to the growing antler. Knipe (1977:13) reported an antler abnormality in a captive Coues whitetail that was caused by a screw worm infestation. A screw worm burrowed into the base of the antler when the velvet was freshly stripped and caused the following year's antler growth to be malformed.

There are many possible causes of antler abnormalities, but injuries and genetic factors account for the majority of those cases observed in the field. Oddities and freaks are not normally tolerated by nature; natural selection quickly removes them from the gene pool. However, an enormous nontypical freak is certainly not at a disadvantage when it comes to battling for dominance.

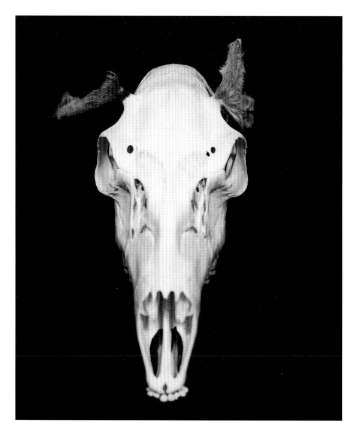

Figure 26. Desert mule deer doe with antlers. This doe was harvested as a buck, but her reproductive tract indicated she had raised a fawn the previous year. *Photo by author*

Antlered Does

The earliest reference to an antlered white-tailed doe comes from a label on an antler in the British Museum of Natural History: "This horn grew on the frontlet of a DOE in New England in America, 1607" (Owen 1853, in Donaldson and Doutt 1965). Since then, antlered does have been reported in European roe deer, fallow deer, red deer, wapiti, whitetails, blacktails, and mule deer (Wislocki 1954). Antlered does are reported occasionally in the Southwest (Mearns 1907:196, Knipe 1977:14) (Fig. 26). An antlered mule deer doe was seen several consecutive years accompanied by fawns on the Kaibab Plateau (Diem 1958).

    Some of the antlered "does" that have been reported in the past were males with malformed external genitalia. Many early reports of antlered does were not accompanied by a discussion of the condition of the *internal* reproductive tract.

Looking at the arrangement of the internal organs is crucial when one inspects an antlered deer of questionable gender.

Three different types of animals are commonly referred to as "antlered does": hermaphrodites, cryptorchid males, and true antlered does (Wislocki 1954).

Hermaphrodites

These are deer that possess both male and female sex organs. They usually have an ovary on one side and an internal testicle on the other (lateral hermaphrodites). These animals are not capable of reproducing and because of the presence of testosterone, usually carry out a normal antler cycle: stripping velvet, polishing, and shedding antlers. These deer normally have polished antlers during the fall hunting season unless the ovary produces enough female hormone to offset the effects of the testosterone.

Cryptorchid Males

These are not really does at all, but rather bucks with testicles that never descended into the scrotum. They remain inside the body cavity encased in fatty tissue. The penis is also inside the body and the buck urinates through an opening that looks very much like female genitalia. These animals clearly look like does but are actually males. Sometimes they are called "pseudohermaphrodites" because they look like a hermaphrodite but lack any female reproductive tissue. They usually have polished antlers in the fall because of the presence of testosterone produced by the internal testicles. Swank's (1958:19) 4×4 mule deer "doe" is most likely a cryptorchid male.

True Antlered Does

Most true antlered does have fully functional female reproductive tracts. These does breed, become pregnant, and successfully raise fawns (Diem 1958) (Fig 26). What actually initiates the antler development in these does is still somewhat of a mystery. It is known that the testicles are not the only source of testosterone to initiate the growth of pedicles; the adrenal gland in both sexes can also produce this male hormone. Castrated males have been shown to have adequate levels of testosterone in their bloodstream to initiate antler development, but it is not known if this is the initiating influence in antlered does. Researchers have been able to grow antlers experimentally in does by giving them a single shot of testosterone. The antlers developed normally, but the velvet was not lost because they lacked the sudden increase in testosterone in the prerut period. When given a second testosterone shot, the does' velvet dried and was promptly stripped off.

Because of the lack of increasing testosterone levels at the end of summer, true antlered does in the wild do not have polished antlers. The antlers never lose their velvet and are often deformed, lacking basal burrs, and permanent (not shed).

Some antlered females have been found to have a cyst or tumor on an ovary or their adrenal gland. Such tumors can disrupt the normal balance of hormones and provide enough male-type hormones (androgens) to initiate antler growth. Physical injury to the frontal bones of the skull can also induce the development of antlers in does. Injuries and tumors may explain some, but not all of the antlered does reported in the wild. An antlered doe in Mesa Verde National Park (southwestern Colorado) produced fawns for several years. Remarkably, one of her female fawns later grew 5-inch antlers when 1.5 years old. This indicates that the physiological cause of that antler was inherited from her mother (Mierau 1972).

Antlered does have been reported in the scientific literature in Arizona, Colorado, Idaho, Iowa, Massachusetts, Michigan, Missouri, New York, Pennsylvania, Texas, Utah, and Washington. Out of 163,000 bucks harvested in Pennsylvania in 1958–61, 31 so-called antlered does were documented (Donaldson and Doutt 1965). Of these, 17 were functional females (one with a tumor), 4 were cryptorchid males, and 1 was a hermaphrodite. All 17 females were still in velvet, while the 4 cryptorchid males had polished antlers.

Chapter 5

Diet and Water Requirements

Important Southwestern Deer Foods

NY DISCUSSION of deer diets must begin with a look at the main food items. Pictured in this section are a few important plants in the diet of southwestern deer (Figs. 27–38).

Deer Diets

Kufield et al. (1973) summarized 99 studies of mule deer diet and reported that 788 different species of plants were eaten by mule deer. Despite the abundance of diet studies throughout the country on whitetails and mule deer, it is still difficult to make generalized statements about what plants deer eat. The species composition of plants consumed by deer varies tremendously by geographical area, elevation, season, and year. Some shrubs that are relished by deer do not appear in any deer diet studies simply because no studies were conducted where those plants were common. Obviously, deer living in different vegetation associations will not eat the same proportions of the same plant species. Mule deer in southern Colorado could not have the same diet as those in Sonora, Mexico. Fluctuating climatic conditions and the resulting effect on the vegetation causes major differences in plant growth and nutritional quality.

With this in mind, it is evident that one short-term study on "what deer eat" could not possibly describe the complicated fluctuations in deer diets. Some studies were conducted during a time when deer were nutritionally stressed, so the most prevalent food items consumed may be those available at the time, rather than the types of food deer prefer. In addition, studies have been conducted using very different methods to determine which foods are important to deer—for example, examining rumen contents or fecal pellet contents, watching wild or tame deer eat, and feeding tame deer different plants. All this complicates our ability to compare different studies. Even studies using the same methods may report the results in different ways. For example, some data are presented as the percentage of sampled deer that ate each plant species, while other studies reported the percentage of the stomach contents that a particular plant amounted to.

In spite of these limitations, however, looking at the major deer foods reported in the more significant studies allows us to piece together a picture of which plants are the most nutritionally important to southwestern deer.

Mule Deer

Through the annual, seasonal, and elevational changes in diet, deer rely heavily on browse for much of the year (Table 6). Although mule deer use browse species heavily in most seasons, it is especially important during times of less rainfall. Depending on precipitation and temperature, diets shift heavily in favor of forbs (broad-leaved weeds) when they become available following the summer or winter rainfall periods. In the Chihuahuan Desert, a higher percentage of the annual rainfall occurs in the summer; the resulting summer forb growth is a more significant factor for mule deer in that part of the Southwest. Cactus and shrub/tree fruits are always used intensively where and when they occur in abundance. Most deer studies have shown that grass makes up a very small portion of deer diets, usually being important only for a short time when it is the first vegetation to green up in the spring.

Other plants, such as mistletoe, can make important contributions to a deer's diet. Mistletoe is known to be relished by deer, so we should not be surprised to learn that it is high in carbohydrates and is easily digestible (Urness 1969). Mistletoe is not normally high in protein, but one of the desert varieties *(Phoradendron californicum)* can run as high as 15–25 percent protein.

The first organized study on deer diets in the Southwest was the pioneering work by A. A. Nichol (1938), one of the first research projects conducted with the new Wildlife Restoration Funds derived from the Pittman-Robertson Act of 1937 (see Table 3). Nichol assembled a collection of seven mule deer from the Kaibab Plateau and one from southwestern Arizona, and held them in captivity on the Santa Rita Experimental Range south of Tucson. Throughout the study, these deer were fed 168 different native plant species to determine which ones they considered palatable. By measuring how much of each plant species the deer consumed, Nichol developed a list of the plants deer preferred. This was a step in the right direction, but the diet of deer in the wild is dependent on what plants are available

TABLE 6. SEASONAL PERCENT (%) COMPOSITION OF MULE DEER DIETS THROUGHOUT THE SOUTHWEST

Location	Spring Browse	Spring Forb	Spring Other	Summer Browse	Summer Forb	Summer Other	Fall Browse	Fall Forb	Fall Other	Winter Browse	Winter Forb	Winter Other	Source
Southeastern California													
Imperial Co.	48–62	41–63	0–8	68–73	25–43	0–3	51–89	8–45	3–24	66–93	7–34	0–10	Marshal et al. 2004
Southwestern Arizona													
Harquahala Mts.	30.7	67.7	1.1 cacti	77.8	22.2	1.8 cacti	88.8	10.7	...	52.6	46.7	0.4 cacti	Krausman et al. 1989
Belmont Mts.	86	12.3	...	72.3	26.3	...	95.5	1.5	3.0 cacti	68.9	25.1	...	Krausman et al. 1997
Picacho Mts.	77.2	21.4	...	97.3	1.3	...	94.7	0.3	...	60.8	38.4	0.4 cacti	Krausman et al. 1997
Kofa NWR	66.1	33.5	0.4 grass	87	6	3 cacti	84.2	14.2	0.9 cacti	64.6	34.8	0.3 cacti	Krausman et al. 1997
Central Arizona													
Three Bar Wildlife Area	...	...	...	32	6	62 shrub fruit	77–92	5–22	5–6 grass	53–85	8–45	2–6 grass	McCulloch 1973
Prescott	90	10	...	...	...	...	...	...	...	99	...	...	McCulloch 1978
Bloody Basin	38	62	...	...	...	...	...	...	...	70	30	...	McCulloch 1978
Texas													
Panhandle	51.4	41.2	...	72.7	14.3	...	65.2	25.6	...	42.7	34.5	...	Sowell 1981
Stockton Plateau	51.7	25	1.5	45.2	29.5	1	36.9	40.4	...	21.2	38.3	9.3	Ratliff 1980
Trans-Pecos (Longfellow Ranch)	56.1	35.5	1.3 grass	56.8	28.5	3.2 grass	34.1	52.7	4 grass	34.7	51.3	5 grass	Keller 1975
Big Bend NP	9.6	3.8	70.6 succulents	33.5	18.1	22.0 succulents	35.2	24.8	26.1 succulents	16.2	17.4	55.0 succulents	Krausman 1978a
Trans-Pecos (Black Gap)	...	...	...	69.4	27.2	3.4 grass, cacti	45.1	27.7	24.7 cactus	67.6	27.1	3.5 cactus	Brownlee 1981
Guadalupe Mt. National Park	74	23	3 grass	85	14	1 grass	79	19	2 grass	70	28	2 grass	Krysl 1979

(*continued*)

TABLE 6. (CONTINUED)

Location	Spring			Summer			Fall			Winter			Source
	Browse	Forb	Other	Browse	Forb	Other	Browse	Forb	Other	Browse	Forb	Other	
Southern New Mexico													
Guadalupe Mts.	68.3	30.7	...	62.1	33.7	...	71.2	28.4	...	80.4	16.5	...	Anderson et al. 1965
Sacramento Mts.	84.1	14.1	1.8 grass	72.8	25.4	1.8 grass	...	...	...	80.9	18.9	1.1	Mahgoub 1984
Southwestern New Mexico													
Ft. Bayard	...	...	...	79.7	18.1	2.2	53.2	39.1	7.7	94.0	6.0	...	Tofoya et al. 2001
Ft. Bayard	58	32	10	50	42	8	87	4	9	94	2	4	Short et al. 1977
Ft. Bayard	60	30	...	56	42	...	86	6	...	94	4	...	Boeker et al. 1972
North-central Arizona													
Beaver Creek	...	...	...	...	...	...	...	...	...	60	34	5	McCulloch 1961
Beaver Creek [Juniper]	66–69	21–24	7–13	51–75	24–47	1–2	60	39	1	53–94	5–39	1–8 grass	Neff 1974
Beaver Creek [Ponderosa Pine]	23	45	32 grass	53	44	3 grass	58	40	2 grass	64	33	3 grass	Neff 1974
Hualapai Tribe	...	...	...	...	...	...	43	33	16 acorn	...	...	...	McCulloch 1978
Northern Arizona													
Kaibab Plateau	...	...	...	21–52	16–49	4–47 grass	...	...	...	...	...	...	Hungerford 1970
Kaibab Plateau	...	...	...	25	66	9 grass	...	...	...	63	22	14 grass	McCulloch 1978
Southwestern Colorado													
Mesa Verde NP	45.0	31.6	21.2 grass	71.9	23.8	3.2 grass	51.8	16.7	29.6	87.0	0.6	10.5 grass	Mierau and Schmidt 1981

Note: Percentages were determined by many different methods.

to them in the habitat they occupy. In the decades since that early work, researchers have assembled dietary information on mule deer herds from many diverse areas of the Southwest (Table 7). An analysis of desert mule deer diets throughout their range found over 185 plants species that make up at least 1 percent of their diet (Krausman et al. 1997).

Southwest Deserts

In areas with higher plant diversity the total number of species eaten would be higher. In the low-elevation southwest deserts of Baja, Sonora, southwestern Arizona, and southeastern California, browse (shrub species) is the staple food source, supplemented by forbs (broad-leafed weeds) and cactus fruits when they are available. The average annual diet of desert mule deer in this region consists of over 75 percent browse and 22 percent forbs (Krausman et al. 1989, 1997). Browse species, such as jojoba, ironwood, desert ceanothus, and palo verde, provide a year-round food source, while forbs, such as globemallow, lupine, filaree, and buckwheat, account for a significant portion of the winter-spring diet because of their availability after the winter rains (Table 7; see appendix for scientific names). Grasses make up less than 5 percent; the use of cactus fruits is sporadic and can be quite high at times.

Sonoran Desert Grassland and Chaparral

Diets of mule deer in the Sonoran Desert grassland and chaparral have been studied most intensively on the Three Bar Wildlife Area and to a lesser extent in a few southeastern Arizona populations. McCulloch (1973) summarized much of the research in this part of mule deer range and reported 106 species of plants eaten by mule deer. In the spring, browse items such as jojoba, fairy duster, desert ceanothus, and mountain mahogany are important because they are high in protein and relatively easy to digest thanks to the low fiber and high water content of the new growth.

Forbs and succulents like buckwheat, deer vetch, and spurge are important when adequate winter rain falls (Swank 1958, McCulloch 1973). As early summer approaches and the range dries out, deer diets shift to shrub fruits (beans and nuts), cactus, cactus fruit, and the leaves of some browse, such as fairy duster (Short 1977). Shrub fruits provide a good source of phosphorus, and browse leaves continue to be attractive and nutritious because they can take advantage of deeper soil moisture than the shallow-rooted forbs.

Summer rains in July offer an increase in forbs from the abundant seeds that lie dormant in the soil. During this time the consumption of browse (range ratany, kidney wood, fairy duster, jojoba) is supplemented with forbs such as ayenia, metastelma, buckwheat, spurge, deer weed, and penstemon, as long as they remain available. As the summer's green forb growth begins to cure in the coming fall, deer switch more heavily to protein-rich browse (jojoba, fairy duster,

TABLE 7. THE MOST IMPORTANT SOUTHWESTERN MULE DEER FOOD ITEMS BY SEASON

Location	Spring	Summer	Fall	Winter	Source
Mexico (Baja del Norte)	oak, ceanothus, chamise, redberry, cactus fruits	→			Leopold 1959
Mexico (Cedros Island, Baja)	elephant tree, jojoba, lentisco, California copperleaf, mission manzanita, boundary ephedra, lemonade berry, buckwheat	→			Perez-Gil 1981
Northern Mexico	ironwood, dogweed, atriplex, palo verde, cactus fruits, sotol	→			Leopold 1959
Mexico (LaMichilia–Durango)	mistletoe, manzanita, ebony, condalia, juniper, oak, madrone, bacharis, dalea, sedge, deer vetch	→			Ezcurra et al. 1980
Southwestern California (Fort Pendleton)	coastal sagebrush, chamise, black sage, white sage, laurel sumac, oak, buckwheat, lemonade berry, monkey flower, filaree	→			Pious 1989
Arizona (captive deer)[a]	aspen, desert hackberry, mountain hackberry, fairy duster, buckwheat, fendler ceanothus, fendlera, mistletoe, careless weed, lamb's quarters, filaree, poppy, purslane, sowthisle, talinum	→			Nichol 1938

Location					Reference
Southeastern California (Imperial & Riverside Co.)	ironwood, mesquite, black willow, galleta, yucca, agave, saltbush, ocotillo, palo verde				Longhurst et al. 1952:120
Southeastern California (Imperial Co.)	palo verde, brittlebush, ironwood, arrowweed, buckwheat	burro-weed, brittlebush, ironwood, mesquite	palo verde, ironwood, arrowweed, prickly pear, buckwheat	palo verde, ironwood, arrowweed, burro-weed, saltbush, buckwheat	Marshal et al. 2004
Southwestern Arizona (Harquahala Mts.)	buckwheat, lupine, globemallow, white sage	turbinella oak, jojoba, buckwheat, lupine, globemallow, desert ceanothus, ironwood	jojoba, globemallow, tidestromia, ironwood, copperleaf, desert ceanothus	jojoba, buckwheat, globemallow, lupine, borage	Krausman et al. 1989
Southwestern Arizona (Belmont Mts.)	ironwood, jojoba, filaree, lupine	lupine, ironwood, jojoba, desert vine, tidestromia	ironwood, jojoba	ironwood, jojoba, catclaw	Krausman et al. 1997
Southwestern Arizona (Picacho Mts.)	desert vine, jojoba, buckwheat, borage	desert vine, jojoba	desert vine, jojoba, globemallow	jojoba, paperflower, globemallow, lupine, borage, buckwheat	Krausman et al. 1997
Southwestern Arizona (King Valley, Kofa NWR)	ironwood, filaree, ratany	ironwood, dalea, mesquite	ironwood, dalea, filaree	ironwood, plantain, dalea, filaree	Krausman et al. 1997
Southwestern Arizona	jojoba, buckwheat, deer vetch	ayenia, deer vetch, prickly pear cactus	barrel cactus, ayenia, metastelma	barrel cactus, ayenia, metastelma	Truett 1971
Southeastern Arizona (Dos Cabeza Mts.)	...	...	whitethorn acacia, mountain mahogany, buckwheat, alligator juniper berries	...	Anthony and Smith 1974
Southeastern Arizona (San Cayetano Mts.)	...	ratany, wire-lettuce, kidney wood, mountain hackberry, fairy duster	...	...	Anthony 1976
Southeastern Arizona (Santa Rita Exp. Range)	jumping cholla fruit, catclaw, eriastrum, spurge	mesquite beans, fairy duster, spurge, prickly pear fruit	catclaw, fairy duster, barrel cactus fruits	cane cholla fruits, barrel cactus fruits, fairy duster	Short 1977

(continued)

TABLE 7. (CONTINUED)

Location	Season				Source
	Spring	Summer	Fall	Winter	
Central and Southeastern Arizona	desert ceanothus, holly-leaf buckthorn, juniper, mountain mahogany	...	...	...	Swank 1958
Central Arizona (Bloody Basin)	desert ceanothus, spurge, white sage	...	...	holly-leaf buckthorn, juniper, white sage	McCulloch 1978
Central Arizona (Three Bar Wildlife Area)	fairy duster, jojoba, ratany, desert ceanothus, buckwheat, spurge	mesquite beans, turbinella oak acorns, fairy duster, jojoba, buckwheat, spurge	fairy duster, jojoba, spurge, deer weed	jojoba, filaree, spurge, white sage, birch-leaf mountain mahogany, fairy duster, grassnuts, buckwheat	McCulloch 1973
Central Arizona (Prescott)	birch-leaf mountain mahogany, skunkbush	...	...	turbinella oak, birch-leaf mountain mahogany, manzanita	McCulloch 1978
Texas Panhandle (Clarendon and Canadian River)	skunkbush, juniper, sand sagebrush, half-shrub sundrop, white sage, bladderpods	skunkbush, mountain mahogany, half-shrub sundrop, white sage	skunkbush, juniper, half-shrub sundrop, white sage	juniper, half-shrub sundrop, bladderpods, skunkbush	Sowell 1981
Texas (Stockton Plateau)	acacia, dalea, Mohr shrub oak, littleleaf sumac, spurge, needleleaf bluets, bladderpod, plantain, milkwort	acacia, juniper, Mohr shrub oak, littleleaf sumac, skeletonleaf goldeneye, dalea, spurge, gumhead, needleleaf bluets	acacia, juniper, Mohr shrub oak, skeletonleaf goldeneye, dalea, gumhead, needleleaf bluets, longstalk greenthread	juniper, Mohr shrub oak, gumhead, needleleaf bluets, longstalk greenthread, sedge, grass	Ratliff 1980

Location				Reference
Texas Trans-Pecos (Del Norte, Sierra Diablo, Black Gap)	littleleaf sumac, yucca, skunkbush, Apache plume	wavyleaf oak, lechuguilla, littleleaf sumac, spurge, ditaxis, Apache plume	Mohr oak, lechuguilla, wavyleaf oak, Apache plume, littleleaf sumac, prickly pear, fourwing saltbush	Mohr oak, wavyleaf oak, lechuguilla, spurge, sotol, yucca, prickly pear, cholla, juniper, Apache plume, littleleaf sumac, ditaxis	Uzzell 1958
Texas Trans-Pecos (Black Gap)	...	lechuguilla, spurge, acacia	acacia, cactus fruits	lechuguilla, guayacan, skeletonleaf goldeneye, spurge	Brownlee 1981
Texas Trans-Pecos (Longfellow Ranch)	acacia, Emory oak, spurge, juniper, dalea	mesquite, dalea, littleleaf sumac, spurge, acacia, skeletonleaf goldeneye	dalea, Emory oak, spurge, littleleaf sumac, skeletonleaf goldeneye	broom snakeweed, bladderpod, dalea, juniper, Emory oak	Keller 1975
Texas Trans-Pecos (Black Gap)	guayacan, mesquite, cenizo, lechuguilla, ephedra, yucca, catclaw, spurge, desert hibiscus, milkwort, dalea, bahia, silktassel, flax				Wallmo 1960
Texas (Big Bend NP)	oaks, lechuguilla, prickly pear cactus	acacia, littleleaf sumac, dalea, spurge, prickly pear cactus	guayacan, spurge, lechuguilla, prickly pear cactus	evergreen sumac, menodora, prickly pear cactus, lechuguilla	Krausman 1978a
Texas (Guadalupe Mt. National Park)	oak, desert ceanothus, mountain mahogany, Apache plume, littleleaf sumac, bladderpod	oak, desert ceanothus, mountain mahogany, Apache plume, skunkbush, globemallow	oak, mountain mahogany, Apache plume, juniper, desert ceanothus, globemallow	oak, mountain mahogany, juniper, bladderpod, leatherweed croton, Apache plume	Krysl 1979
Southern New Mexico (Guadalupe Mts.)	wavyleaf oak, yucca, mountain mahogany	wavyleaf oak, skunkbush, juniper, dogweed, white sage, spurge	wavyleaf oak, mountain mahogany, flax, juniper, skunkbush	juniper, yucca, mountain mahogany, silktassel	Brown 1961
Southern New Mexico (Guadalupe Mts.)	wavyleaf oak, birchleaf mountain mahogany, yucca	wavyleaf oak, juniper, skunkbush, white sage, spurge, dogweed	wavyleaf oak, juniper, skunkbush, flax	juniper, birchleaf mountain mahogany, yucca, silktassel, flax	Anderson et al. 1965

(*continued*)

TABLE 7. (CONTINUED)

Location	Spring	Summer	Fall	Winter	Source
Southern New Mexico (Sacramento Mts., Black Range)	white fir, misc. forbs, lichens, oak, aspen, mountain mahogany, Utah honeysuckle, alligator juniper	oak, skunkbush, mountain mahogany, misc. forbs, alligator juniper	alligator juniper, oak, pinyon pine	alligator juniper, pinyon pine, oak, mountain mahogany, misc. forbs	Taylor 1961 and 1962
Southern New Mexico (Sacramento Mts.)	wavyleaf oak, juniper, skunkbush, pinyon pine, mountain mahogany, buckwheat, globemallow, vervain	wavyleaf oak, juniper, skunkbush, pinyon pine, mountain mahogany, Apache plume, buckwheat	wavyleaf oak, juniper, skunkbush, mountain mahogany, Apache plume, brickellia, buckwheat	...	Mahgoub et al. 1987
Southern New Mexico (Capitan Mts.)	Fendler's ceanothus, Gambel oak	Gambel oak, anemone, Fendler's ceanothus, vetch, lichen, aspen	Gambel oak, Fendler's ceanothus	...	Stewart 1957
Southwestern New Mexico (Fort Bayard)	...	oak, mountain mahogany, juniper, gourd, globemallow	mountain mahogany, silktassel, oak, pigweed	mountain mahogany, oak, silktassel, prickly pear cactus	Tafoya et al. 2001
Southwestern New Mexico (Fort Bayard)	oak, birchleaf mountain mahogany, buffalo gourd, tansymustard, spurge	oak, birchleaf mountain mahogany, fendler ground cherry, dayflower, spurge	oak, birchleaf mountain mahogany, silktassel, skunkbush	oak, birchleaf mountain mahogany, silktassel, skunkbush	Boeker et al. 1972
Southwestern New Mexico (Fort Bayard)	mountain mahogany, gray oak, white sweet clover, silktassel, heart-leaf goldeneye	mountain mahogany, gray oak, dalea, morning glory, bird's bill dayflower, James bundleflower	mountain mahogany, gray oak, dalea, white sweet clover, deer vetch, heart-leaf goldeneye	mountain mahogany, gray oak, silktassel, alligator juniper	Hunt 1978

Location				Reference
Southwestern New Mexico (Fort Bayard)	oak, birchleaf mountain mahogany, tansymustard	oak, birchleaf mountain mahogany, skunkbush, dayflower, James bundleflower	oak, birchleaf mountain mahogany, juniper, silktassel	oak, birchleaf mountain mahogany, juniper	Short et al. 1977
Central New Mexico (Cibola N.F.)	white fir, juniper, pinyon pine, grass, mountain mahogany	oak, misc. forbs, mountain mahogany, purslane	oak, mountain mahogany, misc. forbs, skunkbush	oak, mountain mahogany, pinyon pine, juniper, misc. forbs	Illige 1956
Northern New Mexico (Cimarron Canyon)	bearberry, ponderosa pine, Oregon grape, mountain mahogany, aspen	Oregon grape, misc. forbs, mountain mahogany, Gambel oak, bearberry, aspen	chokecherry, cottonwood, rose, Douglas fir, Gambel oak	chokecherry, Oregon grape, Douglas fir, penstemon, woodrush	Taylor 1963
North-central Arizona (Beaver Creek)	...	...	Utah/alligator juniper, turbinella oak, mountain mahogany, penstemon	...	McCulloch 1961
North-central Arizona (Beaver Creek [Juniper])	mountain mahogany, squirreltail grass, turbinella oak, buckwheat	mountain mahogany, cliffrose, clover	buckwheat, clover, mountain mahogany, cliffrose	mountain mahogany, turbinella oak, buckwheat, catclaw	Neff 1974
North-central Arizona (Beaver Creek [Ponderosa Pine])	Gambel oak, grasses, clover, orchard grass	Gambel oak, deer vetch, buckwheat, slender milkvetch	Gambel oak, buckwheat	Gambel oak, ponderosa pine, grasses	Neff 1974
Northern Arizona (Kaibab Plateau summer range)	...	aspen, white fir, daisy, mountain dandelion, clover, orchard grass	...	...	Hungerford 1970
Northern Arizona (Kaibab Plateau)	aspen, cliffrose, serviceberry, mountain mahogany, elderberry, horsemint, Indian paint brush, deer vetch, lewisia, clover				Julander 1937

(*continued*)

TABLE 7. *(CONTINUED)*

Location	Spring	Summer	Fall	Winter	Source
Northern Arizona (Kaibab Plateau)	...	aspen, white fir, milkvetch, Oregon grape	cliffrose, penstemon, buckwheat, sagebrush	cliffrose, sagebrush, Apache plume, Utah juniper, pinyon pine	McCulloch 1978
South-central Colorado (Fort Garland)	big sagebrush, fringed sagewort, mountain mahogany, bladderpod, pinyon pine	mountain mahogany, sedge, fescue, bluegrass, fleabane, Gambel oak, juniper	big sagebrush, mountain mahogany, pinyon pine, juniper, rabbitbrush	big sagebrush, pinyon pine, fringed sagewort, mountain mahogany, brome	Hansen and Reid 1975
Southwestern Colorado (Mesa Verde NP)	serviceberry, big sagebrush, milkvetch	serviceberry, Gambel oak, common chokecherry, mountain snowberry, milkvetch	serviceberry, antelope bitterbrush, rubber rabbitbrush, mountain mahogany	big sagebrush, rubber rabbitbrush, pinyon pine, Utah juniper	Mierau and Schmidt 1981
Southwestern Utah (Harmony Mt. and Pine Valley)	...	Gambel oak, serviceberry, mountain mahogany, bigtooth maple, aster, deer vetch, water birch	Gambel oak, serviceberry, mountain mahogany, big sagebrush, deer vetch, mushrooms	...	Beale and Darby 1991:16
Central Utah (Fishlake N.F.)	grass	aspen, chokecherry, oak, elderberry, snowberry, lupine, penstemon, clover	...	big sagebrush, cliffrose, bitterbrush	Smith 1952

Note: See the appendix for scientific names of plants. Arrows indicate a span of more than one season in which these items were eaten by deer.

[a] Captive deer were fed 168 species of native plants; those listed are the species deer preferred.

Figure 27. Fairy duster (*Calliandra eriophylla*). *Photo by author*

Figure 28. Buckwheat (*Eriogonum wrightii*). *Photo courtesy of Dan Robinett/NRCS*

Figure 29. Mountain mahogany (*Cercocarpus* spp.). *Photo by author*

Figure 30. Desert ceanothus (*Ceanothus greggii*). *Photo by author*

Figure 31. Jojoba (*Simmondsia chinensis*). *Photo by author*

Figure 32. Spurge (*Euphorbia melenadenia*). *Photo by Clay McCulloch*

Figure 33. Skunkbush sumac (*Rhus trilobata*). *Photo by author*

Figure 34. Holly-leaf buckthorn (*Rhamnus crocea*). *Photo by author*

Figure 35. Cliffrose (*Cowania mexicana*). *Photo by author*

Figure 36. Big sagebrush (*Artemisia tridentata*). *Photo by author*

Figure 37. Gambel oak (*Quercus gambelii*). *Photo by author*

Figure 38. Trembling aspen (*Populus tremuloides*). *Photo by author*

catclaw acacia, mountain mahogany) and the fruits of prickly pear and barrel cactus. When fairy duster loses its leaves in early winter and the yellow fruits of the barrel cactus become scarce, deer become more reliant on cane cholla fruit, browse, and forbs produced by the winter rains (Table 7).

Chihuahuan Desert

Seasonal relationships between the use of browse and forbs in the Chihuahuan Desert grasslands/shrublands of Chihuahua, Coahuila, New Mexico, and West Texas are much the same as in the Sonoran Desert. However, in the Chihuahuan Desert most of the annual rainfall comes during the summer. Summer storms bring precipitation that fuels a late summer greenup and allows mule deer to switch from cactus fruits and browse to vitamin-rich forbs (Table 6).

Oaks play a more prominent role in the Chihuahuan Desert, with wavyleaf, Emory, and Mohr shrub oak being very important in the Texas Trans-Pecos and southern New Mexico (Table 7). In addition to the oaks, littleleaf sumac, skunkbush, Apache plume, and mountain mahogany provide the browse forage during the spring months. Forbs such as spurge, dalea, globemallow, tansymustard, and milkwort are also important when provided by winter moisture.

Although forbs figure prominently in spring, summer, and fall diets, the greenup following the late summer monsoon rains is the most significant. This is reflected in a noticeable shift to summer foods like skeletonleaf goldeneye, spurge, dalea, buckwheat, dayflower, dogweed, and needleleaf bluets. Mule deer diets on Fort Bayard in southwestern New Mexico were reported to be 42 percent forbs during this time (Boeker et al. 1972, Short et al. 1977).

As the range dries and the summer forbs disappear, deer switch to the dependable browse species like skunkbush, juniper, Mohr shrub oak, wavyleaf oak, littleleaf sumac, Apache plume, silktassel, and mountain mahogany. To these browse species are added forbs like buckwheat, spurge, sketelonleaf goldeneye, dalea, gumhead, and longstalk greenthread. The later the summer rains, the larger role forbs may play in deer diets at this time of year.

During winter, mule deer diets shift in favor of browse species, with 80–94 percent of southern New Mexico diets consisting of browse, such as wavyleaf oak, silktassel, juniper, mountain mahogany, and skunkbush (Table 6). In Big Bend National Park in West Texas, Krausman (1978a) reported that mule deer diets throughout the year were 22–70 percent succulents, such as prickly pear and stalks of lechuguilla. In the arid Southwest, succulents are an important source of water for deer.

Mule deer in the Texas Panhandle rely on skunkbush, half-shrub sundrop, sand sagebrush, white sage, and juniper. Panhandle deer have a less diverse list of food items to select from, and their diets are below optimum in terms of protein and phosphorus requirements (Sowell et al. 1985). However, some Panhandle mule deer herds have access to highly nutritious agricultural crops, such as wheat, rye, alfalfa, corn, and sorghum (Cantu and Richardson 1997:9). Sowell (1981) found that mule deer using cultivated fields in the Panhandle were getting nearly half of their digestible energy, crude protein, and phosphorus from this supplement to their diet.

Juniper/Ponderosa Pine

Mule deer in the juniper and ponderosa pine vegetation types in the Southwest (northern Arizona/New Mexico and Southern Utah/Colorado) have a much different assemblage of plant species to choose from compared to their desert counterparts. Many of these mule deer herds migrate to lower elevations because of a lack of available food or to escape heavy snow cover, and return to their high-elevation home ranges as the snow melts. In the spring, mule deer in juniper and ponderosa pine habitats feed on turbinella and Gambel oak, mountain mahogany, sagebrush, buckwheat, spurge, clover, and some tender green grasses that are just emerging (Tables 6 and 7).

In summer, Gambel oak, mountain mahogany, aspen, white fir, serviceberry, mountain snowberry, and common chokecherry are supplemented with abundant forbs, such as clover, buckwheat, milkvetch, Oregon grape, daisy, mountain dandelion, and sedges, providing protein and energy for a burst of summer growth. Aspen is used heavily where available. On the Kaibab Plateau, aspen is the most important summer browse species. The leaves of a small aspen knocked down by logging operations or wind are picked clean by the next day (Hungerford 1970).

After the late summer rains (July and August), mule deer diets shift away from forbs and toward sagebrush, cliffrose, mountain mahogany, Gambel oak, juniper, antelope bitterbrush, rubber rabbitbrush, and serviceberry (Table 7). Forbs that may still be available into the fall months include buckwheat, clover, penstemon, deer vetch, and also a variety of mushrooms (McCulloch and Smith 1991). Mushrooms can be an important part of mule deer diets in the southern Rocky Mountains. Mushrooms are most abundant in late summer and fall, and their high levels of moisture, protein, phosphorus, and potassium are a welcome addition to the diet of deer preparing for winter.

As winter sets in and storms bring snow, deer must concentrate on browse almost exclusively, and many move to lower elevations for the winter. Gambel oak, cliffrose, mountain mahogany, and sagebrush are still important, as are pinyon pine, Apache plume, juniper, holly-leaf buckthorn, and turbinella oak. Grass is not normally an important component in deer diets except in spring when it is just sprouting. However, grass can be a significant part of deer diets at times in sagebrush-pinyon-juniper associations. Sagebrush alone contains compounds that inhibit proper digestive function in the rumen, but dilution with other species creates a nutritious diet. Mutton grass and wheatgrass serve this purpose on the Kaibab winter range (C. McCulloch, personal communication, 1999). Heavy reliance of deer on cliffrose in these vegetation associations may also be linked to its value in diluting the intake of sagebrush.

White-Tailed Deer

Whitetail diets are very similar to those of mule deer in areas they where they coexist (Anthony and Smith 1977). Like mule deer, whitetails are also primarily browsers, with forbs becoming important following the summer and winter rains (Table 8). In some areas tree or shrub fruits (acorns, beans, pods, nuts, berries) are nutritious supplements when available. Grass is taken in small amounts, usually when tender new shoots are the first to green up.

That whitetails and mule deer tend to occupy different habitat structure (amount of brushiness) and differential elevational zones is probably the only reason one might find differences in their diets. There are no known physical or physiological differences that would account for them selecting different food items in the same habitat. As a result, they may compete with one another when resources are in short supply in areas where they both occur.

Nichol (1938) held nine whitetails in captivity with his mule deer at the Santa Rita Experimental Range to determine the relative preference of different deer foods. Of 168 potential deer foods offered to both species of deer, whitetails preferred mountain hackberry, aspen, Emory oak, false indigo, fairy duster, fendlera, mistletoe, and wild grape (Table 9).

No work on the diet composition of Coues whitetails has been conducted in the low deserts of western Sonora, but research on the Organ Pipe Cactus National Monument in southwestern Arizona probably approximates what whitetails in these desert areas are eating. The isolated population of whitetails on Organ Pipe Cactus NM have a diet very similar to that of mule deer in the low desert scrub. Annual diets were 78 percent woody species, with seasonal use of forbs after the winter rains (Henry and Sowls 1980). Major food items included jojoba (over 25 percent of the diet most of the year), desert vine, fairy duster, copperleaf, buckwheat, and Indian mallow (Table 9). Abundance of forbs on the desert floor is so low that these deer rely to a great extent on woody browse throughout the year. Although not dependable, the periodic flush of forbs after summer precipitation events is critical to maintaining a stable population.

Most research on the diet of Coues whitetails has been conducted in southeastern Arizona, which probably yields a very representative idea of what most Coues whitetails are eating because that area is near the geographical center of their distribution. An evaluation of whitetail diets found—to no one's surprise—that browse made up a majority of the year-round diet. Shrubs like fairy duster, ceanothus, velvet-pod mimosa, mountain mahogany, oaks, silk tassel, holly-leaf buckthorn, and ratany are dietary staples and are actively sought out by whitetails wherever they occur (McCulloch 1973). Velvet-pod mimosa is one of the most

TABLE 8. SEASONAL PERCENT (%) COMPOSITION OF WHITE-TAILED DEER DIETS THROUGHOUT THE SOUTHWEST

Location	Spring			Summer			Fall			Winter			Source
	Browse	Forb	Other	Browse	Forb	Other	Browse	Forb	Other	Browse	Forb	Other	
Southwest Arizona Ajo Mts., Organ Pipe Cactus NM	59.1	36.2	...	91	2.2	...	...	...	...	76.4	20	...	Henry and Sowls 1980
Southeast Arizona Florida Canyon, Santa Rita Mts.	...	...	...	80	18	2	98	0.5	1.5 grass	100	...	...	White 1957
Santa Rita Mts.	...	...	...	59.9	25.4	...	...	...	...	...	...	...	Anthony and Smith 1974
Santa Rita Mts.	32–61	37–56	1–7.3 grass	68–80	18–32	1 cacti	64–81	12–32	2 cacti	67–77	11–28	3 cacti	Ockenfels et al. 1991
Central Arizona Three Bar Wildlife Area, 3,000–5,000 ft.	76	10	9 shrub fruit	14	16	70 shrub fruit	45–48	35–48	4–17 shrub fruit	50–68	27–40	5–10 grass	McCulloch 1973
Southwestern New Mexico Ft. Bayard	...	...	...	65.6	34	0.4	49.2	45.1	5.7	87.6	12.4	...	Tafoya et al. 2001
Texas Big Bend NP	26.7	12.3	48.7 succulents	18.3	20.7	24.3 succulents	59.6	8.0	6.9 succulents	31.4	17.9	27.6 succulents	Krausman 1978
Mexico La Michilia, Durango	96.9	3.1	...	82.2	17.8	...	72	28	...	80.6	19.4	...	Gallina et al. 1978

Note: Percentages were determined by many different methods.

TABLE 9. THE MOST IMPORTANT SOUTHWESTERN WHITE-TAILED DEER FOOD ITEMS BY SEASON

	Season				
Location	Spring	Summer	Fall	Winter	Source
Arizona (Ajo Mts., Organ Pipe Cactus NM)	jojoba, fairy duster, mallow, grassnuts	jojoba, desert vine, fairy duster, white sage, agave, Indian mallow	...	jojoba, desert vine, fairy duster, copperleaf, malvaceae, grassnuts	Henry and Sowls 1980
Central Arizona (Three Bar Wildlife Area, 3,000–5,000 ft.)	desert ceanothus, fairy duster, ratany, sugar sumac, holly-leaf buckthorn, skunkbush, white sage, nightshade	turbinella oak acorns and leaves, spurge, fairy duster, holly-leaf buckthorn, mesquite beans	morning glory, holly-leaf buckthorn, fairy duster, spurge, white sage	desert ceanothus, holly-leaf buckthorn, buckwheat, birchleaf mountain mahogany, spurge, grassnuts	McCulloch 1973
Southeastern Arizona (Santa Ritas, Chiricahuas, Canelo Hills, 4,000–5,500 ft.)	...	velvet-pod mimosa, ocotillo, fairy duster, ratany, morning glory, barrel cactus, buckwheat, Fendler's ceanothus, tall larkspur	↑		White 1957
Southeastern Arizona (Santa Ritas, Chiricahuas, Canelo Hills, 8,000–9,500 ft.)	...	Fendler's ceanothus, oaks, tall larkspur, spikebent, lupine, fairy duster, alligator juniper (berries and leaves)	↑		White 1957
Southeastern Arizona (Santa Ritas)	...	ratany, dalea, globemallow, oak, fairy duster	...		Maghini 1990
Southeastern Arizona (Santa Ritas)	velvet-pod mimosa, ratany, oak, globemallow, dalea, mesquite, fairy duster, buckwheat	↑			Ockenfels et al. 1991
Southeastern Arizona (Santa Ritas)	...	fairy duster, prickly poppy, kidney wood, silktassel	...		Anthony and Smith 1974

(*continued*)

TABLE 9. (*CONTINUED*)

Location	Spring	Summer	Fall	Winter	Source
Southeastern Arizona (San Cayetanos)	...	kidney wood, fendlera, ratany, wire lettuce, buckwheat	...	...	Anthony 1976
Southeastern Arizona (Whetstones)	...	desert ceanothus	desert ceanothus, goldeneye, crown-beard, meadow rue, brickellia, green sprangletop	desert ceanothus	Barsch 1977
Southeastern Arizona (Chiricahuas)	...	...	mountain mahogany, juniper leaves/berries, acorns, mearns sumac berries, silktassel	...	Day 1964
Arizona	mountain mahogany, fendlera, silktassel, Emory oak, desert ceanothus	mountain mahogany, skunkbush, grape, kidney wood	mountain mahogany, skunkbush, grape, kidney wood, desert hackberry	mountain mahogany, silktassel, Emory oak, madrone, Arizona white oak, one-seed juniper	Knipe 1977
New Mexico	mountain mahogany, live oak, ceanothus, various sumac, silktassel, Apache plume, aspen, pinyon nuts, acorns, mushrooms, vetch, pea vines				Raught 1967

Location						Reference
Southwestern New Mexico (Ft. Bayard)	...			oak, mountain mahogany, juniper, gourd, globemallow	mountain mahogany, oak, silktassel, prickly pear cactus	Tafoya et al. 2001
Texas (Big Bend NP)	oak, acacia, evergreen sumac, littleleaf sumac, spurge, buckwheat, lechuguilla, prickly pear	mountain mahogany, littleleaf sumac, acacia, spurge, prickly pear, lechuguilla, grama grass	mountain mahogany, silktassel, oak, globemallow	guayacan, acacia, oak, silktassel, mesquite, spurge, prickly pear, lechuguilla	evergreen sumac, acacia, spurge, deer vetch, fleabane, lechuguilla	Krausman 1978
Mexico (La Michilia, Durango)	alligator juniper, madrone, oaks, mistletoe, manzanita, huisache, lupine	→				Gallina et al. 1978
Mexico (Chihuahua)	mountain mahogany, ceanothus, oak, manzanita and juniper berries	→				Leopold 1959
Mexico (Coahuila)	madrone, oak, kidney wood, wild rose, snowberry, mountain mahogany, morning glory tree oak (leaves and acorns)	→				Leopold 1959
Mexico (Sierra Madre Occidental)	mesquite, mountain mahogany, manzanita, ceanothus, mimosa, dalea, astragalas, lupine	→				Galindo-Leal and Weber 1998
Mexico (Nuevo Leon)	cenizo, acacia, desert yaupon, guajillo, yellow ground cherry, leatherstem, malva	→				Martinez et al. 1997

Note: See the appendix for scientific names of plants. Arrows indicate a span of more than one season in which these items were eaten by deer.

[a] Captive deer were fed 168 species of native plants; those listed are the species deer preferred.

important browse species in many mountain ranges (for example, the Santa Ritas, in Arizona); however, ceanothus and mountain mahogany assume greater importance in other ranges, such as the Chiricahuas and Whetstones (Ockenfels et al. 1991). In fact, Barsch (1977) found that the distribution of ceanothus was the most important factor determining the distribution of whitetails. Other browse plants like kidney wood are very important whitetail browse in some areas, but have been largely missed by many diet studies.

In August or September, after the summer rains, there is a pronounced dietary switch to forbs like spurge, morning glory, buckwheat, tall larkspur, lupine, dalea, globemallow, goldeneye, grape, and vetch (Gallina et al. 1978) (Table 9). Other plants may make sporadic contributions, depending on the timing of rainfall and plant phenology. For example, ocotillos at lower elevations leaf out very quickly after a rain, and being the only tender green leaves available, are used heavily by whitetails for a short period. Shrub fruits, nuts, and beans from plants like jojoba, oaks, mesquite, and manzanita can add considerable body fat and are eaten whenever available. Whitetails also consume mistletoe and the stalks of lechuguilla and other agaves, and they relish the ripe fruits of prickly pear, cholla, and barrel cactus. With cactus fruits, deer rarely eat more than a few fruits before moving on to select a different food item. During the fall, it is very common to find the tiny black seeds from barrel cactus fruits throughout a whitetail's rumen.

Some diet information available from Mexico and New Mexico supports the results gathered in Arizona (Raught 1967, Gallina et al. 1978). Littleleaf sumac, prickly pear, and lechuguilla play a larger role in whitetail diets in New Mexico and the Big Bend region of West Texas. In Chihuahua, Coahuila, Durango, and the rest of the Sierra Madres most of the same plants are available, though in different combinations. In addition to the food items listed here, additional browse species like manzanita, madrone, huisache, and snowberry become important locally across northern Mexico (Gallina et al. 1978, Galindo-Leal and Weber 1998).

Water Requirements

Water is the most important and abundant substance in the bodies of mammals; it makes up 63–73 percent of the body weight of mule deer (Anderson 1981). Without water all body functions shut down rapidly. "Free water" is available to deer in the form of natural seeps, springs, creeks, rock basins, stock tanks, windmills, and artificial water catchments. Water is also contained in food (called "preformed water"), and this is an important source of water for deer. When deer foods are very succulent (high water content), deer may satisfy the majority of their water needs solely from their forage. Nichol (1938) monitored the water consumption of whitetails and mule deer in captivity, and reported that deer drank 1.2 quarts per day in the winter and 2.3 quarts in summer. Water consumption rose 25–65 percent when deer were fed dry forage. Another, less-known source of water is that which is released during digestion as a result of the metabolism of carbohydrates contained in plant parts. This "metabolic water" supplements the other two sources of water and contributes to the overall water intake.

Animals living in the desert have adaptations that help them cope with dry environmental conditions. Many of these adaptations are behavioral, such as reducing daytime activity or shifting to foods with higher water content, such as cactus. Some animals also have physiological adaptations that allow them to function better in an arid environment. These include more water-efficient ways of voiding waste (like concentrating urine) and the ability to function in a state of partial dehydration.

Natural water sources in the Southwest are usually least abundant in June, which corresponds to the period of greatest physiological need for deer because of high temperatures and dry forage conditions. Rosenstock et al. (2004) recorded the most visits to water by mule deer in May, June, and July, when temperatures were the highest and relative humidity lowest. In addition, does are in their later stages of pregnancy at that time and have a higher demand for water. Many studies show that pregnant does use habitat closer to reliable water sources (Clark 1953, Hervert and Krausman 1986). Fox and Krausman (1994) found that desert mule deer fawning sites were unrelated to distance from water, but fawns are usually born after the onset of the summer rains when deer disperse and are not found close to water.

Human activities over the last 130 years have caused the lowering of the water table in many areas, which has resulted in the disappearance of some springs, cienegas, artesian wells, and even entire rivers. Concurrent with this disappearance of natural water sources was the rapid development of artificial sources by the ranching community. These developments provide water for wildlife, such as deer, where natural sources have been depleted. Thousands of artificial water sources were established (and more continue to be created) throughout the desert Southwest; however, in some cases, water is turned off when cattle are not grazing that particular pasture.

The effect of the increased distribution of water on southwestern deer populations is unclear, in part because we lack good information on historical deer abundance and in part because of a multitude of other factors that occurred during the same period (reduction in livestock numbers and the beginning of hunting restrictions). The Arizona Game and Fish Department alone has built over 800 water catchments since 1946 to provide alternative sources of water for wildlife populations (AGFD 1997). Originally many of these catchments were designed for small game. However, research in the 1960s showed that they provided little benefit to small game, so older catchments were modified and new catchments were designed with a larger capacity for use by all wildlife species.

Not much research has been conducted to determine the effects of water catchments on deer abundance, but several studies have documented deer altering their habitat use and movements in accordance with water distribution. Determining the long-term value of artificial water sources is no easy task. Aldo Leopold (1933: 288) aptly described the difficulty: "The watering habits or preferences of game when water is plentiful, and its real requirements when water is scarce, are two different things. To see a species drinking is not proof that it must drink. To prove that it must drink under one condition of food and weather is not proof that it must drink under any and all conditions."

Broyles (1995, 1997) expressed concern that adding water sources to desert environments might have negative consequences. For example, if predators are attracted to water sources, then more water may result in more predation, which would counterbalance at least some of the benefit the water provided. Research conducted in response to those concerns found predator (primarily coyote) sign near water sources seven times more abundant, but little evidence indicated that predators frequented water sources specifically to ambush prey (DeStefano et al. 2000, Rosenstock et al. 2004). Researchers speculated that predators were there to drink and that any incidental predation events were trivial to deer population dynamics.

If a new water source becomes available to cattle in an area formerly lightly grazed, the new water may result in heavier grazing in that area and lead to a reduction in deer cover and forage. If the water is not solely for wildlife use, sometimes cattle stocking rates are increased with the addition of more water sources. If stocking rates result in inappropriately heavy grazing in dry periods, this would result in a net decrease in deer habitat quality.

Deer may not benefit from water catchments during most of the year (and not at all during wet years). However, in dry months deer often concentrate around water sources (Brownlee 1979, Welch 1960, Wood et al. 1970) and may travel long distances outside their home range to drink (Hervert and Krausman 1986, Rautenstrauch and Krausman 1989). These shifts in distribution are an indication that water sources are important to deer. Well-distributed water sources throughout otherwise suitable habitat will distribute deer better, thereby allowing them to occupy previously unused areas. This effectively increases the overall carrying capacity of the habitat and reduces the need for long-range movements out of normal home ranges, which could increase deer survival. Even if deer do not shift their areas of use, the availability of open water allows them to use a greater variety of foods, including very dry forage. If this results in a better overall nutritional intake for deer, their health and survival would be improved over deer with less access to water. This has been shown in domestic sheep (Hutchings 1946), and the metabolic use of water by deer is no different than for sheep (Knox et al. 1969).

Still, most of these benefits remain speculative and theoretical. Definitive, population-level information is lacking on whether the addition of hundreds of wildlife water catchments throughout the Southwest has resulted in a wider distribution or higher densities of deer than before their installation (Krausman and Etchberger 1995, AGFD 1997, Rosenstock et al. 1999) (Fig. 39).

Mule Deer

Amount and Frequency

Hervert and Krausman (1986) reported that desert mule deer came to drink most often during the hours surrounding sunset (7 to 10 PM), with 38 percent of the does watering during daylight hours. This timing was reconfirmed by research in southwestern Arizona which found that visits to artificial water catchments peaked between 8 PM and midnight (Chantal O'Brien, personal communication, 2005). However, deer visit water sources mostly at night during the hot summer months (Rogers 1977). Tricia Cutler (personal communication, 1998) recorded 242 instances of desert mule deer drinking from water catchments in southwestern Arizona. Across all seasons, only 10 percent of the observations were made during daylight hours. During the spring/summer period, only 4 percent of the visits occurred during the day, but 42 percent of the observations were made during the day in the winter period.

Desert mule deer drink water daily during hot, dry summer months (Clark 1954, Elder 1956). Three radiocollared mule deer bucks were monitored for a period of ten days in the driest part of the year (June 22–July 2) and the bucks visited water catchments once per day (Hazam and Krausman 1988). In the Belmont and Picacho mountains of south-central and southwestern Arizona, desert mule deer does visited water once a day during the summer (Hervert and Krausman 1986). Bucks are frequently found farther from water than does, and they visited the water catchments unpredictably every one to four days.

When there is very little moisture in the deer forage plants, mule deer may consume 4–10 quarts (an average of 6.3 qts.) per day (Elder 1954, Hervert and Krausman 1986). Clark (1953) found that pregnant does remained within a quarter-mile of water and sometimes came to drink up to four times a day.

Water-Related Movements and Habitat Use

Mule deer in chaparral appear to move 1–1.5 miles to water (Hanson and McCulloch 1955, Swank 1958). Much, but certainly not all, mule deer habitat is well watered, in part because of the artificial sources provided for livestock use. However, during dry years many of these water sources dry up.

Although mule deer may not depend on free water daily, they do shift their area of activity within their home range, or even move out of their home range when water sources dry up (Rogers 1977). Hervert and Krausman (1986) found that when water sources within the home ranges of several mule deer does were rendered inaccessible, those does left their home range and traveled 1–1.5 miles to other water sources to drink. Once they drank, they immediately returned to their home range.

In the Sonoran Desert of southwestern Arizona, mule deer use habitat closer to water during the dry summer months, but not during other seasons (Rautenstrauch and Krausman 1989). Interestingly, Rautenstrauch and Krausman docu-

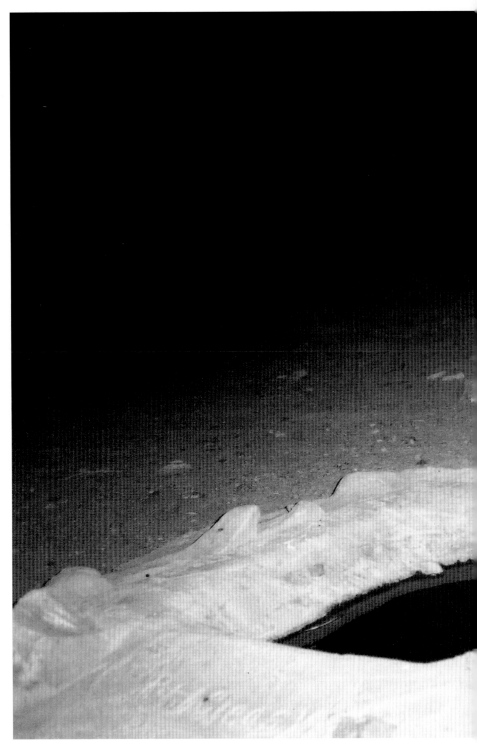

Figure 39. Artificial water developments for wildlife in the desert may not be needed during some times of the year or even during some years, but they are critical during extended dry periods. *Photo by Dennis Segura/USFWS*

mented a desert mule deer population that made long-distance movements to water. Some individuals in this population moved up to 20 miles in early summer to areas with freestanding water. Within a few weeks after the arrival of summer rains, the deer returned to the areas they occupied during the remainder of the year.

Deer throughout the Southwest concentrate within one-half or one mile from water sources during the dry period of the year (Ordway and Krausman 1986, Sánchez-Rojas and Gallina 2000). Habitat use by radiocollared male and female desert mule deer was studied in the Picacho Mountains in southern Arizona. Both sexes were found significantly closer to water sources during the May–October summer period than in other seasons (Ordway and Krausman 1986). Krysl (1979) found that 85 percent of mule deer in the Guadalupe Mountains of southern New Mexico were within one mile of permanent water. During the dry months in southern California 97 percent of mule deer observations were within about a half-mile of water (Bowyer 1986a).

In 1989 extremely dry conditions resulted in most natural and artificial water sources becoming dry on the Kaibab Plateau in Arizona and on west-side winter range. By early winter, mule deer appeared to be heavily dependent on the few remaining water sources, and water was trucked to the water catchments. At that time it was not uncommon to observe up to 200 mule deer in close proximity to a water source (T. Britt, personal communication, 2000).

This congregation near water sources results in a higher level of browsing in the vicinity of water. During exceptionally dry years or during an extended dry period, browse near water sources can be severely affected by overbrowsing. If drought conditions prevail or intensify, deer survival may be threatened (Swank 1958, Truett 1971).

That mule deer shift their movements in relation to water availability implies that in those cases their survival would be negatively affected if they remained in waterless areas. Still lacking is solid research on watered and unwatered areas clearly showing a long-term increase in deer density with the addition of water, or that deer moved into a recently watered area and stayed there year-round for several years.

On Fort Stanton in southern New Mexico, Wood et al. (1970) reported that deer densities in the well-watered West Pasture were higher than in the rest of the area. As water sources were added to the East Pasture, deer densities increased. Since there was no area left "waterless" for comparison, it is difficult to determine what role range conditions, precipitation, and deer movements played in that increase. It appeared throughout the study that deer densities fluctuated along with the availability of water each year, possibly because of movements onto and off the area. However, overall deer densities on Fort Stanton did not appreciably increase through the five years of adding water sources (Wood et al. 1970:fig. 7).

Researchers in West Texas started monitoring deer densities on the Black Gap and Sierra Diablo areas and then closed several water sources to see what effect that had on the desert mule deer. Densities quickly plummeted when water became unavailable and stayed low, then increased again when the water was reestablished three and a half years later (Brownlee 1979). The change in deer density was probably due to deer movement, since sharp increases were noted at a time when fawns were not being born. Also, temporary increases in deer density on the Black Gap area occurred immediately following periods of rainfall that filled potholes with water.

Recommended Water Distribution

Desert mule deer will readily move one and a half miles to water, but are found in decreasing densities as one moves away from a water source (Wood et al. 1970). At a minimum, water sources should not be more than three miles apart so all mule deer habitat is within one and a half miles of a permanent water source (Brownlee 1979, Dickinson and Garner 1979). Because does during fawning season and all deer during dry times are found to congregate around water sources, the optimum spacing would be one mile between waters to provide for the times of highest water use.

White-Tailed Deer

Amount and Frequency

The frequency and timing of drinking in any area depends on many things, including temperature, moisture in forage, abundance of water sources, disturbance near waters, and availability of cover nearby. Coues white-tailed deer have also been shown to drink primarily at first light and after dark to avoid the midday heat (Welch 1960). During the summer, Knipe (1977) reported that whitetails watered in the morning (8:00 AM–noon) and the late afternoon (6:30–8:30 PM), with only 15 percent watering after dark. However, McMichael (1972) used a time-lapse camera at a water source and found that whitetails in central Arizona chaparral visit water in late afternoon between 2 PM and 7 PM; any deer watering at night would not have been recorded. Coues whitetails watered mostly (64 percent) after dark in a dry year, but only 32 percent did so the following wet year (Maghini 1990). In Maghini's study, deer visited the water sources most between 2:30 PM and 11:30 PM, with very little visitation during midday.

At the extreme western edge of Coues whitetail range in the Organ Pipe Cactus National Monument, deer came to water every one to four days in the summer, averaging 2.1 days between drinks (Henry and Sowls 1980). In the cooler winter months deer made little use of permanent water, either because they did not need water at this time, or because it was available throughout their home range in small, temporary basins. These whitetails watered at any daylight hour, but preferred 6:00–10:00 AM and sunset, with no indication of nighttime watering. These diverse results clearly show that there is great variation in when deer use available water.

Water-Related Movements and Habitat Use

Knipe (1977:54) reported that whitetails frequently travel as much as one and a half miles to water during the extremely dry late May and June period. In a study of seasonal movements in the Chiricahua Mountains in extreme southeastern Arizona, Welch (1960:33) found that deer moved closer to permanent water in June when water became scarce. During the study a doe moved over two miles to available water and stayed there, returning only after the summer rains came. Welch also reported that browse in the vicinity of these permanent sources showed evidence of heavy past and current use.

Whitetails in the Santa Rita Mountains southeast of Tucson selected habitat within one-half mile of permanent water, with an even stronger affinity for areas within one-quarter mile (Ockenfels et al. 1991). Very few deer were located more than three-quarters of a mile from open water. The 32 whitetail home ranges identified in that study contained an average of 7.5 water sources, with males having significantly more water sources than females because of their larger home ranges.

White (1957:49) found that drinking water was not a major limiting factor for southern Arizona whitetails in the lower, more accessible portions of the Santa Rita Mountains and nearby Canelo Hills. He did, however, note that deer concentrations were near available drinking water during his study.

The need for water sources varies between years and even between seasons of the same year. Using radiocollared Coues whitetails, Maghini (1990) monitored deer movements in relation to water sources during a dry summer and a wet summer. Only one of the seven deer had an identified source of permanent water in the home range during the dry summer, but six of the seven had access to open water the following wet year. Additionally, the use of permanent water sources declined drastically fourteen days after the arrival of the summer rains.

Recommended Water Distribution

Because water distribution is generally better in the higher elevations occupied by whitetails, this may be of less concern for deer management than in the case of mule deer. However, managers should still evaluate water distribution in whitetail habitat and improve upon it when needed. Water sources would have to be closer in rugged country to be readily available because it is more difficult to move long distances on rough terrain. Also, whitetails are inherently less desert-adapted and may require a closer distribution of water throughout their habitat than do mule deer. Carmen Mountains white-tailed deer in West Texas are always found associated with free water and thick brush, yet desert mule deer in the area are more widely distributed (Krausman and Ables 1981:40).

Whitetail movements and habitat use in the Whetstone Mountains of Arizona are more dependent on the distribution and quality of browse than the distribution of water (Barsch 1977). Study plots located closer to water did not receive significantly higher deer use. In that study all plots were within 0.8 miles of permanent water. This provides some evidence that habitat within three-quarters of a mile from water is used evenly; thus a spacing of one and a half miles between water sources may represent an adequate density of water in whitetail habitat. This agrees with other research showing few whitetails were found farther than three-quarters of a mile from water in the Santa Rita Mountains (Ockenfels et al. 1990). Others have recommended a distribution of two and a half water sources per square mile (Maghini 1990), which would result in a spacing of one-half to three-quarter mile between sources.

Chapter 6

Density, Home Range, and Movements

Density

DEER DENSITY is defined as the number of deer per unit area at a point in time, such as 12 deer per square mile (mi^2) in June. This sounds straightforward: survey a square mile completely with a helicopter, count all the deer, and you have it. Unfortunately, the situation is not that simple. If deer were evenly distributed throughout the habitat, if all habitat looked the same, if the land was flat, and if you saw all the animals in that square mile, your measurement would come close to the actual density of deer. But there are many factors that complicate the estimation of deer density.

Estimating Deer Density

In the Southwest, deer density can fluctuate wildly through time. It is important that estimates of deer density be expressed for a specific time. Deer density changes throughout the year as animals are born, die, or move. Density can almost double when fawns are born, but it will then decrease throughout the year to a low point shortly before the next fawning period. Besides this annual cycle, deer populations increase and decrease through time as a result of many factors. For the Southwest, the most important factor is rainfall pattern. This results in fluctuations in deer density from year to year and precludes the use of one standard density estimate all the time.

In mountainous terrain, one square mile, or "section" on the map, may contain two square miles of surface area because of ridges, hills, and canyons. If one could physically pull outward on the corners of a mountainous section until it was flat, it would be much more than a square mile in size. Even if deer were evenly distributed throughout the habitat (exactly 100 yards apart in all directions) you would have many more deer in a mountainous section than in a flat one because of the greater surface area.

Deer are not evenly distributed 100 yards apart in all directions. Deer habitat contains some areas that are better than others, and deer are concentrated in these areas. Additionally, some aspects of deer behavior contribute to an uneven distribution of deer in the habitat. For example, in late summer, bucks run together in bachelor groups and does are found with last year's fawns and other does. During rut, bucks are found in association with does, and then they are found singly in the postrut period.

Even within one area, no habitat is uniform in structure or quality. Because of this, deer prefer some areas (those that contain important resources for survival) and avoid other areas (those that lack an important habitat component). How your sample areas rate in terms of habitat quality has a lot to do with how many deer you will observe there. Deer may be concentrated in recently burned areas or near water sources in late summer, or they may avoid areas that have received intense grazing pressure. On a larger scale, habitat quality varies from one geographic area to another. Mule deer densities are obviously not the same in the Sonoran Desert as on the Kaibab Plateau.

Research on the percentage of animals seen during aerial surveys shows that surveyors rarely see more than 75 percent of the animals present in the survey area. To complicate matters, the percentage seen varies widely from survey to survey, even when the method of observation, time of day, time of year, and identity of observer are the same. The observation rate (percentage actually observed) for white-tailed deer from a helicopter can vary from 17 percent to 65 percent in the same study (Leon et al. 1987).

If survey efforts are designed carefully with all these problems in mind, it is possible to estimate the average number of deer for a particular area (Koenen et al. 2002). Sample units are often divided into different topographic classes (flat, gently rolling hills, rugged) to allow the comparison of rugged vs. flat terrain. A related issue is the different vegetation associations, such as creosote flats, mesquite washes, ocotillo foothills, and oak woodlands. With enough sample units, one can obtain a minimum density estimate (deer/mi^2) for each habitat type and topographic class. These are always minimum estimates because only a portion of the deer present in the sample unit will have been observed. These density estimates (deer/mi^2) can then be applied to the actual number of square miles of that vegetation type or topographic class in the area of interest to get an overall population estimate at a given time. More important, population estimates generated the same way year after year can provide a useful index to actual population fluctuations.

The problem of determining deer density becomes very complicated when one considers the details behind the estimates. One must take into account the assumptions used when calculating these estimates, along with the weaknesses, and exercise caution when using or reporting deer densities.

Southwestern Deer Densities

In spite of the difficulties of measuring density, several research projects have yielded estimates of deer density in various habitats. Density has been estimated primarily by observing individually identifiable deer, by making pellet counts, or by using helicopters to count deer. Before any better methods for estimating deer populations were developed, some biologists recorded observations of deer with unique markings (split ears, for example) and did their best to estimate the number of different deer using the area.

As the field of wildlife management became more scientifically based, fecal pellet counts became popular as a method of estimating populations. With this method, randomly selected areas of soil are cleared of existing deer fecal pellets. After a specified time period the biologist returns to the cleared areas and counts the number of deer pellet groups that have been deposited during that time interval. Since these cleared areas are of a known size, one can estimate the number of deer pellet groups deposited per day per square mile. Although the figures used vary widely, most researchers have used figures ranging from 12 to 21 pellet groups deposited per deer per day. Using this information a biologist can calculate how many deer must have been present to deposit the estimated number of pellet groups per square mile.

Although this method was initially popular, it is used much less now for estimating the actual population size. This is because pellet deposition rates vary dramatically with diet, deer activity, and other factors. To estimate deer density using this method, one has to guess what the deposition rate is for that area at that time. Guessing correctly is extremely difficult, and using a slightly different defecation rate in the calculation can change the population estimate dramatically and lead to inaccurate density estimates (Collins and Urness 1981, Galindo-Leal 1992). Even using a consistent pellet deposition rate, estimates of deer density can vary widely on repeated sampling of the same area (Raught 1960). This method can, however, provide a rough index to deer abundance through time.

Helicopter surveys can yield an estimate of density in several ways. Besides the survey quadrants mentioned above, transects can also be flown to count deer over a subset of the total area of interest. Having marked (ear-tagged or radiocollared) deer in the population allows biologists to estimate the percentage of the animals they are seeing when surveying. For example, if there are 50 radiocollared deer on a study area and only 25 of them are seen during a survey, we can assume that only

50 percent of all the deer present were observed. Aerial surveys with fixed-wing aircraft or helicopters also allow biologists to fly transects and count deer within 200 yards of each side of the flight line. Since the length and width of the transects are known, one can calculate the total area (mi^2) surveyed and derive the number of deer seen per mi^2. Again, not all deer within the transect are seen, so counts have to be adjusted for the percentage of deer missed.

Estimates of deer densities for mule deer and white-tailed deer in the Southwest are summarized in tables 10 and 11. These estimates vary considerably because of fluctuating deer populations, differences between vegetation associations, and many other factors. These densities were derived from past research and apply only to a particular area at that particular time. As such, they are not intended to be used as a basis to calculate the number of deer in any area at a given time. However, the figures do give us a feel for the boundaries of variation around the Southwest and through time.

Deer densities in the Southwest are generally less than fifteen deer per square mile in most situations. Deer populations in the high-elevation, wetter areas of the Southwest have the capacity to increase faster and reach higher densities (the Kaibab Plateau, for example). Still, southwestern deer populations do not reach densities seen in the eastern whitetail, which can exceed a hundred deer per square mile in agricultural areas.

Home Range

A *home range* is simply the area used by an animal as it meets its needs for food, water, cover, and social interactions. It may refer to an annual home range (year-round) or to a seasonal home range. For example, an individual deer's summer home range may be very different from its winter home range if it migrates. Within the annual home range there is usually a core area where the animal spends a disproportionate amount of time, presumably because conditions are optimal in that area of the home range.

Biologists have estimated home range by a variety of methods; initially, the preferred method was repeated observations of individually identifiable deer (ear tags or natural marks). This provided a crude estimate of home range, but not until the advent of radiotelemetry technology could home ranges be accurately estimated. Even then, several different methods were used to summarize the radio locations and estimate an individual's home range. The methods generally differ in the way they treat locations outside the periphery of most other locations. One method of delineating the actual boundaries of a deer's home range is to simply draw a line to connect all the outermost locations (this is called the "minimum convex polygon" method). Another method draws a line around most locations so that 90 percent are contained within the perimeter. Many computer software packages can now easily calculate home ranges using this method or many other, related meth-

TABLE 10. SOUTHWESTERN MULE DEER DENSITIES REPORTED BY VARIOUS METHODS

Location	Years	Deer/mi^2	Source
Baja Sur, Mexico			
Sierra de La Laguna[a]	1987–93	49.2	Alvarez-Cárdenas et al. 1999
Durango, Mexico			
San Ignacio Area, Mapimi Biosphere Reserve[a]	1996–98	1.8–10.9	Sánchez-Rojas and Gallina 2000
Coronas Area, Mapimi Biosphere Reserve[a]	1996–98	2.0–7.1	Sánchez-Rojas and Gallina 2000
Arizona			
Three Bar Wildlife Area[b]	1952–53	12	McCulloch 1955
Prescott[b]	1954	5–15	Hanson and McCulloch 1955
Mingus Mountain[b]	1954	8	Hanson and McCulloch 1955
Chaparral in general	…	10	Hanson and McCulloch 1955
Three Bar Wildlife Area[a]	1969–70	6	Horejsi 1982
Three Bar Wildlife Area[a]	1977–81	7.2–12.8	Horejsi 1982
Three Bar Wildlife Area[a]	1970–80	6–13	Horejsi and Smith 1983
Kaibab summer range[a]	1972–79	8–17	McCulloch and Smith 1982
Buenos Aires NWR[c]	1996	2.3–6.5	Koenen 1999, Koenen et al. 2002
New Mexico			
Fort Stanton[a]	1964–68	8.4–18.5	Wood et al. 1970
Fort Stanton[a]	1964–66	14.4–17.8	Howard 1966
Fort Stanton[a]	1970–73, 1986	3.8–69.4	Rosier 1987
SW quarter of state[a]	1956–60	7–26	Raught and Frary 1961
Statewide[a]	1967–69	10–12.3	Raught 1969
Texas			
Longfellow Ranch[a]	1973–74	39–82	Phillips and Hanselka 1975
Black Gap WMA[d]	1982–89	2.5–7.3	Hobson 1990
Black Gap WMA[d]	1963–68	3–24 (ave.=9.8)	Brownlee 1979
Trans-Pecos	2001–2002	4.3–4.7	Bone 2003
Panhandle	2001–2002	5.9–6.7	Bone 2003
Panhandle (Canadian River and Clarendon)[e]	1978	13	Koerth 1981
SW Colorado			
Mesa Verde NP[a]	1968–69	10–20	Mierau and Schmidt 1981

[a] Estimates based on surveys of fecal pellet groups.
[b] Intensive observation of individually identifiable deer.
[c] Point sampling.
[d] Spotlight surveys.
[e] Helicopter line-transect or block surveys.

TABLE 11. SOUTHWESTERN WHITE-TAILED DEER DENSITIES REPORTED BY VARIOUS METHODS

Location	Years	Deer/mi^2	Source
Durango, Mexico			
La Reserva La Michilía[b]	1976–86	54.4	Gallina 1994[c]
Arizona			
Chiricahua Mountains[b]	1958–59	12–37	Welch 1960:39
Coronado National Monument	1959	>8	Welles 1959
Aravaipa Canyon (both deer species)[a]	1991–93	9.6–13.7	Cunningham et al. 1995
West Texas			
Chisos Mountains[b]	1969–73	18–36	Krausman and Ables 1981[c]

[a] Helicopter transect surveys.
[b] Estimates based on surveys of fecal pellet groups.
[c] May be overestimates because of the sampling strategy used (Krausman and Ables 1981, Galindo-Leal 1992).

ods. Unusual trips and temporary forays outside of the normal area of use are always problematic. For instance, the researcher has to decide how to treat the cases where a deer stays within three square miles for two years and then moves ten miles away for a week before returning to his home range. Is this excursion part of his home range?

Like estimations of deer density, the estimation of home range is subject to wide variation depending on the species, gender, terrain, water distribution, season, quality of forage, and methods used for estimating. Deer living in the arid southwest deserts have larger home ranges than those in the chaparral, simply because it takes more area to supply all the needed resources (Table 12). More important, home ranges can change somewhat from year to year depending on the condition of the habitat. Home ranges are generally smaller in wet years and larger in dry years. Even under similar habitat conditions, individual deer show variations in how much habitat they use. When studying home ranges and movements, researchers assume the deer they provide with radiocollars are representative of the deer in the population. This is more likely to be true if a lot of deer are radiocollared to better sample the range of variation among individuals.

In general, home ranges are smaller where resources (food, water, shelter) are abundant and well distributed. However, under certain circumstances, home ranges may shrink seasonally as food or water becomes less available (for example, limited winter range or near water sources during the dry months). As a rule, home ranges for bucks are much larger than those of does. Females travel in groups of related does and their offspring. The home ranges of all does and fawns in these matriarchal groups overlap almost completely because they travel together. The larger home ranges of bucks overlap those of several female groups. During rut, the dominant bucks expand their home ranges and travel over a larger area to

TABLE 12. HOME RANGE ESTIMATES FOR SOUTHWESTERN MULE DEER
(MINIMUM CONVEX POLYGON METHOD)

Location	Season	Sex	Home range (mi^2)	Min.–max. home range (sample size)	Source
Nuevo Leon & Coahuila, Mexico					
Rancho San Francisco	Annual	M	0.79	(7)	Gallina et al. 1998
		F	0.80	(6)	
Arizona					
Kaibab Plateau	Summer	M	10.8	...	Haywood et al. 1987
		F	5.2	...	
	Winter	M	2.9	...	
		F	2.8	...	
Three Bar Wildlife Area and Tonto Basin	Annual	M	3.7	...	Horejsi et al. 1988
		F	2.0	...	
Santa Rita Exp. Range	Annual	M	14.9	2.7–36.5 (9)	Ragotzkie 1988
		F	4.8	2.7–7.0 (10)	
Santa Rita Exp. Range	Summer	M	3.0	1.8–4.8 (3)	Rodgers et al. 1978
		F	2.8	1.5–4.1 (4)	
Belmont and Big Horn Mts.	Annual	F	25.5	9.7–63.5 (8)	Krausman and Etchberger 1993
Belmont and Big Horn Mts.	Annual[a]	M	35.5	(11)	Krausman and Etchberger 1995
		F	13.2	(30)	
Belmont and Big Horn Mts.	Winter	F	12.2	(8)	Fox and Krausman 1994
	Spring	F	11.1	(8)	
	Summer	F	8.3	(8)	
	Fall	F	8.7	(8)	
Belmont and Big Horn Mts.	Annual	F	12.5	(5)	Hayes and Krausman 1993
Belmont and Big Horn Mts.	Annual	M	37.1	22.9–54.6 (5)	Krausman 1985
		F	33.2	4.9–66.5 (8)	
Picacho Mts.	Annual	M	27.7	7.6–50.3 (6)	Krausman 1985
		F	4.6	2.3–7.3 (15)	
Granite Wash, Harquahala and Little Harquahala	Annual	M	62.0	5.5–150.3 (5)	Krausman 1985
		F	53.8	25.8–84.5 (4)	
King Valley	Annual	Both	56.0	18.1–218.7 (22)	Rautenstrauch 1987
West Texas					
Longfellow Ranch, Pecos Co.	Annual	M	3.0	2.3–3.6 (3)	Wampler 1981
		F	2.3	2.3 (1)	
Longfellow Ranch, Pecos Co.	Annual	M	1.9	0.4–3.5 (5)	Dickinson and Garner 1979
		F	1.1	0.9–1.4 (5)	
Brewster County	Annual	M	4.3	3.8–4.9 (10)	Lawrence et al. 1994
		F	2.2	2.0–2.4 (22)	
Elephant Mt., Brewster Co.	Annual	M	5.3	5.2–5.4 (37)	Relyea et al. 2000
		F	2.2	1.9–2.5 (36)	
Texas Panhandle					
Oldham and Donley Co.	Annual	F	9.8	4.4–21.2 (6)	Koerth et al. 1985
Oldham Co.	Annual	M	26.5	24.8–28.1 (2)	Koerth and Bryant 1982

[a]Does not include movement between summer and winter ranges.

monitor as many doe groups as possible for estrus females (Koerth 1981:50). Conversely, females restrict their home range to 1.5–2.5 mi^2 for the two weeks before and after giving birth to their fawn (Fox and Krausman 1994).

Mule Deer

Radiotelemetry data from migratory mule deer on the Kaibab Plateau in the high elevations of northern Arizona show that home ranges are much smaller where deer are concentrated on winter range (2.9 mi^2), then expand during the summer while deer occupy the higher elevations (Haywood et al. 1987). Home ranges of males and females are similar when they are occupying limited winter range, but with resources more widely distributed on summer range, bucks (10.8 mi^2) have home ranges that are twice that of does (5.2 mi^2).

In the chaparral habitat, home ranges are smallest during the early summer dry period in May and June (Horejsi et al. 1988). In this environment, the summer period represents the time when food is limited in quality and quantity, but deer cannot range far from water sources that are also limited. Home ranges of bucks in this habitat (3.7 mi^2) are nearly twice the size of does' (2.0 mi^2).

In the desert grasslands home ranges are slightly larger because the lower average productivity of this environment forces deer to travel over a larger area to obtain all the things they need. In desert grassland south of Tucson, Arizona, annual home ranges for mule deer bucks ranged from about 3 to 37 mi^2 with an average near 15 mi^2 (Ragotzkie 1988). Home ranges for females were much smaller, ranging from 3 to 7 mi^2 and averaging just less than 5 mi^2. Even within this single study area, deer home ranges were significantly larger at lower elevations where precipitation was less and productivity lower.

Deer in the arid deserts of southwestern Arizona, Sonora, Baja, and southeastern California must maintain very large home ranges to obtain the necessities for survival. Most estimates of desert mule deer home ranges in this area are 30–50 mi^2. Rautenstrauch (1987) documented average annual home ranges up to 218 mi^2 in the King Valley on and near the Kofa National Wildlife Refuge. Most studies reinforce the early observations that bucks have larger home ranges than does (Table 12). Limited water sources force desert mule deer to restrict their movements, but this all changes when the summer or winter rains arrive. The rainfall initiates browse and forb growth, and deer are then free to travel throughout a much larger area (Celentano and Garcia 1984).

In the Chihuahuan Deserts of New Mexico, West Texas, and northern Mexico, home ranges average much smaller than those in the arid Sonoran Desert. Several research projects in West Texas involving more than 100 radiocollared mule deer consistently estimated home ranges of 2–5 mi^2 (Table 12). Less work has been done in the Texas Pandhandle, but radiotelemetry research there showed deer had larger home ranges (4–28 mi^2), primarily because of long-distance movements in fall and winter to areas with agricultural crops (Koerth et al. 1985).

White-Tailed Deer

There have been very few efforts to estimate home ranges of southwestern white-tailed deer. The best estimate of Coues whitetail home ranges comes from the research done by Ockenfels et al. (1991) in typical whitetail habitat in the Santa Rita Mountains. A total of 35 (14 male and 21 female) Coues whitetails were captured and instrumented with radiotelemetry collars. These deer were then located repeatedly for three years; some were located several hundred times. The overall area they used during that time, and a more intensive "core area," was identified by analyzing these locations. Does had home ranges averaging 2 mi^2 (range = 0.2–7.0 mi^2) with a core area of more intensive use of 0.7 mi^2. As is typical, bucks had larger home ranges, averaging 4.1 mi^2 (range = 1.7–6.5 mi^2) with a core area of 1.7 mi^2.

Movements

Seasonal Movements

Deer evolved migratory patterns as a survival strategy to reduce overall energy expenditure while meeting body needs in the face of seasonal changes in the environment. Deer migrate seasonally to escape unfavorable conditions and return as soon as their original habitat is once again able to supply their needs. Some southwestern deer herds are migratory; however, in much of the Southwest habitat conditions do not fluctuate to the extremes necessary to induce migration. In these cases, deer occupy the same areas year-round. Some populations may migrate only in certain years or even adopt a "mixed strategy," where only some members of the population migrate and others do not (Nicholson et al. 1997).

Although most southwestern deer do not migrate, they do make seasonal movements in response to water and food distribution, segregation of the sexes, fawning, weather, human disturbance, heavy livestock use, and rut. Any deer home range consists of a mixture of different plants, many providing a source of food only during a particular part of the year. The distribution of deer throughout the year depends on the distribution and arrangement of food, water, concealment cover, and all other factors influencing deer movements.

Mule Deer

The deer on the Kaibab Plateau in northern Arizona are one of the nation's best known migratory deer herds. From 1978 to 1984, researchers radiocollared 115 mule deer on the plateau and located them 3,912 times to study their migration patterns and habitat use (Haywood et al. 1987a). They found that migration off the high plateau and down onto winter range occurred in November (October 29–December 9). The return trip up to summer range took place in May (April 23–June 3). Deer spent an average of 52 days in migration, but many completed migration in 10–20 days. Twenty-nine percent migrated less than 3 miles, and over half (57 percent) traveled between 3 and 18 miles. This is similar to migration distances of 6–13 miles reported in Mesa Verde National Park in southwestern Colorado (Mierau and Schmidt 1981).

As the time of fall migration approaches, deer move to "staging areas" closer to the lower elevation winter range, with bucks migrating at the same time as does. Fall migration starts as a trickle in October, regardless of snow cover, as the summer forage items dry out and wither from frost (Clay McCulloch, personal communication, 1999). This slow, dispersed migration can turn into an intensive mass movement when an early heavy snowfall covers much of the available forage on summer range.

In spring, when conditions allow the return to summer range, deer return to the same summer home range they occupied the previous summer. Upward spring migration of Kaibab deer coincides with seasonal maturation of muttongrass, when its digestibility and protein content normally decreases on winter range (Clay McCulloch, personal communication, 1999). Of the 115 deer tracked, only 6 deer reversed which side of the plateau (east or west) they wintered on in different years. Deer summering on the plateau winter mostly to the west (65 percent), but also on the lower elevation areas to the east (25 percent), south (5 percent), and north (5 percent) (McCulloch and Smith 1991).

Of the 93 deer captured on Kaibab National Forest or Bureau of Land Management winter range, 77 (83 percent) entered the Grand Canyon National Park at the south end of the plateau at least once in the summer. However, 78 percent of those left the park for lower elevations by October 25, making them available during the hunts and allowing for proper management of the herd to protect the habitat from overuse.

In 1995, 83 mule deer were radiocollared in southern Utah and northern Arizona to study the interstate movements between the Paunsaugunt (Utah) and the Kaibab Plateau (Arizona). Timing of migration for Kaibab deer was nearly the same as that reported from the earlier study. Some Kaibab deer share winter range around and in the Buckskin Mountains with deer from the Paunsaugunt (Carrel et al. 1999). Also, a small number of Utah bucks migrate to winter range in northern Arizona before the Arizona deer seasons open in late October.

Migration patterns of deer along the high-elevation areas of central Arizona and New Mexico are variable depending on the weather; this is a transitional area between migratory and nonmigratory deer populations. High-elevation areas of the White and Gila Mountains along the Arizona/New Mexico border have migratory deer herds, while many other areas show variation by year and area. Some deer summer in ponderosa pine and winter in pinyon-juniper, while some spend all year in the pinyon-juniper. In these transitional zones, small-scale movements from north-facing to south-facing slopes are often sufficient to satisfy the food and cover requirements of deer.

Deer are present year-round throughout the chaparral region in central Arizona, but areas adjoining coniferous forests also receive an influx of wintering deer from higher elevations (Hanson and McCulloch 1955). Likewise, mule deer habitat in southern Arizona does not receive snow depths sufficient to prompt migration during the winter months. On the contrary, a winter with high precipitation in the southern ranges produces an abundance of forbs and results in higher adult deer survival and an increase in fawn recruitment the following summer.

Mule deer demonstrate homing ability when translocated to new areas. In the 1960s, 248 mule deer were trapped in north-central New Mexico and released in seven places around the state to restock areas of low deer density. Of the 129 deer tagged during that operation, 14 were later seen and 7 of those returned to the area from which they were trapped (Eberhardt and Pickens 1979). The returning deer traveled 7–31 miles back to the capture site; however, no deer translocated more than 31 miles returned.

When water sources dry up, mule deer sometimes shift their areas of activity. Several studies have clearly shown desert mule deer abandoning areas when water sources dry up and using other, well-watered portions of their home range (Clark 1953, Rodgers et al. 1978, Bellantoni et al. 1993). Conversely, when water was added to formerly dry range on Fort Stanton in southern New Mexico, deer begin to use that area (Wood et al. 1970).

In the most arid portions of the Southwest, deer make significant movements in relation to water availability. In the King Valley of southwestern Arizona, where there are no permanent water sources in about 463 mi^2, mule deer move during the summer dry season to the surrounding areas, which have dependable sources of water (Rautenstrauch and Krausman 1989). Deer then return to the valley as soon as the summer rains cause the first greenup; they apparently have the ability to detect and react to distant rainfall when summer rains return to the valley. In Rautenstrauch and Krausman's study, one doe returned a distance of 19 miles just three days after rain fell in the valley, even though it did not rain where the doe was. Of the 15 deer radiotracked for over one year, 10 moved to new areas near water, traveling an average of 8.5 miles between these seasonal ranges. All the bucks monitored during this study made these seasonal shifts to well-watered areas.

It is typical for desert mule deer to use foothills and riparian areas in the dry summer months to stay closer to reliable water sources. However, as soon as the summer rains arrive, deer disperse much farther from perennial water sources and take advantage of the high moisture content in the new layer of forbs and also the ephemeral water sources that show up throughout their habitat. In the deserts of southeastern California, deer use the Colorado River in summer because that riparian corridor provides water and shade. When winter rains come, these desert mule deer disperse into the mountains because of the abundance of forbs there and a lack of browse along the river when the deciduous vegetation drops its leaves (Celentano and Garcia 1984:20). Bowyer (1986a) found that mule deer in San Diego County, California, showed no difference in seasonal habitat use, but preferred open meadows and stands of oak.

As with most members of the deer family, mule deer males and females occupy different habitats at different times of the year. Mule deer does in the Sonoran Desert use mountain and foothill areas all year long. The bucks also use these areas seasonally (especially the northern slopes), but also spend time in the flats away from the mountains (Ordway and Krausman 1986). The use of these foothills may be related to the greater diversity of shrubs and forbs usually found there, as well as to a greater abundance of natural water sources. Bucks use the mesquite bosque (thickets) and washes running through creosote flats extensively, except during the dry season (June) when they are confined to the foothills.

Sánchez-Rojas and Gallina (2000) found mule deer in the central Chihuahuan Desert primarily use areas near water with a higher shrub diversity and uneven terrain. This is similar to the habitat that is important to mule deer in the Texas Panhandle. In the summer, deer use riparian areas and then move to agricultural fields and juniper breaks in the fall and winter (Koerth et al. 1985). Mule deer use the north- and east-facing slopes of juniper breaks with light livestock and human use, but travel between these breaks and agricultural crops when available. Koerth and colleagues found that although agricultural fields made up only 2 percent of the study area, they held 22 percent of the deer locations in fall and winter. Cultivated fields are not as widely available to mule deer in the Southwest, but this source of nutrition is important when and where it occurs.

Without agricultural areas, habitat use by mule deer in the northern Chihuahuan Desert is heavily influenced by cover and browse diversity/distribution. Mule deer in West Texas use the plant communities dominated by skeletonleaf goldeneye, catclaw, javelina bush, and juniper (Dickinson and Garner 1979, Wampler 1981). Germany (1969) found that mule deer on Fort Stanton in southern New Mexico prefer a diverse mixture of browse rather than vegetation associations dominated by a single species.

Similar to their counterparts in the Sonoran Desert of Arizona, male and female desert mule deer in the Trans-Pecos of West Texas also use slightly different habitat throughout the year. Compared to does, bucks use areas that are lower in elevation, less steep, closer to water, and with higher productivity, but having lower amounts of preferred forage (Lawrence et al. 1994). Unlike the Arizona findings, West Texas mule deer appeared to select south- and west-facing slopes.

One of the biggest seasonal movements comes during the rut, when mule deer bucks roam wide areas to locate receptive does (Relyea and Demarais 1994). Some bucks may move over six miles during rut (Ragotzkie 1988), sometimes leaving their annual home ranges (Rodgers et al. 1977, Koerth and Bryant 1982).

Near the time of the fawn drop, does separate themselves from other deer and typically move to specific areas to have their fawns. These areas are usually more mountainous and brushier than habitat used at other times of the year. Fox and Krausman (1994) found that does showed no preference according to the different vegetation types available to them during most of the year, but shifted to the upper portions of steep slopes in mountains where they were able to find greater hiding cover and shade.

White-Tailed Deer

Coues whitetails do not migrate in the traditional sense, but like other nonmigratory deer, they do show seasonal shifts in habitat use induced by the seasonal availability of food, water, and cover, and the effects of weather, hunting, breeding, grazing, and other factors (Welch 1960, Knipe 1977). After periods of heavy snowfall in the higher elevations, whitetails will move down to lower elevations or to alternative slopes where travel is easier and food more available (Welch 1960). These deer return to their original ranges as soon as habitat conditions allow.

Even without the need to escape snow cover, some whitetails in southeastern Arizona use the lower portions of their home ranges in winter and higher elevations in the summer (Welch 1960). This is mostly in response to the distribution of the browse during the winter months and forbs in the summer. Barsch (1977) determined that the presence of desirable forage plants was the most important factor affecting white-tailed deer distribution on his study area in the Whetstone Mountains.

An intensive investigation of seasonal movements of whitetails in the Chiricahua Mountains showed that these deer were generally most abundant on the southern and eastern exposures at the 6,000–7,000 feet elevation during the winter (Welch 1960). This coincided with the distribution of mountain mahogany, which is the most important browse species in the area. As spring approached, deer moved to areas where forb growth was available—mostly above 7,500 feet where moisture was more abundant. With the coming of higher summer temperatures, deer gradually spent more time on brush-covered northern exposures, where it was shaded and cooler. These areas also provided excellent hiding cover for fawns at this time of the year. During the dry periods of May and June, deer congregated around permanent water sources, causing overutilization of browse plants in the area. As temperatures lowered in the fall, northern exposures were less favorable, and deer moved to the warmer, browse-covered southern and eastern slopes where forbs and Gambel oak acorns became available. White-tailed deer in West Texas and northern Durango, Mexico, preferred areas with denser vegetation where shrub and forb diversity was greatest (Krausman and Ables 1981, Galindo-Leal et al. 1994, Gallina 1994).

As with mule deer, the rut drastically changes the activity patterns of whitetail bucks. By mid-December, bucks are easily seen at all times of the day as they search for receptive does. These large-scale movements may take them out of their normal home range (Welch 1960, Koerth 1981:50).

Daily Activity Patterns

Daily movements occur in response to weather, disturbance by humans/predators, and the distribution and juxtaposition of cover, food, and water. These movements are driven by the daily requirements of deer not only to stay alive, but also to maintain a level of comfort in their environment.

Deer are active primarily in the early morning and late evening hours. After a morning feeding session, they find a secluded place to bed down and ruminate (chew their cud) undisturbed. By late afternoon they are up feeding until dark or for several hours afterward. Deer usually bed down a few hours after sunrise; however, in the winter they may remain active most of the day.

Mule Deer

In summer, mule deer on the Kaibab Plateau select areas with a diversity of pine, mixed conifer, meadow, and aspen (Haywood et al. 1987a). Within this mixture of habitat types, deer select feeding areas in relation to what foods are available at that time. Areas with important browse plants show high levels of deer use, as do exposed benches and meadows at times of high forb growth (Swank 1958).

Desert mule deer on the Buenos Aires National Wildlife Refuge in southern Arizona spent an average of 39 percent of their daylight hours foraging, 29 percent traveling, 26 percent standing, and 6 percent bedded (Koenen and Krausman 2002). Radiocollared mule deer in West Texas and southern Arizona made daily movements ranging from 100 yards to three-quarters of a mile (Rodgers 1977, Dickinson and Garner 1979). Desert mule deer actively feed an average of about two hours after sunrise. In a study by Truett (1971:26), mule deer did not bed down any sooner after sunrise in the warmer summer months; however, they moved to sunny slopes on cold mornings and to shaded slopes when the day began to warm up. Bed sites are selected to maximize comfort; sunny beds are used on cool days and dense shade is sought during hot weather (Tull et al. 2001).

Deer use desert washes heavily and avoid areas that have been cleared of mesquite and other native shrubs. Desert washes provide a favorable microclimate for a greater diversity of plants used as food and cover, and also provide shelter from inclement weather (Phillips and Hanselka 1975, Krausman et al. 1985). Deer affected by high winds coupled with cool temperatures seek shelter on the protected side of hills and ledges, or in washes. Light rain does not affect deer activity substantially, but heavy rain will cause deer to bed down. High summer temperatures result in less daytime activity and more nighttime activity (Leopold and Krausman 1987, Hayes and Krausman 1993).

Moon phase is an oft-discussed factor in relation to deer activity and movements. Little has been done to determine the extent of this relationship in mule deer. Buss and Harbert (1950) reported more mule deer visiting a salt lick during a full moon than at new moon. These observations were recorded over a one-month period, in one location, in one year, with no accounting of other factors that may have affected deer activity during that period. Truett (1971) compared the length of time desert mule deer were actively feeding in the morning among the various moon phases and found no difference in morning activity following the different moon phases. Studying desert mule deer nighttime activity, Hayes and Krausman (1993) found no difference in deer movements among three periods of moonlight. In California and Colorado, researchers also found no difference in mule deer activity between new and full moons (Linsdale and Tomich 1953, Kufeld et al. 1988).

Human disturbance, especially during the hunting season, may cause changes in deer movements, although supporting data for mule deer are not as well established as for whitetails. Rodgers et al. (1978) reported movements onto the unhunted Santa Rita Experimental Range south of Tucson during the hunting season, but this was based on the movement of one radiocollared buck and a report of an "influx of deer" by other people. Mule deer bucks are more elusive during the hunting season, but much of this may be due to behavioral changes at that time of year, rather than to the effects of hunting. Swank (1958) found a reduced observation rate that began several weeks before the season and also occurred on the Three Bar Wildlife Area, which was entirely closed to hunting. More recent data from the Three Bar Wildlife Area and Tonto Basin provide some evidence of altered behavior of mule deer during the hunting season. One of the study areas (Tonto Basin) was subjected to heavy hunter densities, and 38 percent of the radiocollared bucks (all of them mature) made what appeared to be evasive movements during the hunting season (Horejsi et al. 1988). These movements consisted of distinct treks to steep, brushy, inaccessible portions of the study area, but did not exceed the boundaries of the normal home range. These findings mirror the results of a similar study of mule deer does in Colorado (Kufeld et al. 1988).

The presence of cattle can influence deer movements in some cases. Rodgers et al. (1978) reported that moderately grazed (40 percent utilization) areas did not affect deer movements in desert grassland habitat of the Santa Rita Experimental Range in southern Arizona. However, on the same study area, later research showed that radiocollared deer used ungrazed areas more than grazed areas within their home range (Ragotzkie 1988, Ragotzkie and Bailey 1991).

After nearly twenty years of no grazing, cattle were introduced into part of the Chevelon Wildlife Management Area of eastern Arizona in 1980; mule deer numbers in the grazed portion dropped significantly (McIntosh and Krausman 1982, Wallace and Krausman 1987). Heavy grazing that removes most of the green herbaceous vegetation undoubtedly affects the use of those areas by deer. If excessive grazing is widespread, or continues for several years, deer populations will decline due to substandard nutrition and lack of fawn security cover. Truett (1971) reported that during parts of the year, very little ground cover remained and cattle were browsing the brush heavily. Many of the important deer browse species were heavily used by cattle, forcing deer to move to rough terrain not grazed by cattle or to stand on their hind legs to reach blossoms or fruit. More than twenty-five years later, this condition still exists in too many areas throughout the Southwest. Not surprisingly, the resident desert mule deer populations suffer more during extended dry periods in these areas where the herbaceous layer is perpetually missing.

White-Tailed Deer

Like mule deer, whitetails are active in the early morning hours and again in late afternoon, with more activity at midday during the cooler months. Most southwestern whitetails are active in the winter months between 8 am and 10 am and again between 4 pm and 6 pm (Knipe 1977:33, Krausman and Ables 1981). During the summer months, 70 percent of the deer observed by Knipe (1977) between 9 am and 5 pm were bedded. Summer bed sites are usually under a few large trees that provide mostly shaded ground (Ockenfels and Brooks 1994).

During the summer, grasslands invaded by mesquite shrub and mixed grasslands are heavily favored by bucks, but not does, in Arizona's Santa Rita Mountains (Ockenfels et al. 1991). Both bucks and does prefer slopes of 20–29 degrees for resting and feeding, usually on the upper third of major and minor ridges. Oaks within 400 yards of drainages provide excellent habitat, with deer using northerly facing slopes more than others for bedding and foraging. Whitetails also avoid areas within 400 yards of a graded road (Ockenfels et al. 1991).

Based on deer observations, Knipe (1977:50) felt that moon phase had little to do with whitetail activity patterns. Beier and McCullough (1990) analyzed a large amount of data on movement and habitat use of whitetails and concluded that moonlight did not affect deer activity, or that its effect was minimal at most. Many researchers have investigated the effects of lunar cycles on deer movement, but so far no one has presented any compelling evidence for such a relationship. Those selling lunar-based hunting and fishing charts seem to be the only ones claiming such a relationship.

High hunter density and disturbance has been reported to cause seasonal movements in white-tailed deer. Welch (1960) used pellet counts to estimate the seasonal movements of deer and reported that deer vacated areas of high hunter activity. After a heavy snow, he tracked deer to determine the effect of hunters on deer movements and found that some deer made short circles of less than a quarter-mile, while others fled to canyon bottoms or brushy northern slopes. Deer became nocturnal and bedded in thick, brushy areas all day, but did not leave their home range unless hunter density was abnormally high. Toward the end of the hunt, when hunter activity was low, the deer resumed their normal activity patterns.

There is no evidence that hunting during the rut disrupts deer reproduction in any meaningful way. Bristow (1992) simulated high hunter densities and intensive deer disturbance during rut and monitored the effects on radiocollared deer in the Santa Rita Mountains of southern Arizona. He found that even at unusually high levels of disturbance, deer did not leave their home ranges. Additionally, reproductive success and the date of breeding was no different between disturbed and undisturbed does. It is highly unlikely that the current density of hunters in the Southwest creates anything more than an occasional local disturbance of breeding activity.

Presence of cattle in whitetail habitat may induce temporary avoidance movements (Brown 1984). However, competition with livestock is probably less of an issue for whitetails than for mule deer because much of the whitetail range is not grazed as heavily as the lower elevation areas that mule deer occupy. Additionally, the rugged terrain limits the use of many areas by cattle, reserving more of the browse and forbs for deer. Drawing on his experience in Mexico and southern Arizona, Welles (1959) reported that whitetails actively avoided cattle. He stated that competition with livestock is the most important limiting factor on the white-tailed deer population because they share the same sources of forage during a few critical times of the year (early summer and winter). In some parts of the Southwest overgrazing in dry years is still a problem for whitetails and their habitat (Galindo-Leal et al. 1994). However, during years of average or above-average precipitation, grazing probably affects white-tailed deer very little if at all (Ockenfels and Lewis 1997).

Chapter 7

Reproduction

Rutting Behavior

THE DRIVE of self-perpetuation is a powerful force in nature that shapes the breeding behavior of every species. Individuals benefit from having the healthiest mate to help produce their offspring because this assures that their own genes have the best chance of persisting in the future. This drive has given rise to an interesting repertoire of rutting behavior in deer. Although whitetails and mule deer show similarities in much of their breeding behavior, they have developed different courtship mannerisms and signals that influence gene flow between the species.

Prerut

During the summer, bucks travel in bachelor groups and occupy home ranges separated from those of doe-fawn groups. During this time, bucks may begin to establish a dominance hierarchy (pecking order) while their antlers are growing and still in velvet; they do this through flailing hooves and aggressive body posturing. This allows the bucks to assess the strengths and weaknesses of their competitors, thereby reducing the amount of energy expended in costly fighting later in the season. Bucks experience a rise in testosterone levels by late summer that causes the shedding of antler velvet. Once the velvet is shed, mature bucks expand their home ranges to overlap those of doe-fawn groups.

Sparring

Observers frequently confuse sparring with fights to determine dominance. Sparring is not a serious fight, but rather a contest between two bucks that are familiar with each other. Some have likened it to an arm-wrestling match—a good analogy because sparring represents more neck-twisting and shoving than an all-out clash of antlers. Sparring often occurs between bucks of unequal size (Kucera 1978). If there is a great size difference between the two bucks, the larger of the two may offer only one of his antlers to the smaller buck (Hirth 1977, Geist 1981). These matches are initiated when one buck approaches and presents his antlers to another buck. The second buck may then engage by carefully placing his antlers in contact with the antlers of the other buck. The two bucks then begin clicking their antler tines together and pushing mildly. In the case of two different-sized bucks, either may initiate the sparring. There is very little aggression on the part of either buck, but the level of aggressiveness may increase somewhat as the rut approaches. Bucks may break from sparring matches simultaneously and resume feeding, or one buck may jump back but not be pursued by the winner.

Sparring allows bucks to test their strength against other bucks, and probably serves to fine-tune their position in the hierarchy. Young sparring partners of a dominant buck may remain near him during the rut. This can benefit the mature buck if they intercept unfamiliar bucks looking for does in estrus. These previous sparring partners can save the dominant buck from having to constantly defend does he is courting. In return, the younger bucks may have the opportunity to breed an occasional doe when the dominant buck is preoccupied with other does.

There is little difference in sparring between mule deer and white-tailed deer, although whitetails appear to be more cautious, and more effort is exerted by younger bucks in maintaining a subordinate attitude toward a larger buck (Geist 1981). As the peak of rut approaches, a continual rise in testosterone levels results in increasing aggression and intolerance of other bucks. Sparring among mature bucks becomes uncommon, although yearling bucks may continue to spar into rut.

Dominance Fights

By the time the does come into estrus, or "heat," the dominance hierarchy is well established. Dominance is enforced with vocalizations, scent marking, body postures, and threat gestures. When two closely matched mature bucks meet, if neither yields to the threats of the other, a short but violent fight ensues. Violent dominance fights are rare in most populations, but they are intense brawls when they do occur. Serious injuries, and sometimes death, can result. However, one buck most often breaks contact and runs, leaving the victor to breed does in that area.

Antler Rubbing

Mule Deer

Antler rubbing on small trees serves a different purpose in mule deer than in whitetails. Mule deer use this activity as an auditory expression of dominance (Geist 1981). In wooded areas, this communicates dominance status to other bucks in the area. Mule deer vigorously rub or "horn" bushes for long durations and then stop to listen for a buck horning in response. If another buck is detected horning in reply, the first buck will move in that direction to investigate the challenger. Unlike blacktails and whitetails, mule deer spend less effort rubbing their foreheads on the bush they are horning. Mule deer lack a dense glandular area that whitetails have in their forehead skin, indicating that horning is not important in this species for the deposition of scent (Quay and Müller-Schwarze 1971).

White-Tailed Deer

Whitetails begin rubbing their antlers as soon as their velvet begins to dry. This strips the velvet off and helps strengthen a buck's neck muscles, but more important, it serves as a visual and scent-laden signpost. A dense concentration of glandular tissue under the forehead hair of whitetails is the source of scent that may serve to identify individual bucks to other deer in the population (Atkeson and Marchington 1982). The sight of a rub draws male and female deer nearer to investigate the information held by the scent. Rubs inform other deer which bucks are in the area, and probably also communicate dominance status. Rubs occur wherever suitable saplings of the right size and species are found. Rubs do not signify boundaries of a buck's rutting territory or area; rather, their distribution is influenced by type of vegetation, food sources, and topography. The density of rubs is related to the nutritional level and density of bucks 2.5 years old and older (Miller et al. 1987). Generally, mature bucks make rubs that are larger and twice as numerous as those of yearling bucks.

Rub-Urination

Rub-urination is a rutting behavior in which bucks urinate directly onto the tarsal glands, leaving scent on the ground and on the animal itself. As urine passes over the tarsal tufts, a deer shifts its hips from side to side, rubbing the urine into the hair tufts. Urine deposition in this manner has been documented in whitetail does and fawns, but it is most often done by mature bucks during rut (Kucera 1978). Whitetails usually do this behavior in scrapes (see following). At the peak of rut, tarsal glands turn black from a combination of urine, dirt, and glandular secretions. Sometimes, mature bucks have urine and mud encrusted on the inside of their legs from the tarsal glands to the hooves.

Scraping

Mule deer have not been documented to make scrapes, but scraping is an integral part of rutting behavior for whitetails. Scrapes seem to serve two primary purposes for breeding whitetails: to communicate presence to rival males and to attract breeding females. The term *scrape* describes a circle 1–3 feet in diameter that is scraped down to bare soil. Most scrapes (80–90 percent) are located under a low-hanging overhead branch that is sometimes called a licking branch. A buck approaches the licking branch and mouths it, probably sensing and/or leaving scent messages. Bucks also may rub their preorbital and forehead glands on the branch, then paw a bare area in the soil, and step forward to rub-urinate in it. This results in scent deposited at the site from the preorbital and forehead glands (licking branch), interdigital gland between the toes (from pawing), and possibly the prepupital and tarsal glands (rub-urination).

Most scrapes are made by mature bucks before or during the rut, and scent left in the scrape conveys a buck's presence and possibly dominance status. Researchers at the University of Georgia used automatic cameras to study scraping, and this method revealed several interesting facts: (1) although bucks visit scrapes made by other bucks, they may or may not return to a scrape they themselves established; (2) young bucks may perform the full scraping sequence; (3) most scraping activity (85 percent) occurs at night; and (4) does rarely urinate in a buck's scrape (Alexy et al. 2001). Females have also been documented to exhibit some scraping behavior, though only rarely. Does frequently visit scrapes, but they do not carry out the full scrape sequence because they usually do not mark an overhead limb. Establishment and maintenance of this form of scent post peaks right before rut. Once does start coming into estrus, scraping activity by bucks is reduced as bucks turn their attention to tending does.

Courtship

Mule Deer

As rut arrives, bucks become less tolerant of one another and begin to move around alone, monitoring does for signs of estrus. Dominant bucks roam the home ranges of doe-fawn groups, while other, "floater" bucks wander about in an unpredictable fashion (Geist 1981). Mule deer bucks technically do not gather harems; females are not actively "rounded up" and maintained as a group by a buck. Rather, does begin to gather near dominant bucks. Some speculate that this provides does protection against the constant harassment of younger bucks.

Bucks monitor a doe's readiness to breed through very ritualized courtship behavior. Geist (1981) identified two forms of courtship: juvenile mimicry courtship, and rush courtship. In *juvenile mimicry courtship,* a buck approaches a doe with a slow, submissive posture, head held low and neck outstretched. A doe typically ignores the advance initially and continues feeding and walking slowly forward; the buck will try to head her off, but she continues to walk away from him. Finally, the doe squats and urinates. The buck then moves forward and takes some urine from the doe into his mouth and performs a lip-curl, or flehman. The purpose of this lip-curl is to draw the urine into a specialized organ, called the vomeronasal, in the roof of his mouth. The vomeronasal was once thought to play a direct role in deciphering how close to estrus the doe is (Geist 1981), but more recent research on whitetails indicates it may be more important in synchronizing the reproductive readiness of a buck (K. V. Miller, personal communication, 1997). *Rush courtship* occurs less frequently than juvenile mimicry courtship, and usually when a doe is not responding to the latter. As the name implies, a buck rushes the doe, stomping and vocalizing. The rush may cover a few yards or they may run over 100 yards, but in the end the doe urinates for the buck, which stops his advances unless she is ready to breed.

When a mule deer doe is ready to be bred, she will stand for a buck as he approaches and allow him to lick and nuzzle her hindquarters. He then may lift a leg onto her back, testing her willingness to stand for him. The doe will stand still for a few seconds and then step or hop forward from underneath the buck. These precopulatory mounts progress gradually as the buck puts more and more weight on the doe's back, culminating in copulation. Copulation involves one thrust by the male, with the buck's hind feet often leaving the ground (Kucera 1978). A buck and doe may copulate several times over a day or two before the buck moves on.

White-Tailed Deer

Compared to mule deer, whitetails have more active courtship behaviors. Whitetails do not use the juvenile mimicry courtship (Geist 1981), but use the rush courtship exclusively. Running is a large component of whitetail courtship. Bucks travel among doe groups in search of receptive females, resting and feeding very little. Bucks may lose more than 20 percent of their body weight during the rigors of rut. As estrus (female receptiveness) approaches, bucks run does relentlessly, with the doe about 10 yards ahead of the pursuing buck (Hirth 1977). A doe's fawns will stay with her when she is only trotting away from the buck, but when the serious chasing starts, fawns are left behind. Sometimes a young buck runs with the buck and doe, keeping back a safe distance so the mature buck ignores him. As for mule deer, getting the doe to urinate is an important part of the whitetail's courtship.

When a buck locates a doe that is in estrus, there is much less chasing as they enter the tending phase. While tending, bucks closely follow and bed quietly with the does until they will allow the buck to mount them. During a period of receptiveness that lasts twenty-four to thirty-six hours, a doe will not run from the buck, and she will eventually allow him to rest his chin on top of her rump. This is a sign that she is ready; the buck then mounts the doe and the complete act of copulation occurs in less than fifteen seconds (Marchington and Miller 1994). If the doe is not bred during her first estrus cycle, she will cycle back into heat in three or four weeks.

Timing of Rut

The timing of rut is of interest because it is a period when deer activity is dramatically increased, and breeding behavior can be more easily observed. Exactly when rut reaches its peak and what dictates its timing is a source of much discussion. With the rise in several male hormones starting in summer, bucks begin producing sperm and are physically capable of breeding for at least five and perhaps as many as nine months (Bubenik et al. 1982). Does, however, will not allow a male to breed them until they reach the peak of their estrus, which lasts only twenty-four to thirty-six hours. The peak of the rut, then, is simply the period when most of the does are in estrus.

Timing of the breeding season is influenced by genetic factors and changes in the relative amounts of daylight and darkness in each twenty-four-hour period (photoperiod). A pea-sized gland near the center of the brain, called the pineal gland, receives input from the eyes and triggers the release of a hormone called melatonin every night when it becomes dark. Nighttime levels of melatonin may reach five to thirty times daytime levels (Bubenik and Smith 1987). As days become shorter in autumn, the pineal gland secretes more melatonin, which acts on the pituitary gland and regulates the release of an array of hormones. This seasonal fluctuation in the timing of the melatonin increase is critical to many seasonal changes, including antler cycles, winter/summer coat changes, food intake, and the rut (G. Bubenik 1990a). If the pineal gland is removed or damaged, the timing of these natural seasonal cycles disintegrates. Also, by artificially administering melatonin in the spring to simulate the normal autumn increase, some researchers have made whitetail bucks shed velvet, harden their partially grown antlers, molt into a winter coat, and begin rutting behavior in mid-summer (G. Bubenik 1983, Bubenik et al. 1986).

Photoperiod triggers physiological changes that lead up to the rut; however, a strong genetic factor predetermines a certain breeding time for deer in a particular location. In northern climates, any fawn born in late summer may not grow large enough to survive harsh winters and then will not pass on genes that are programmed for late breeding. Offspring of animals with late breeding dates would then be removed from the population, causing optimal breeding times to develop through natural selection. Indeed, deer populations in northern states have a narrower window of breeding times and a more predictable, intense rut than those in southern parts of their range. South American whitetail populations contain animals that are in breeding condition (and have hardened antlers) during all months of the year. The genetic contribution to rut timing was demonstrated when researchers transported captive deer from Michigan to Mississippi (and vice versa) and found that the deer retained their breeding dates from their state of origin for at least three years (Jacobson and Lukefahr 1998). In addition, offspring of Michigan × Mississippi matings bred at a time that was intermediate to those of their parents.

Because the actual peak of rut is governed by does coming into estrus, the nutrition and subsequent body condition of individual does in the months leading up to rut can play a role in fine-tuning the timing of rut (Verme 1965, McGinnes and Downing 1977). Factors such as weather and temperature can also play a part, in any given year. Once photoperiod has prepared a deer herd physiologically for breeding, a period of cold weather, or especially a cold front with falling barometric pressure, may increase deer activity. Unseasonably warm weather seems to prolong the rut and suppress rutting activity, especially during the day. Cool weather and storms do not influence the physiological readiness of breeders, but simply provide a comfortable temperature range that allows deer to move more.

Recently, Alsheimer (1999) challenged the traditional theory that timing of rut is based primarily on photoperiodism. He postulated that the lunar cycle acts in concert with changes in day length to set the reproductive clock each year. The lunar cycle theory argues that the peak of the rut corresponds to the "rutting moon" each year, defined as the second full moon after the autumnal equinox. If this theory is correct, the rut should occur at slightly different, but predictable, times each year. Support for this theory is largely anecdotal, and currently lacks scientific corroboration. Moreover, nineteen years of research in Michigan involving 503 recorded breeding dates of white-tailed deer indicated no connection between the timing of the moon and the onset of breeding activity (Ozoga 2000).

Mule Deer

Throughout the Southwest, research has provided insight into the actual timing of the rut in various areas. In northern, high-elevation areas of the Southwest, the rut starts in late November and trails off into January, peaking in early to mid-December. These dates appear consistent throughout southern Colorado/Utah and northern Arizona/New Mexico (Table 13). Illige (1954) reported Kaibab deer rutting from November 20 to December 19, indicating that breeding occurs primarily on winter range and transitional areas.

The rut is somewhat later for desert mule deer in areas of southern California, Arizona, New Mexico, West Texas, eastern Sonora, and northeastern Mexico. Necks begin to swell in late November or early December, and some rutting occurs during early December through early February (Table 13). The peak of rut throughout the southwest deserts, grasslands, and central chaparral habitats

TABLE 13. PEAK BREEDING AND FAWNING DATES FOR MULE DEER IN THE SOUTHWEST

Location	Breeding dates	Fawning dates	Source
California			
San Jacinto Mts.	Oct. 28–Nov. 22	May 21–June 14	Schaefer 1999
Cuyamaca State Park, San Diego Co.	Late Nov. to late Dec.	...	Bowyer 1986b
Cuyamaca State Park, San Diego Co.	...	Last half of June to early July	Bowyer 1991
Southeastern Deserts	Jan. to Mar.	Early Aug. to mid-Oct.	Celentano et al. 1984
Colorado			
Mesa Verde Nat. Park	Nov. 23–Dec. 10	June 14–July 19	Mierau and Schmidt 1981
Arizona			
Kaibab Plateau	...	Late June to early July	Hall 1925
Kaibab Plateau	Mid-Nov. to mid-Dec.	June 15–July 1	Illige 1954
Kaibab Plateau	December	Late June to early July	Russo 1964
Chaparral	Dec. 15–Jan. 30	Late June to early Aug.	Illige 1954
Chaparral	Late Dec. to early Jan.	Last week July to first week Aug.	Hanson and McCulloch 1955
Chaparral	Dec. 1–Feb. 20	July 15–Aug. 15	Swank 1958
Southern Arizona, Tucson Mts.	Mid-Dec. to mid-Feb.	Aug. 1–Sep. 15	Clark 1953
Southern Arizona, north of Tucson	Late Dec. to early Jan.	July 1–Aug. 31	Truett 1971
Southern Arizona, Santa Rita Exp. Range	Mid-Dec. to mid-Feb.	Late July to early Aug.	Rodgers 1977
Southern Arizona, Santa Rita Exp. Range	Late Dec. to early Jan.	Mid-July through Aug.	Ragotzkie 1988
Southwest Arizona	...	First week Aug. to first week Oct.	Krausman and Etchberger 1993
Colorado River Delta	...	August	Stone 1905
Southwest Arizona, Belmont/Bighorn Mts.	...	First week Aug. to first week Oct.	Fox and Krausman 1994

Location	December	January	June	July	Reference
New Mexico					
Northern New Mexico			June		Humphreys and Elenowitz 1988
Southern New Mexico				July	Humphreys and Elenowitz 1988
Texas					
West Texas, Longfellow Ranch	Dec. 25–Jan. 25		Mid-Aug. to early Sept.		Phillips and Hanselka 1975
West Texas, Longfellow Ranch	Mid-Dec. to mid-Jan.		...		Ratliff 1980
West Texas, Big Bend Nat. Park	Dec. 15–Jan. 15				Kucera 1978
West Texas, Black Gap Wildlife Area	Mid-Dec. to mid-Feb.		Mid-July to mid-Sept.		Wallmo 1960
West Texas, Black Gap Wildlife Area	Late-Nov. to early Jan.		Late June to mid-Sept.		Brownlee 1971
West Texas, Black Gap Wildlife Area	Early Dec. to early Feb.		Mid-July to mid-Sept.		Desai 1962
West Texas, Pecos Co.	Mid-Dec. to mid-Jan.		...		Dickinson 1978
West Texas	Mid-Dec. to early Jan.		...		Pittman and Bone 1987
Panhandle	Dec. 1–Dec. 30		...		Pittman and Bone 1987
Mexico					
Baja California	Last week in Dec. to mid-Jan.		...		J. Vazquez-Pineda, Pers. Comm., 2003
Baja California	Nov. to Dec.		Late June to July		Leopold 1959
Sonora and Chihuahua		January	Late July to Aug.		Leopold 1959
Cedros Island		Sept. to Nov.	April		Perez-Gil Salcido 1981

TABLE 14. PEAK BREEDING AND FAWNING DATES FOR WHITE-TAILED DEER IN THE SOUTHWEST

Location	Breeding dates	Fawning dates	Source
Arizona			
Southeastern Arizona	Late Dec. to mid-Feb.	Mid-July to mid-Sept.	White 1957
Chiricahua Mts.	Late Dec. through Jan.	Mid-July to mid-Aug.	Welch 1960
Santa Rita Mts.	Mid-Jan.	...	Nichol 1938
Coronado Nat. Monument	January	August	Welles 1959
Statewide	January	July and Aug.	Knipe 1977
Santa Rita Mts.	January	Early to mid-Aug.	Ockenfels et al. 1991
Chiricahua Mts.	Jan. 19–Feb. 16[a]	Aug. 7–Sept. 4	Day 1964:81
Santa Rita Mts.	...	August	Smith 1984
New Mexico			
Southwestern New Mexico	Jan. and Feb.	Aug. and Sept.	Lang 1959
Southeastern New Mexico	Dec. and Jan.	July and Aug.	Lang 1959
Texas			
Trans-Pecos	Early Dec.	...	Russ 1984
Big Bend Nat. Park	Mid-Dec. to mid-Jan.	July and early Aug.	Krausman and Ables 1978:52
Mexico			
Chihuahua, Rio Gavilan	Mid-Jan.	Mid-Aug.	McCabe and Leopold 1951
Sierra Madres	Mid-Jan. to late Feb.	Aug. and Sept.	Galindo-Leal and Weber 1998
Chihuahua/ Coahuila	January	August	Leopold 1959
Durango, La Michilia	Mid-Jan. to mid-March	Late Aug. to late Sept.	Weber et al. 1995
Durango, La Michilia	...	June to Aug.	Ezcurra and Gallina 1981

[a] Deer were nutritionally stressed during this study, which may have delayed the breeding season.

occurs during the last week in December and the first week in January. Desert mule deer in more arid deserts of southwestern Arizona, southeastern California, and western Sonora, Mexico, rut even later in January.

White-Tailed Deer

The peak of the whitetail rut is a little later, on average, than that of mule deer in the same areas. This is yet another way hybridization between these two species is minimized. The necks of whitetail bucks begin to swell in November, but breeding does not generally start until late December and may continue into late February (Table 14). The peak in rutting activity generally occurs during early to mid-January throughout most of Coues whitetail range, but rut may occur a little earlier in higher elevations of the Sierra Madres of Sonora and Chihuahua, Mexico.

Some of the best information on breeding dates of Coues whitetails comes from Ockenfels et al. (1991), who monitored thirty-six radiocollared whitetails in the Santa Rita Mountains of southeastern Arizona. The earliest rutting behavior during that study occurred on November 7 and the latest on March 16. Many of these extreme observations were likely inexperienced yearling bucks. The average peak of the rut was mid-January based on observations of rut behavior between 1988 and 1991. Bristow (1992) examined twenty pregnant does in this same region and found the average date of conception was during the first week of January.

Gestation and Fawning

The length of gestation, or pregnancy, in deer is about 200–207 days, just short of seven months (Nichol 1938, Robinette et al. 1977). The weight of the fawn or fawns appears to be the major factor triggering birth, but the timing of birth can vary slightly from year to year based on the timing of conception and the nutrition of the does. Very poor nutrition may cause not only a delay in breeding, but also a lengthening of the gestation period (Verme 1965). Smith (1984) found a two-week delay in the whitetail fawning period in the Santa Rita Mountains of Arizona during a dry year (1970) as compared to a wet year (1983).

As the time of parturition (fawning) approaches, a doe disassociates itself from other does and finds a secluded spot to give birth (Michael 1964). Evidence suggests that older does return to the same fawning locations year after year (Ozoga et al. 1982). The actual birth takes place with a doe standing or lying, and labor may take up to two hours. Following birth, the doe consumes the afterbirth and fetal tissues and licks the fawn or fawns clean. Twins are usually born about ten to thirty minutes apart. After only ten to twenty minutes the fawns may stand up and begin nursing.

Shortly after birth the doe leaves the fawns and forages on her own, returning only two or three times per day to nurse for the first few days. Twin fawns are usually separated in different bed sites, which reduces the chance of a predator finding both. The doe may leave the fawns all day (Truett 1971), perhaps explaining why many well-meaning people collect "abandoned" fawns from the wild and bring them in for "rehabilitation." In reality, does probably observe people taking their fawns and thus lose an immense investment in time and energy.

A doe's milk contains twice as much fat as cow's milk, and this provides the sustenance fawns need to grow rapidly. After a few weeks, fawns start to sample green forage, and they are basically dependent on green feed by five weeks of age (Short 1964). After the first two months following birth, fawns begin to move about with their mother and are completely weaned at two to four months of age (Dixon 1934, Short 1964, Knipe 1977).

Mule Deer
Since the rut in the northern and high-elevation areas of the Southwest occurs earlier than in desert areas, it follows that the peak of the fawn drop will also

be earlier. Mule deer does in northern areas drop their fawns from June to August, peaking in the last week of June and first week in July (Table 13). Desert mule deer have their fawns during a later time window, from July to September (Hoffmeister 1986). In chaparral and desert grassland areas of West Texas, New Mexico, Arizona, and northeastern Mexico, most fawns are born from mid-July through mid-August (Table 13). Research in the deserts of southwestern Arizona, southeastern California, and western Sonora, Mexico, however, revealed an even later fawning peak, during August and September.

Mule deer does in desert areas reduce the size of their home ranges and move to mountainous areas to have their fawns. In the Belmont and Bighorn mountains of southern Arizona, Fox and Krausman (1994) found that does selected the upper one-third of steep (more than 30 percent) slopes to hide their fawns, and always used palo verde trees and other vegetation to conceal fawns and keep them cool. Healthy mule deer fawns weigh 7–8 pounds at birth, but their weight doubles in two weeks and quadruples in thirty to forty days (Nichol 1938). A fawn's spotted coat helps camouflage it for about two and a half months (Nichol 1938). Fawns born near the end of the July/August fawning period may still have visible spots in late October.

White-Tailed Deer

In southeastern Arizona, most whitetails drop their fawns from mid-July through mid-September, with fawning reaching a peak in late July through August (Table 14). Knipe (1977) reported births as early as June 30 and as late as October 7. During a time when the deer population was under nutritional stress, Day (1964) used measurement of fetuses to estimate the peak of fawning in the Chiricahua Mountains of southern Arizona as August 7 through September 4. Ockenfels et al. (1991) observed radiocollared and unmarked whitetails in the Santa Rita Mountains of southern Arizona for four years, and reported a peak fawn drop during mid- to late August. On the same study area, but in a different year, Bristow (1992) examined eleven fetuses and estimated an average birth date during the last week of July. Coues whitetails are born weighing 4–5 pounds, but gain weight very rapidly, as is the case for mule deer. Whitetail fawns retain their spots for forty to eighty days (Welles 1959, Knipe 1977), and late-born whitetail fawns can sometimes retain spots into mid-November (O'Conner 1939).

Reproductive Rate

Deer population fluctuations are a function of additions to, and losses from, the population. Additions result from reproduction and immigration into the area; losses are caused by mortality and emigration from the population. Because movements into and out of a population are usually constant, changes in deer abundance are caused primarily by the relationship between reproduction and mortality

(White 1984). If mortality in any year exceeds recruitment into the population, deer abundance declines. Conversely, if recruitment exceeds mortality, a population increases. Understanding deer population dynamics requires an acquaintance with the factors that affect reproduction and mortality.

Fertility

Under good habitat conditions, deer fertility is consistently high. Low fawn:doe ratios are not caused by infertile deer, yet barren or "dry" does are still frequently discussed among deer enthusiasts. The fertility of old does was illustrated by Robinette et al. (1977:99), who reported an average of 1.78 fetuses per doe for 139 Utah mule deer does eight years old and older. Fawns die from many factors, from the time of conception to weaning. Some does may be better at raising fawns than others, but "dry" does are simply those who have lost their fawn(s) that particular year.

Measures of fertility in deer come from a variety of research projects conducted throughout the Southwest. Analysis of female reproductive tracts can reveal the number of ova (eggs) fertilized and the number of fetuses carried (Tables 15 and 16). During a normal female reproductive cycle, one to four ova appear on the surface of the ovaries in a tiny, fluid-filled sac. At the time of ovulation, these sacs rupture spontaneously and release ova into the oviduct and eventually the uterus, where one or more is fertilized (Zwank 1976). The empty sacs from which ova erupted then become hormone-producing glands called *corpora lutea*. Corpora lutea disappear rapidly if no ova are fertilized, but upon fertilization of any ova, corpora lutea are altered by hormonal changes and are referred to as *corpora albicantia* or *corpora rubra*, depending on the pigmentation. Generally, one corpus albicans represents a fertilized egg. Biologists can dissect ovaries from dead does and count corpora albicantia to study the recent reproductive activity of that doe. Some prenatal mortality occurs (about 5–16 percent loss from the egg stage to mid-pregnancy), making corpora albicantia counts a slight overestimate of actual number of fawns born. Older does actually ovulate more eggs than middle-aged does, but prenatal loss is greater (Robinette et al. 1977).

In the Southwest, females rarely breed as fawns, so it is very uncommon to see yearling does with fawns at their side (Tables 17 and 18). In some parts of the country, especially those with extensive agricultural areas, as many as half of the female fawns will breed their first fall when they are six months old. Good physical condition is a prerequisite for breeding as fawns. Southwestern deer are born much later than northern deer and do not have the time and nutritional input necessary to regularly achieve breeding condition as fawns.

Likewise, a low percentage of yearling does (1.5 years) become pregnant, and those that do carry fewer fetuses. During their third rut (2.5 years old), does reach maximum productive potential and will remain reproductively active throughout their lives. The lower reproductive rates in younger does mean that consecutive

TABLE 15. AVERAGE NUMBER OF FETUSES PER DOE BY AGE CLASS FOR MULE DEER IN THE SOUTHWEST

Location	Ave. number of fetuses per doe by age class			Doe sample size by age class (0.5, 1.5, 2+)	Source
	0.5 yr.	1.5 yrs.	2+ yrs.		
California					
North Kings River	0.0	1.33	1.71	3, 21, 81	Salwasser 1974, Salwasser et al. 1978[a]
Camp Pendleton	0.0	0.85	1.73	20, 14, 68	Bischoff 1958[a]
San Jacinto Mts.	...	0.50	1.36	0, 4, 11	Schaefer 1999[a]
Utah					
Oak Creek	0.0	1.06	1.67	77, 236, 863	Robinette et al. 1977[b]
Colorado					
Mesa Verde NP	0.0	1.0	1.79	10, 3, 28	Mierau and Schmidt 1981
Arizona					
N. Kaibab	0.01	0.80	2.04	206, 116, 739	Swank 1958[b]
N. Kaibab	0.0	2.0	1.86	13, 3, 14	Pregler 1974[b]
Mingus Mt.	0.0	0.88	2.5	8, 8, 8	Swank 1956[b]
Bill Williams	0.05	0.90	1.9	21, 20, 90	Swank 1956[b]
Alpine	...	...	1.88	0, 0, 8	Webb 1971[a]
Winslow	0.0	1.05	2.61[c]	15, 42, 124	Swank 1956[b]
Three Bar	...	...	1.25	0, 0, 12	McMichael 1967[a]
Santa Rita Exp. Range	0.0	1.0	1.30	5, 2, 10	Short 1979[a]
Kofa NWR	0.0	1.67	2.56	3, 3, 25	Johnson and Swank 1957[b]
Statewide	0.01	1.00	1.96	348, 430, 1893	All Arizona studies
New Mexico					
Guadalupe Mts.	0.01	0.88	1.53	100, 105, 196	Anderson et al. 1970[b]
Guadalupe Mts.	...	0.90	1.50	0, 33, 164	Humphreys and Elenowitz 1988
MacGregor Range	...	0.61	1.43	0, 16, 124	Humphreys and Elenowitz 1988
Mount Taylor	...	0.91	1.61	0, 14, 141	Humphreys and Elenowitz 1988
Dry Mesa	...	0.95	1.63	0, 15, 148	Humphreys and Elenowitz 1988
Colfax Co.	...	0.00	1.00	0, 5, 27	Bender, unpublished data
Texas					
Black Gap WMA	0.81	2.04	1.84	101	Desai 1962
Black Gap WMA	0.20	0.58	1.51	5, 19, 59	Brownlee 1971
Black Gap WMA	0.0	0.28	1.95	6, 18, 44	Wallmo 1961
Stockton Plateau	...	0.75	1.05	0, 4, 22	Ratliff 1980
Trans-Pecos	...	0.60	1.32	0, 13, 124	Pittman and Bone 1987[a]
Panhandle	...	0.80	1.74	0, 6, 50	Pittman and Bone 1987[a]

Note: Averages include nonpregnant does.
[a] Based on counts of actual number of fetuses present in uterus.
[b] Based on counts of corpora albicantia representing number of eggs fertilized.
[c] Swank reports 41 corpora lutea present in 9 does in 1954. It is possible this includes some corpora lutea from eggs not fertilized (corpora lutea of estrus) or eggs fertilized the previous year, thereby inflating this estimate (W. Swank, personal communication, 1998).

TABLE 16. AVERAGE NUMBER OF FETUSES PER DOE BY AGE CLASS FOR WHITE-TAILED DEER IN ARIZONA

Location	Ave. number of fetuses per doe by age class			Doe sample size by age class (0.5, 1.5, 2+)	Source
	0.5 yr.	1.5 yrs.	2+ yrs.		
Chiricahua Mts.	...	←—1.15[a]—→		0, 123	Day 1960[b]
Chiricahua Mts.	0.0	0.7	1.73[a]	6, 3, 21	White 1957[b]
Huachuca Mts.	0.0	1.17	1.42	3, 6, 12	Johnson and Swank 1957[b]
Huachuca Mts.	...	←—1.05—→		0, 38	Day 1960[b]
Baboquivari Mts.	0.14	0.60	1.11	7, 10, 9	Johnson and Swank 1957[b]
Santa Rita Mts.	0.0	0.0	1.53[a]	1, 1, 15	Smith 1984[b,c] [1971]
Santa Rita Mts.	0.25	1.0	1.9	4, 7, 10	Smith 1984[b,c] [1983]
Santa Rita Mts.	0.0	0.5	1.35	1, 2, 17	Bristow 1992
Statewide	0.09	0.80	1.53	22, 274	All Arizona studies

Note: Averages include nonpregnant does. In some cases, data were originally recorded according to broader age categories than those indicated by a single column in this table. In such cases, arrows indicate the span of years to which a particular value applies.
[a] Data collected when deer physical condition was very poor.
[b] Based on counts of *corpora albicantia* representing number of eggs fertilized.
[c] Based on counts of actual number of fetuses present in uterus.

years of good rainfall are important to build a robust deer population. Good nutrition is a prerequisite to healthy does dropping healthy fawns (year 1). Above-average precipitation the next year (year 2) will help a large percentage of these fawns survive to be yearlings (although they will not yet produce fawns). The third year of above-average rain (nutrition) then allows most of the yearling does to breed and creates a reproductive pulse for the population.

When deer populations decline, some question whether all does are becoming pregnant. Time and again, investigations have shown that low numbers of fawns seen in the fall is not caused by does failing to become pregnant (Andelt et al. 2004). Does collected early in pregnancy invariably show that fetal and pregnancy rates are near the biological maximum, except in cases of substandard nutrition (Tables 13 and 14). Deer herds in the Southwest can increase with only 0.5 fawns per doe present in January (five months after the fawn drop). Female mule deer and white-tailed deer commonly carry more than 1.5 fawns per doe during pregnancy (Tables 15 and 16). This potential reproductive surplus is nature's way of allowing populations to increase rapidly when habitat conditions are favorable. Clearly, at least two-thirds of the fawns born are in excess of a level needed to maintain stable populations.

Mule deer pregnancy rates are usually reported to be less than those of whitetails, but it is doubtful whether there is a significant difference in does more than

TABLE 17. AVERAGE PERCENTAGE (%) OF FEMALES PREGNANT BY AGE CLASS FOR DESERT MULE DEER IN THE SOUTHWEST

Location	Ave. % of females pregnant by age class			Doe sample size by age class (0.5, 1.5, 2+)	Source
	0.5 yr.	1.5 yrs.	2+ yrs.		
California					
North Kings River	0	90	95	3, 21, 81	Salwasser 1974, Salwasser et al. 1978[c]
Camp Pendleton	0	73	94	20, 14, 68	Bischoff 1958[a]
San Jacinto Mts.	…	50	91	0, 4, 11	Schaefer 1999[a]
Utah					
Oak Creek	2.6	82.7	93.7	77, 263, 863	Robinette et al. 1977[b]
Colorado					
Mesa Verde NP	0	100	100	10, 3, 28	Mierau and Schmidt 1981[b]
Arizona					
N. Kaibab	1	53.4	86.9	206, 116, 739	Swank 1958[c]
Mingus Mt.	0	62.5	100	8, 8, 8	Swank 1956[c]
Bill Williams	5	75	92.2	21, 20, 90	Swank 1956[c]
Winslow	0	66.7	88.7	15, 42, 124	Swank 1956[c]
Alpine	…	…	100	0, 0, 8	Webb 1971[c]
Three Bar Wildlife Area	…	…	83	0, 0, 12	McMichael 1967[b]
Kofa NWR	0	100	96	3, 3, 25	Johnson and Swank 1957[c]
New Mexico					
Guadalupe Mts.	1	66.7	86.7	100, 105, 196	Anderson et al. 1970[c]
Colfax Co.	…	0	67	0, 5, 27	Bender, unpublished data
Texas					
Stockton Plateau	…	75	81.8	0, 4, 22	Ratliff 1980
Elephant Mt. WMA	7.9	20	92.3	38, 10, 52	Lawrence et al. 2004
Black Gap WMA	20	47.4	93.2	5, 19, 59	Brownlee 1971[c]

[a]Data collected when deer physical condition was very poor.
[b]Based on counts of actual number of fetuses present in uterus.
[c]Based on counts of corpora albicantia representing number of eggs fertilized

two years old (McCullough 1987). Analysis of differences in pregnancy throughout the range of both species indicated that differences result from variations in habitat quality, rather than inherent fertility (Robinette 1956:416, Beasom and Wiggers 1984). In the Southwest, ratios of about 35–50 fawns per 100 does in January are required to maintain a stable population over the long term. This is much lower than the 60–70 fawns per 100 does needed in the Rocky Mountain states to compensate for the high winter mortality of fawns in that region (Unsworth et al. 1999).

TABLE 18. AVERAGE PERCENTAGE (%) OF FEMALES PREGNANT BY AGE CLASS FOR COUES WHITE-TAILED DEER IN ARIZONA

Location	Ave. number of females pregnant by age class			Doe sample size by age class (0.5, 1.5, 2+)	Source
	0.5 yr.	1.5 yrs.	2+ yrs.		
Chiricahua Mts.	...	←—75.6[a]—→		0, 123	Day 1960[b]
Chiricahua Mts.	0	66.7	100[a]	6, 3, 21	White 1957[c]
Baboquivari Mts.	14.3	40	77.8	7, 10, 9	Johnson and Swank 1957[c]
Huachuca Mts.	0	66.7	83.3	3, 6, 12	Johnson and Swank 1957[c]
Santa Rita Mts.	0	0	100	1, 1, 15	Smith 1984[b,c] [1971]
Santa Rita Mts.	25	85.7	100	4, 7, 10	Smith 1984[b,c] [1983]
Santa Rita Mts.	0	50	100	1, 2, 17	Bristow 1992

[a]Data collected when deer physical condition was very poor.
[b]Based on counts of actual number of fetuses present in uterus.
[c]Based on counts of corpora albicantia representing number of eggs fertilized.

Nutrition

The first step in the continuum of producing a fawn is ovulation—the release of eggs from the ovaries. Nutrition preceding estrus (rut) plays a role in ovulation in mule deer (Zwank 1976) and whitetails (Verme 1962, Smith 1984).

Not all eggs ovulated result in a fetus. Some 5–16 percent of the eggs do not result in mid-term fetuses, due to lack of fertilization, flushing of eggs, or early intrauterine mortality of the embryo (Verme 1962, Rhodes et al. 1992). Fawns may even be resorbed in the womb or aborted if nutritional conditions worsen during pregnancy; the doe's survival is paramount over reproduction. However, embryo resorption and spontaneous abortion of fawns rarely occur. When females are nutritionally stressed, fawns are more often stillborn, or born alive but grossly underweight and unthrifty (Verme 1962, Robinette et al. 1977). These weakened fawns quickly succumb to myriad interrelated factors, including inadequate quantity of milk from the doe, abandonment, predation, disease, parasites, dehydration, and other factors.

Yearling does are most susceptible to nutritionally induced variations in fertility. This is significant because under normal conditions, yearling does represent the largest class of breeding-age females. Fertility of adult does fluctuates with nutritional level, but not as drastically as that of yearling does. This relationship was illustrated by research on whitetails in the Santa Rita Mountains of southern Arizona. Smith (1984) collected does in a year of poor nutrition (1970) and in a year of good nutrition (1983), and found that 100 percent of the adult does (more than two years old) were pregnant in both years. However, there was a drastic difference in the proportion of yearlings that were pregnant in a dry year (0 percent) versus a wet year (85.7 percent) (Table 14). Likewise, the average number of fetuses per doe was similar between wet and dry years in adult does, but differed for yearlings between the dry (0 fetuses/doe) and wet years (1 fetus/doe) (Table 13). Under excellent nutritional conditions (1983), one of four fawns was found to be pregnant. This same nutritional relationship for yearlings holds true for mule deer on the Kaibab Plateau and elsewhere in the Southwest (Swank 1956, Lawrence et al. 2004). The effects of nutrition on herd productivity then, are largely manifested in the proportion of yearlings breeding and the average number of fawns produced.

The nutritional level of a doe during the last third of pregnancy and during the early stages of lactation is of utmost importance to her fawns' survival. Verme (1962) investigated the effects of poor versus good winter/spring diets on pregnant does. Results were dramatic: 92 percent of fawns born to malnourished does died within two days. In contrast, does receiving good nutrition during the last half of pregnancy gave birth to fawns that weighed twice as much, and only 5 percent of their fawns died within a few days. The nutritional condition of pregnant does clearly is the most important factor influencing the size of a fawn crop. The quality of the milk remains consistent regardless of the doe's diet, but the quantity produced by a malnourished doe may be insufficient to provide for the needs of her offspring (Youatt et al. 1965). Also, malnourished does are more likely to lose their maternal instinct and abandon their fawns.

In the Southwest, nutritional status of deer is heavily dependent on rainfall and the intensity of grazing by domestic livestock. Since nutrition before rut and throughout pregnancy is vital to healthy fawns and good fawn survival, winter rains should be closely correlated with deer fawn recruitment. Smith and LeCount (1979) analyzed nine years of mule deer fawn:doe ratios, winter rainfall totals, and deer forage abundance in central Arizona. They found a high correlation between October-April rainfall and forage (forbs and browse species) available to deer in mid-pregnancy (April). Rainfall after February, however, added little to forage production.

Fawn:doe ratios for mule deer in January also are correlated with the amount of forbs produced the previous spring. Winter rainfall is most important to fawn recruitment, but in the absence of winter rains, above-average summer rainfall can buffer the negative effects by providing cover for fawns and nutrition for lactating does (Smith and LeCount 1979). Summer rainfall seems to play a more important role in fawn recruitment in western Arizona below 3,000 feet elevation (A. Fuller and B. Henry, personal communication, 2000). A combination of above-average winter and summer precipitation will result in optimal conditions for fawn production and deer population growth or maintenance.

In the Southwest, woody browse plants that deer prefer appear to have inadequate or marginal levels of protein and phosphorus except during the winter growing season. After annual growth of browse stops in early spring (after April), protein and phosphorus drop below a level recommended for satisfactory growth for the remainder of pregnancy (Urness et al. 1971). To supply rapidly developing fetuses, pregnant females supplement their diet with forbs, which are extremely important because they are highly digestible and supply a disproportionate amount of nutrients like vitamins, protein, and phosphorus.

In high-elevation areas of the Southwest, however, winter precipitation comes as snow, and cold temperatures limit the availability of early spring forbs. McCulloch and Smith (1991) did not find a correlation between winter precipitation (October–May) and pre- or posthunt fawn:doe ratios on the Kaibab Plateau. Winter precipitation was not related to deer population trends, but summer (June–August) rainfall and the cumulative amount of precipitation for the preceding two to three years was related to the trend of the deer population in that part of Arizona.

Brown and Henry (1981) used a Palmer Drought Severity Index (PDSI) to evaluate factors affecting whitetail reproduction. They found that the occurrence of drought in June and November explained about one-third of the variation in fawn:doe ratios. Early summer and fall droughts were thought to be major factors in determining the distribution and density of whitetails in southern Arizona. Rainfall does not explain all the variation in fawn crops, but it is certainly the main factor driving the reproductive engine of southwestern deer (McKinney 2003).

Although precipitation is the most important factor affecting deer nutrition and fawn survival, range conditions, as affected by ungulate stocking density, determine how much of that nutrition is available to deer. High levels of ungulates (cattle, elk, deer, burros, sheep) reduce the amount of forage available, and can negatively affect deer reproduction. Overpopulations of deer occur occasionally in the Southwest. These temporary occurrences of overabundance are eventually reduced to more appropriate levels through management (doe harvest) or natural factors (high natural mortality and reduced reproduction) resulting from depleted forage resources.

Livestock grazing can affect deer nutrition if not properly managed. Conservative grazing (less than 40 percent of forage by weight [Holechek and Galt 2000]) does not affect deer nutrition in most cases, but overuse of forage in arid environments removes much of the herbaceous cover crucial for doe nutrition and fawning cover (Loft et al. 1987, Galindo-Leal et al. 1994). When herbaceous cover is scarce, livestock browse the twigs and leaves of shrubs, exacerbating poor conditions created by removal of forbs (Hanson and McCulloch 1955). Livestock sometimes browse the most important deer forage plants intensively (Swank 1958, Knipe 1977, Heffelfinger et al. 2006). Stocking rates of cattle on some grazing allotments in the Sonoran Desert are based on browse, because herbaceous material is scarce or nonexistent in most years. Some "browse allotments" for cattle in the Sonoran

Desert north of Tucson, Arizona, allow cattle to use 50 percent of the browse over large areas. Cattle in these areas mostly browse jojoba and fairy duster, which are the two most important components of desert mule deer diets in the area.

Deer avoid areas occupied by large numbers of cattle, and they are more abundant in areas ungrazed by cattle (McIntosh and Krausman 1981, Wallace and Krausman 1987). Horejsi (1982) reported that livestock grazing negatively affected fawn survival, but only during drought years. Gallizioli (1977) stated that higher productivity and better habitat conditions in ungrazed areas equate to higher overall mule deer densities. Inappropriately high populations of elk could have a similar effect, although the competitive relationship between deer and elk is not consistent from area to area (Lindzey et al. 1997).

The introduction of some exotic plants, such as Lehmann's lovegrass, red brome, and buffel grass, has been detrimental to southwestern deer (Heffelfinger et al. 2006). The spread of Lehmann's lovegrass and red brome throughout large areas of desert grasslands and Sonoran Desert areas has reduced the natural diversity of native grasses and forbs (D'Antonio and Vitousek 1992). This dense and foreign ground cover provides little benefit to native wildlife, and competes with natural herbaceous plants that once provided a nutritional base for deer (Heffelfinger et al. 2006). Although this knee-deep grass may look good, it has reduced the overall amount of nutrition available to deer herds. In addition, consistent presence of abundant grass cover in the Sonoran Desert increases the frequency of fire, which historically was not a natural component of the Sonoran Desert ecosystem. Subtle changes in deer habitat, such as encroachment by exotic plants, do not cause acute declines in deer abundance, but incrementally reduce habitat quality and, in turn, deer abundance.

Buck:Doe Ratios

Reproductive success is sometimes thought to be directly related to buck:doe ratios, or the number of mature bucks available to breed does. Research in this area has failed to support a biologically meaningful relationship. A low number of bucks per hundred does has been found to be unrelated to fawn recruitment the following year (Horejsi et al. 1988; McCulloch and Smith 1991:39).

Studies of white-tailed deer populations in the southeastern United States have suggested that making buck:doe ratios more equal and letting more bucks mature caused a shorter, earlier, and more intensive rut (Jacobson et al. 1979, Guynn et al. 1988). Changes in the buck:doe ratio and buck age structure seemed to have contributed to this shift in the timing of rut. However, the deer population was reduced dramatically coincident to the changes in buck:doe ratios. This lower density may have improved the nutrition available to the population (Jacobson 1992), and confounded the results. Studies have shown that nutritional deficiencies and social stress caused by high deer densities suppress reproduction (Teer et al. 1965, Mc-

Ginnes and Downing 1977), and that the rut occurs earlier and more intensively when nutrition is high (Verme 1965, Robinette et al. 1973, McCullough 1979).

Michigan whitetail does kept in close confinement (65-foot-square pens) with bucks were found to come into estrus earlier (Verme et al. 1987). The authors of that study hypothesized this may have been the result of constant confinement with a buck, but cautioned that crowded pen conditions were an unnatural situation. Miller et al. (1991) theorized that the presence of a buck induced an earlier estrus through priming pheromones (biostimulation), as shown in domestic ungulates.

Research was conducted on a herd of whitetails in a one-square-mile enclosure in Michigan to determine if a total lack of mature bucks affected reproductive performance. Ozoga and Verme (1985) stocked the enclosure with mature bucks and does for three years, then removed all mature bucks and retained only yearling bucks for another three years. Results showed that with only yearling bucks in the population, rutting activity was more disorganized, and young bucks displayed less stereotyped courtship behavior. However, there were no differences in mean breeding date of mature does, and all adult does were bred. Researchers concluded there was "no evidence that severe exploitation of antlered white-tailed bucks adversely influenced herd productivity" (Ozoga and Verme 1985).

Investigations of low-density deer populations in the Southwest have failed to find any consistent relationship between buck:doe ratios or buck age structure and herd productivity. Haywood et al. (1987b) conducted an extensive exploratory analysis to detect relationships among hunt design, weather, and deer productivity. Most of the variation in fawn:doe ratios was shown to be due to weather. On the Kaibab, the lowest year of buck abundance (1979) was followed by a multi-year increase in the population (McCulloch and Smith 1991:39), providing further indication of the comparative unimportance of buck:doe ratios in deer recruitment. Horejsi et al. (1988) reported a decrease in the number of mature mule deer bucks and the number of bucks per doe with heavy hunting in central Arizona, but no effect on recruitment of fawns.

In comparison, analysis in Colorado indicated a slight relationship between fawn:doe ratios and the previous year's buck:doe ratio. Fawn:doe ratios in that state have declined over the last four decades, but the buck:doe ratio has remained relatively constant for the last fifteen years. White (2001) analyzed twenty years of data from mule deer surveys and found that increasing the sex ratio from 10 to 40 bucks per 100 does results in only 7.4 additional fawns per 100 does. Although the relationship between buck:doe ratios and fawn production was statistically significant, effects were relatively minor and did not account for the long-term decline in recruitment. Interestingly, no lower threshold in buck:doe ratios was detected below which recruitment dropped off rapidly. This also fails to support the theory that fewer bucks per doe dampens recruitment to any meaningful degree.

Some biologists have expressed concern that a later rut could result in fawns entering the winter months at a lower body weight and therefore having a lower overwinter survival rate in cold climates. Overwinter losses are not important for most of the Southwest, but this topic should be investigated in more northern portions of deer range throughout the West.

Human Disturbance

Disturbance of deer during breeding season is often raised as a factor of concern in affecting the reproductive potential of deer herds, yet many states hunt deer during the rut with no discernible negative effects. Bristow (1992) compared pregnancy rates, fawning dates, and reproductive performance of two groups of Coues whitetail does: those in an area where harassment during rut was intensive, and those in another area where deer were not intensively disturbed. Researchers in this case intentionally harassed and fired rifle shots near whitetails they encountered in the disturbance area throughout December and January. Individual deer were followed until they left the study area or could no longer be located, with many deer being disturbed several times. The disturbance level in the treatment area was twice that of the undisturbed area and averaged 28 deer disturbed and 18 shots fired per day in less than six square miles. There was no difference between disturbed and undisturbed areas in pregnancy rate, number of fetuses per doe, or date of conception. There is no evidence that limited and low-density hunts as currently offered during the rut negatively affect reproduction.

Some people have expressed concern that having hunting seasons running from August (archery) through January is detrimental to deer or other wildlife. There is no biological support for this concern. In most hunted areas, the actual number of hunters per square mile is not high enough to disturb deer with the frequency required to cause an energetic drain or to preclude the opportunity for deer to breed. The only exception to this may be places where deer are concentrated on limited winter range, and thus much more vulnerable.

Chapter 8

Mortality

DEER POPULATIONS increase or decrease depending on the sometimes delicate relationship between reproduction and mortality. Just as recruitment is expressed as a rate or ratio (fawns:doe; see chapter 7), so is mortality—usually as the percentage dying per year (annual mortality rate). The annual mortality rate of a deer population can be estimated in several ways. The traditional way, before radiotelemetry, was to collect a large number of deer jaws that reveal the age at which the animals died. These jaws would then be examined and placed in age categories to construct what is called a "life table." The relative numbers of deer in each age category represent the age structure of the population and allow an estimation of the rate of mortality from one age class to another. With the advent of radiotelemetry collars for deer, researchers were able to monitor individual deer remotely. Many of these radiocollars come equipped with a "mortality sensor" that alerts the researcher when the deer remains motionless for a period of time and is presumed dead. This technology allows biologists to estimate mortality rates based on the total amount of time survived by a radiocollared sample of the population.

Estimating the annual mortality rate is an attempt to describe a dynamic process. The mortality rate is constantly changing in response to myriad population-regulating factors. The mortality rate may fluctuate greatly from one year to the next, from one area to another, and it even changes from month to month throughout the same year in the same population. For example, mortality may be much lower in April after the winter rains have produced an abundance of nutritious forbs than in June when the range conditions are very dry (Smith and LeCount 1979). Any estimate of mortality rate applies only to a particular population during the time period measured. Still, looking at the various studies that have been done gives an overall view of the rate at which deer are lost from a population.

A nearly endless list of agents cause deer mortality. These various mortality factors are always acting in concert. For example, a malnourished doe may give birth to a very weak and unthrifty fawn that is very susceptible to bacterial infection or predation. In other cases, disease or parasites may predispose an animal to predation by impairing its eyesight or hearing, or by simply weakening it.

The main factor suppressing a deer population's growth is called the *limiting factor*. Traditional wildlife management principles hold that wildlife managers should work to eliminate or moderate the oppressive effects of the limiting factor so that the population will increase. Unfortunately, there are times where several factors interact to varying degrees, so that when one mortality factor is alleviated, another increases and becomes the limiting factor. For example, Bartmann et al. (1992) removed coyotes from an area in Colorado for three consecutive winters. The predation rate decreased, but the number of fawns that starved increased so that there was no change in overall fawn mortality. In this case, the removal of one source of mortality did not affect the overall fawn recruitment or deer abundance. This phenomenon is termed *compensatory mortality* because less mortality from one cause is compensated for by increased mortality from another cause.

Sometimes a mortality factor causes population losses that add to other causes of mortality; this is called *additive mortality*. These two concepts, additive and compensatory mortality, are sometimes discussed as if they were mutually exclusive, but they can more accurately be thought of as opposite ends of a spectrum. The complex interrelationship of mortality factors usually operates in a partially compensatory fashion.

A single factor, such as the harvest of does by hunters, varies as to its place on the compensatory-additive mortality scale, depending on habitat conditions, weather, timing, and extent of other mortality sources. For example, the killing of a large number of does preceding a harsh winter with heavy snow in northern New Mexico may be almost entirely compensatory because there will not be enough available forage on winter range to feed even a reduced deer population. The does harvested would have been removed from the population anyway. In contrast, a doe harvest in a desert mule deer population preceding the same wet winter (rain) may be additive mortality because those does that were removed would have remained in the population during the lush winter and spring forage growth.

Estimating the proportion of the population that dies due to various causes is useful in determining how most animals are dying; however, this information is sometimes misunderstood or misused. Ninety percent of the mortality can be caused by predators, but this is not a problem if only 10 percent of the population is dying each year. The important figure is the percentage of the population that is lost to various mortality sources on an annual basis. This information is rarely available, but some estimates have been made from radiotelemetry studies.

Also relevant to estimates of mortality is determining the age structure of a deer population. Determining the age of deer serves a greater purpose than merely satisfying a general curiosity. The age structure of the population tells biologists a lot about average mortality rates and the appropriateness of harvest intensity. Biologists can use two methods to estimate the age of deer. Both involve looking at the teeth. The first method, called "field ageing," involves looking at changes in tooth replacement and wear of the teeth in the lower jaw, which is correlated with the age of the deer (Robinette et al. 1957). The second method is based on the fact that like many other animals, deer lay down annual rings in the cementum layer of their teeth. Deer teeth can be decalcified, sliced very thin, and stained so that these rings, called *cementum annuli,* can be counted like tree rings. It is not entirely clear what causes these annual layers, but it must be related to annual hormonal cycles rather than to dietary changes because captive animals fed the same diet year-round can still be aged this way.

Counting the cementum annuli is generally more accurate than tooth eruption and wear (field ageing) for deer over 3.5 years old, but field ageing is more accurate for animals that are 3.5 years of age or younger (DeYoung 1989). More recent studies have shown that placing animals in one-year age classes with field ageing has a high error rate, but lumping deer into multi-year age categories allows for an acceptable accuracy in estimating age.

Deer living in captivity, afforded protection and good nutrition, will commonly live fifteen to twenty years (Hosley 1961:167; H. A. Jacobson, personal communication, 2000). It is much rarer to find cases of *wild* deer living more than fifteen years. Robinette et al. (1977:44) reported a hunter-harvested buck from Utah that was estimated to be nineteen years old by cementum annuli. After ageing 10,000 deer by cementum annuli, Matson's Laboratory in Montana reported no mule deer older than twenty years and no whitetails older than nineteen (Matson 1988). The Arizona Game and Fish Department has used cementum annuli to age wildlife for many years, and the oldest age recorded is thirteen years for mule deer and nine for whitetails (W. Carrel, personal communication, 1999). Even in unhunted herds, wild deer rarely live past fifteen years; at this age, the teeth are worn to the gumline and body condition declines noticeably. Antler development of bucks declines after seven or eight years, which is another indication the deer is having trouble obtaining the excess nutrients it did in its prime.

Mortality Rates

Mule Deer

Many states use some kind of computer population model, along with historical survey and harvest data, to estimate average mortality rates of deer (Green-Hammond 1996, Heffelfinger and Piest 1996). The results from these modeling efforts are usually similar to the mortality rates gathered by radiotelemetry. The Arizona computer model estimates for natural (excluding legal hunting) annual mortality rates (1967–2005) average 21.6 percent (range 16–35 percent) for bucks and 15 percent (range 11–24 percent) for does (A. Munig, personal communication, 2005). In New Mexico a different computer model estimates an annual mortality rate of 21 percent (range 14–32 percent) for bucks and 13 percent (range 3–33 percent) for adult does (Haussamen 1995, Green-Hammond 1996). A higher natural mortality rate in males is typical of deer species. Even unhunted populations frequently have fewer males because of the greater risks involved in rutting activity and their use of different habitat that makes them more vulnerable for much of the year.

Several studies using radiotelemetry collars in the Southwest have provided estimates of mule deer mortality. A large study in Brewster County, West Texas, included 180 radiocollared desert mule deer monitored for three years, 1990–93 (Lawrence et al. 1994). Researchers found that deer mortality differed by season, by year, and by age/sex class. Average annual mortality rates (including hunting of bucks) were 30.3 percent (range 20–46 percent) for bucks and 20.7 percent (range 9–41 percent) for does. The most stressful period for bucks was February and March because of the rigors of rut and lower nutrition at that time. Does, however, died at a higher rate in late spring or during summer when the physical demands of pregnancy and lactation are highest.

During a long-term study in southwestern New Mexico, researchers monitored radiocollared mule deer for ten years. They found that mortality rates were relatively low during a period of average or above-average rainfall (bucks, 12.4 percent; does, 11.7 percent). However, when a drought occurred in the study area between 1992 and 1995, the number of deer the habitat could support (carrying capacity) plummeted. Under these nutritionally stressful conditions, mortality rates rose to an average of 33.8 percent for bucks and 21.3 percent for does (Logan et al. 1996:202).

Bender (2003) documented a relatively high mortality rate (37 percent) for mule deer in northeastern New Mexico during a period of record dry conditions (January–December 2002). During that time period almost one-quarter (23 percent) of the radiocollared does died of malnutrition. Clearly, extended droughts affect not only fawn recruitment, but also survival of adult deer in the population.

On the Arizona's Three Bar Wildlife Area and Tonto Basin, sixty-nine mule deer were radiocollared to study mortality rates and habitat use. Overall mortality rates (hunting and natural) were higher for males (34–36 percent) than females (10–19 percent). The main causes of mortality in that study at that time were predation, illegal kill, and legal harvest (Horejsi et al. 1988).

Eighty-three mule deer were radiocollared along the Arizona-Utah border between 1996 and 1998 (Carrell et al. 1999). Mortality rates and causes were not the primary purpose of the study, but the researchers did report mortality information they collected. Of the eighty-three collared deer, thirty-one died during the study. Mortality rates were estimated as 25.5–40.4 percent for does and 48.8 percent for bucks. These rates were uncharacteristic overestimates of the actual long-term averages because the population should have declined dramatically, but it did not (Carrell et al. 1999). The causes of death were varied: unknown causes (14), mountain lion predation (6), road kills (4), legal hunting (3), poaching (2), and other predators (2). Mortality was highest in October–November, which coincided with the hunting season and autumn migration.

In contrast to this study, thirty-four radiocollared mule deer (mostly females) in the San Bernardino Mountains of southern California had an annual mortality rate of 19 percent. Of the nine mortalitities examined, five were due to mountain lions and one to a domestic dog, two were the result of legal harvest, and one was due to an unknown cause (Schaefer et al. 2000).

White-Tailed Deer

The best radiotelemetry data for Coues whitetails come from a study in the Santa Rita Mountains of southeastern Arizona in the late 1980s. Ockenfels et al. (1991) radiocollared thirty-six whitetails, of which twenty-two died during the four-year study. Annual mortality rates fluctuated widely, but the overall rate was 46 percent (range 16–67 percent) for males and 19 percent (range 8–35 percent) for females. Major causes of mortality were legal hunting and mountain lion predation for bucks and predation (mountain lion and coyote) for does. Male losses occurred mainly during the hunting season and also during the postrut period. Legal harvest was a major component of buck mortality because the study area had excellent access, moderately easy terrain, and good visibility for hunters (Ockenfels et al. 1991). Of the 46-percent annual buck mortality, more than half (28.8 percent) was due to legal hunting; the average *natural* annual mortality rate for bucks was 17.2 percent. Female mortality was higher in midsummer (July–August) before fawning, than during the winter/spring period. This pattern of male and female mortality is typical of deer populations in the arid Southwest.

The estimates of annual natural mortality rates in the study by Ockenfels et al. (1991) are similar to those generated by the Arizona statewide white-tailed deer population model (1978–2005), which averages 21.4 percent (range 13–32.2 percent) for bucks and 11.3 percent (range 6.8–16.9 percent) for does (A. Munig, personal communication, 2005). These mortality rates are also very similar to those of mule deer.

Understanding deer population fluctuations requires an understanding of not only the rates, but also the factors that affect recruitment and mortality. We discussed recruitment in chapter 7, so here we will consider the factors that contribute to mortality.

Malnutrition/Starvation

Malnutrition is the lack of adequate nutrient intake, which if serious and prolonged, can lead to weakening and death. Deer rarely die of starvation per se; even badly malnourished animals usually have food in their stomachs. The problem is lack of nutrients necessary to generate glucose, an important substance that is metabolized to produce energy. Excess glucose that is not needed at the moment is stored as body fat. A healthy layer of body fat enables deer to successfully survive periods of low food availability.

An animal that is malnourished begins to use glycogen stored in the liver and then, as conditions worsen, it uses body fat and then muscle tissue. Fat deposits first disappear from under the skin of the rump and brisket. As conditions worsen, fat deposits are utilized from the heart, kidneys, and then finally the bone marrow. After all fat reserves are used, the animal begins to burn protein from the muscles and begins a slow slide toward death. The amount of fat on the heart provides an index to a deer's physical condition, since it is one of the last internal organs to lose fat deposits. In healthy deer, fat can be found surrounding the top of the heart, all along the major surface blood vessels, and at the tip. In malnourished animals, the heart has no fat, and bone marrow in the leg bones appears red and Jello-like in contrast to the fatty white paste found in the marrow of healthy animals.

In high-elevation and northern portions of the Southwest, nutritious foods are most often lacking in mid- to late winter because of snow cover. This causes deer to concentrate on their winter range, where they are forced to subsist on lower quality browse. During winters with early snowfall, deer are forced to move to wintering areas earlier than usual, which increases the number of days they have to spend on substandard winter nutrition. Also stressful are the years with above-average amounts of snowfall, which forces the animals down to the lowest portion of winter habitat where even browse plants may be in short supply. The nutritional stress associated with such winters can cause disproportionately high mortality in fawns and serves as the major limiting factor to herd growth in many western deer populations (Unsworth et al. 1999).

Outright mortality, however, is not the only way inadequate nutrition can affect deer populations. Malnutrition suppresses a deer population in more subtle ways by reducing reproductive success and increasing a deer's susceptibility to other mortality factors. Lower quality food resources available to does reduces fawn weights at birth, resulting in weaker fawns and lower fawn recruitment.

In lower elevation deserts of the Southwest, winter precipitation benefits the deer populations because of the abundance of forbs produced the following spring. Range conditions typically become progressively drier until the summer rains come, usually in early July. The period of highest nutritional stress for desert deer is the month of June, or early July if summer rains are delayed. With abundant winter rain, deer are able to store body fat to help them get through the period of summer nutritional stress. It is particularly important for pregnant does to have sufficient nutrition to allow sustained growth of the fetuses and, subsequently, adequate lactation.

Fat deposits of deer killed during the fall hunting season may serve as a useful index to herd health on northern deer ranges because it is important for deer to enter the winter period with a good layer of fat in reserve. In southern deserts, fat deposits during the fall are not a good indication of the health of the deer herd, because winter is not normally a nutritional bottleneck. Day (1964) found that bucks killed in October–November in Arizona's Chiricahua Mountains were usually in good condition, but those killed between February and April were in poor condition. It would be more informative on these desert ranges to measure fat reserves in April or May when deer are entering the dry summer period. Over the long term, malnutrition is the most important regulating factor of western deer populations by affecting reproduction (chapter 7) and by increasing susceptibility to other mortality factors.

Legal Harvest

There is a vast difference between the unregulated killing that occurred before game laws were in place and the current regulated harvest. Too often the decline of wildlife populations in the late 1800s is blamed on "hunting." It is important to distinguish between the market hunting of yesteryear and today's highly regulated hunting.

Hunting can affect sex ratios, age structures, and densities of deer populations, depending on what sexes are hunted and in what proportion (Horejsi et al. 1988). More bucks are naturally present in a deer population than are needed to impregnate all the does. Therefore, a buck-only harvest may alter sex ratios and lower the average age of bucks, but will not decrease long-term deer abundance because just as many fawns are born the next year after the annual removal of bucks. Hunting bucks and does in the same proportion in which they occur in the population may not affect sex ratios, but might decrease deer abundance.

Legal harvest is one of the few sources of mortality that can be estimated with some degree of accuracy. All states estimate harvest in some way, usually by game management unit (see chapter 9). The trends in legal harvest do not directly track population fluctuations because of changes in harvest strategies, but they do provide a general indication of deer abundance over historic time periods.

Using data sets of long-term trends in survey and harvest data, some wildlife agencies have constructed computer models that simulate deer population trends and, in some cases, generate population estimates. Although the population estimates themselves may not be entirely accurate, these models are very useful for management decisions. In addition to overall population trends, the models allow biologists to glean additional information, such as the percentage of males harvested and mortality rates due to nonhunting causes. This helps wildlife managers evaluate the appropriateness of the current level of harvest. The Arizona statewide mule deer model simulation for 1967–2002 estimates that an average of 32 percent of the available bucks were harvested each year (range 13–44 percent) (A. Munig, personal communication, 2004). Harvest rates of white-tailed deer are much lower because of the rugged terrain they occupy and the limited access to their habitat. Differences in escape behavior between the two species also result in whitetails being less vulnerable to hunter harvest. As expected, the model estimates that an average of only 19 percent of the available whitetail bucks are harvested each year (range 14–25 percent). This statewide percentage includes many remote areas and thus is understandably lower than the average mortality rate due to legal harvest (28.8 percent) estimated for whitetails in a heavily hunted area of the Santa Rita Mountains (Ockenfels et al. 1991). In the latter study, legal harvest was the leading cause of mortality for bucks, as is commonly the case in hunted areas.

Predation

Predation is a complex and contentious subject that frequently causes friction between wildlife managers and the public. A review of past predation research and management experience in the Southwest is important to understanding the effects predators have on deer populations and the effectiveness and appropriateness of controlling predators.

Ample deer research and historical data show that in the Southwest precipitation affects the quality and quantity of nutrition available to deer, which in turn has an impact on their year-to-year population fluctuations (Smith and LeCount 1979, Marshal et al. 2002). Other factors, such as overgrazing, long-term habitat changes, predation, poaching, and human encroachment, play secondary, but important roles. Widespread declines in deer populations during the 1960s and early 1970s, and the 1990s brought to the forefront the topic of predation and its relationship to deer population dynamics.

When deer populations decline, predators become a focus of concerns because they can kill a considerable number of deer. In the Southwest, mountain lions are the most important predators of adult deer, whereas coyotes can kill a large number of young fawns (Ballard et al. 2001). These two predator species are the focus of most concerns over predation on deer.

In poor-quality habitat, on islands, or after harsh winters in some northern states and Canadian provinces, wolves can also reduce deer densities below what the habitat might support in the absence of predators (Mech and Nelson 2000). Wolves released in Arizona and New Mexico might play a role in deer population dynamics if they are successfully restored to appreciable numbers. Population modeling indicates that restored wolf populations would reduce deer densities by less than 10–15 percent in high-elevation mixed conifer areas, such as the Blue Range on the Arizona–New Mexico border (Green-Hammond 1994). This assumes that wolves are eating mostly elk, which appears to be the case from analysis of wolf diets (Reed 2004). This impact might be substantially higher if wolves were to colonize lower elevations free of elk (Green-Hammond 1994). Not many historical records exist of wolves below 4,500-foot elevation, so they are not expected to become established in high densities below that level (Brown 2002). Other predators, such as black bears (*Ursus americanus*), bobcats (*Lynx rufus*), golden eagles (*Aquila chrysaetos*), and possibly even an occasional transient jaguar (*Panthera onca*), kill deer. However, population-level effects of these predators are likely minimal, because of low densities or inability to exert intensive predation pressure on deer populations.

Impact of Predators on Deer Populations

Whether predators affect deer populations is at the core of often-heated discussions among managers and members of the public. Some believe predators kill only excess deer in populations, animals that would die of other causes such as disease or malnutrition (the sick and the weak), while others blame predators for every reduction in deer abundance. The truth likely lies somewhere in between.

Predators certainly can and do affect deer populations, but abundance of deer in any area depends on interrelationships among different mortality factors. Multiplying a statewide lion population estimate by the number of deer eaten annually by a single lion can be an alarming exercise, but provides little insight into how predators affect deer populations. There is no doubt that predators remove deer. However, just because they *affect* deer populations does not mean they *regulate* them. Predators have been shown to accelerate deer declines associated with poor habitat quality and to delay the recovery of deer populations following declines (Logan and Sweanor 2001). However, coyotes and mountain lions do not *cause* declines in deer populations. Habitat quality likely is the overriding factor driving southwestern deer population fluctuations (Ballard et al. 2001).

Coyotes are important predators on deer fawns because they have wide distribution and high densities, and are efficient predators of small to mid-sized prey. These qualities allow coyotes to prey on a large portion of the annual fawn crop when fawns are born underweight with below-average hiding cover. It is not uncommon for research on causes of fawn mortality to reveal that most fawns are lost to coyote predation (Cook et al. 1971). This sounds alarming, but the important thing to consider is the percentage of the annual fawn production that is dying, rather than the percentage of mortality attributable to predation. For example, if 4 out of 100 fawns die and 3 of those are killed by coyotes, then 75 percent of the mortality is due to coyote predation, but the mortality rate itself is quite low.

Arizona established a 700-acre predator exclosure on the Three Bar Wildlife Area to determine the causes of fawn mortality in mule deer. The survival of fawns was found to be significantly higher inside the exclosure (80 fawns per 100 does), where the deer were protected from predators, than outside (50 fawns per 100 does) (LeCount 1977). The protected deer population increased until it was over the carrying capacity of the habitat and started affecting the food available. As the population peaked, fawn survival decreased because of nutritional stress caused by the high deer density. Even though predators were responsible for a significant loss of fawns outside the exclosure, that part of the deer herd still increased by almost 40 percent from 1971 to 1976, illustrating the predators' inability to keep deer populations from recovering when habitat conditions are favorable. This study was repeated thirty years later, and the deer population again responded vigorously to the lack of predators.

These kinds of experiments excluding coyotes from an area are usually followed by an increase in the deer population, indicating that coyotes depress deer abundance. If you remove all predators from an area, and if the habitat can support more deer, the deer population will likely increase. However, what can be done in a small research exclosure is impossible to replicate cost-effectively on a statewide level or even in a single game management unit.

Several long-term research projects have brought us closer to understanding the relationship between mountain lions and mule deer in the Southwest. In southern California, researchers tracked deer populations for sixteen years and lion populations for ten years. During that time, the deer population declined from 5,500 in 1984 to fewer than 1,000 in 1990, presumably because of below-average rainfall (B. Pierce and V. Bleich, personal communication, 2004). The lion population also declined in response to the decreasing prey base, but the low point of the lion population was not reached until eight years after the deer population reached its lowest point. During this lag time, lion predation was the major cause of deer mortality and may have exerted proportionately greater pressure on the remaining deer. However, forage availability and quality subsequently improved with several years of normal rainfall and fewer deer on the range, and the deer population rebounded to more than 2,200 by 2004.

During a ten-year study in New Mexico, lion predation did not affect a deer population that was stable or increasing during years of adequate rainfall. However, when the study area had a multi-year drought, deer recruitment declined drastically, and lion predation became the major cause of mortality for mule deer, which accelerated the decline (Logan and Sweanor 2001). Studies like this reinforce the belief that the condition of the habitat is what regulates deer populations.

Another question for biologists is what limits or regulates predator populations. Coyotes do not eat deer fawns year-round, so coyote populations vary in response to the density of other prey and food items, which are related to habitat conditions. Early research indicated that mountain lion populations are regulated by very low prey densities, or by social factors such as mutual avoidance or aggression, including killing one another (Hornocker 1970). Given self-regulation, lion densities can remain stable as deer densities rise, because a point is reached where resident lions will not tolerate more lions establishing home ranges or territories (Lindzey et al. 1988, Lindzey et al. 1994).

If self-regulating mechanisms are operating, lion populations would not increase to very high levels and drive deer populations down. However, prey availability—together with, or independent of, social strife—might be a key factor limiting or regulating mountain lion populations (Pierce et al. 2000). In the Southwest, prey (primarily deer) populations can fluctuate dramatically in response to environmental conditions, and lion populations might not respond for several years to increasing or decreasing prey abundance. In general, it is very difficult to evaluate whether prey densities or social interactions exert more influence on mountain lion abundance.

Multi-year lags in the response of mountain lion or other predator populations to changes in prey availability can also create an imbalance in the ratio of predators to prey. Shaw (1977), for example, studied mountain lions in central Arizona in the early 1970s during the depths of a mule deer decline. He concluded that mountain lions were a major source of mortality at reduced deer densities, and the predation contributed to low deer abundance, but lion predation alone did not prevent the deer population from increasing. If alternate prey are available, lions seem to be very adept at switching from predominantly deer to other food items such as cattle, javelina, or rabbits (Leopold and Krausman 1986). Conversely, if a lion population remains high in the face of a declining deer herd, it might exert a greater influence on overall deer mortality (Logan et al. 1996).

Studying predation for just a few years provides only a snapshot of long-term relationships. Effects of predation vary from place to place and among years because of differences in predator species, predator densities, prey species, predator: prey ratios, vegetation type, weather, learned behavior, and other variables. There are also differences in the same area over time. Diversity of predator/prey relationships creates a complex and confusing body of information and opinions that fuel never-ending discussions about the effect of predators on deer abundance.

Elton (1930:17) expressed well the complexities of species interactions within ecosystems: "The balance of nature does not exist, and perhaps has never existed. The numbers of wild animals are constantly varying to a greater or lesser extent, and the variations are usually irregular in period and always irregular in amplitude. Each variation in the numbers of a species causes direct and indirect repercussions on the numbers of the others, and since many of the latter are themselves independently varying in numbers, the resultant confusion is remarkable."

Effectiveness of Predator Control

Focusing on whether predators affect deer populations distracts from the real question of what actually might be done to improve deer abundance in specific instances. Once the debate moves past whether "lions eat deer," it becomes increasingly complex. There are few realistic options for reducing predator populations effectively over a wide area. Discussion of the effectiveness of predator control should focus on how control actions will reduce overall deer mortality significantly over a large area. In other words, what will the actions realistically do to increase deer abundance?

Research in Arizona and southern Utah showed that when a lion was removed, the vacant territory could be filled rapidly by other lions if the removal area was contiguous with occupied lion habitat (Cunningham et al. 1991, Laing and Lindzey 1993). Knowledge of lion behavior and movements indicates that removing adult male lions with large territories may result in more young lions with smaller territories, as younger animals appear to be more tolerant of each other. Because of this, some mountain lion researchers have even theorized that a light to moderate removal of lions may actually *increase* the density of lions in an area.

Coyotes are extremely resistant to population reduction and show considerable resilience. Computer modeling indicates that coyotes can sustain annual losses up to 60–75 percent and maintain a long-term stable population through increased reproductive rates and immigration from surrounding areas (Connolly and Longhurst 1975, Pitt et al. 2001). Clearly, removing predators from an area does not mean the predator population is reduced enough to have any population-level impact on the prey species of interest.

Poisons such as compound 1080 were used extensively to control coyotes from the 1940s to 1971. It is assumed that deer abundance was greater during that time as a direct result of lower predator populations. Use of these predicides definitely decreased predator populations in some areas, but it is not entirely clear whether their application was widespread enough to explain the multi-state peak in deer numbers that occurred during those years. Deer populations later rose to high levels in the 1980s without the use of poisons when increased precipitation improved habitat conditions considerably.

High fur prices in the 1980s generated more interest in harvesting predators, and this coincided with deer populations that were increasing. Predator and furbearer harvest did increase significantly during that time, but how much that higher level of harvest contributed to the increasing deer populations is debatable. Based on past deer research and management, it is more likely that the series of very wet years during that time was the main factor allowing deer populations to increase. Concurrent with this rise in deer populations throughout the Southwest were some of the highest rainfall years ever recorded. The second and third highest summer rainfall years on record in Arizona were 1983 and 1984, with the winters of 1982–83 and 1984–85 about 50 percent above normal precipitation levels.

Given what we know about the response in predator reproductive rates, this level of trapping is unlikely to have significantly reduced the coyote and bobcat densities during the fawn drop. Sport trapping occurs primarily during winter, and predators have their young in spring before the next deer fawning period. Indeed, Hamlin (1997) showed that very high coyote harvest levels induced by high fur prices in Montana did not affect the number of breeding pairs or overall coyote population levels. If sport hunting and trapping were effective enough to cause declines in predator and fur-bearer populations, bag limits and more restrictive seasons would be necessary to protect these animals, but such is not the case.

From 1982 to 1989, twenty-five mountain lions and sixty-two coyotes were removed from Black Gap Wildlife Management Area in West Texas to study the effect of predator control on the desert mule deer population (Hobson 1990). Most of the lions were removed in the first two years of the study, and an increase in the deer population occurred in 1982–86. However, this was the same time period of record high rainfall throughout much of the Southwest, including West Texas, and Hobson (1990) concluded that there were too many factors changing at the same time to be able to draw any conclusions about the effectiveness of the predator removal.

New Mexico implemented a coyote removal program in five game management units from 2000 to 2003. After three years and with 1,082 coyotes removed, the coyote removal effort showed no discernible benefits to the game populations so it was discontinued (B. Hale, personal communication, 2004).

In 1996 Utah began a similar large-scale predator control program in fifteen deer management units to improve the condition of those deer herds. Following this predator removal effort, some deer populations increased, some decreased, and some remained stable with no apparent relationship to the number of predators removed. To supplement this effort, the Utah legislature appropriated $200,000 for predator control in 2000; half of this amount was to go to a bounty program in which hunters were paid $20 for each pair of coyote ears turned in. Bartel and Brunson (2003) reported that this Utah bounty resulted in probably less than 1 percent of the coyote population being harvested in participating counties. Including all predator removal efforts statewide, it was estimated that less than 10 percent of the coyote population was removed. This is far below what would be required to elicit a positive response in deer survival or fawn recruitment.

The Kaibab Plateau has long served as a predator-prey laboratory. An intensive predator control program implemented during 1905–31 resulted in 781 lions, 4,849 coyotes, 30 wolves, and 554 bobcats being removed (Mann and Locke 1931). During these years, the deer population increased dramatically, peaking in 1928, before crashing due to overuse of forage and destruction of habitat by deer and domestic livestock. However, a ban on doe hunts, above-average precipitation, and a dramatic reduction in livestock occurred concurrently with removal of predators. There were reportedly 20,000 cattle and 200,000 sheep grazing the Kaibab deer habitat in 1889; by 1924 grazing pressure had been reduced to 4,000 cattle and 3,500 sheep (Mann and Locke 1931). If these numbers are accurate, the predator control campaign on the isolated Kaibab Plateau undoubtedly contributed to the increase in the deer population, but it is difficult to separate effects of decreased predation from other changes that occurred at the same time.

Another intensive control effort on the Kaibab Plateau took place in the 1950s using poisons to target coyotes (McCulloch 1986). Another increase in the Kaibab deer population followed, until the deer herd was again destructively high and causing damage to important forage plants. Aldo Leopold (1966:140), the founder of modern wildlife management, once wrote regarding the Kaibab: "A buck pulled down by a predator will be replaced in 3 years, deer habitat destroyed by too many deer may fail to replace itself in as many decades."

Concerns over effects of lion predation on the Kaibab herd following a mule deer decline in the late 1960s spurred a study of mountain lion predation on deer. Using radiocollars on deer and lions, researchers documented a decline in lion numbers on the Kaibab (Shaw 1980). Later, the deer population increased, although again this took place during a wet period with abundant forage growth and following a reduction in livestock numbers. Deer populations throughout the region were also increasing in response to better weather patterns at this time.

As part of a fifteen-year study in Utah, researchers reduced a mountain lion population by more than 50 percent, yet saw no measurable difference in mule deer numbers or in the proportion of deer killed by lions (Robinette et al. 1977). They concluded that lion control made only a trivial contribution to the number of deer available to harvest during the hunting season.

Intensive coyote removal can increase deer populations if habitat can support more deer. Data on the removal of lions are less compelling, and depend on the intensity of removal. When a deer population is as high as the habitat can support, there is no room for more deer. In this case, deer saved from predation will die from other mortality factors or cause the deer population to exceed the carrying capacity of the habitat—to the detriment of deer and habitat. If a deer population is below carrying capacity, then well-timed, intensive predator control might be able to reduce overall mortality enough to maintain higher deer densities or accelerate recovery of a population. The problem lies in effecting significant reduction in predator densities over a wide enough area using the methods currently available. Traps are no longer legal on public land in some states, predicides are not registered for general use, aerial gunning is not practical in brushy or forested deer habitat, and many areas are not suitable for effective predator hunting because of rough, remote, and inaccessible terrain.

Sociobiological Concerns

In the early days of game management, predators were targeted with no protective limits in place. Since those early days, the public has learned to appreciate predators as part of the natural system. Some believe this appreciation has gone too far, so that predators are now seen as untouchable. When the topic of predator control comes up, many people still recall mass slaughters of yesteryear, and do not realize how common, prolific, and resilient predators are today. An alarming number of people think mountain lions are rare or endangered.

The Utah study mentioned above showed that a 50 percent reduction in a lion population did not increase a deer population (Robinette et al. 1977). Greater reduction in a lion population would be very difficult to achieve because of the expertise required for that level of effectiveness in removing mountain lions. If a deer population is not below habitat carrying capacity, there is no room for more deer. This is illustrated by the New Mexico lion study, where a reduction of more than 50 percent in the lion population was followed by a severe drought; mule deer fawn recruitment and adult survival decreased drastically because of reduced habitat carrying capacity (Logan and Sweanor 2001). If mountain lion population reduction had been done in either of these cases for the purpose of deer management, it would have been a waste of money. Annual fluctuations of deer habitat carrying capacity in the Southwest add to the difficulties in prescribing well-timed predator control.

Wildlife managers should implement predator reduction programs only when there is a high probability of success in accomplishing a clearly defined goal (such as increasing fawn production or overall population density). Killing predators in a vague attempt to blindly "help the deer" is not likely to accomplish anything meaningful in terms of deer abundance. No predator control should be done without adequate long-term monitoring of prey population demographics so that any benefits of control can be detected. Documenting under what conditions predator control does or does not work will lead to its more effective use as a wildlife management tool and contribute to better management decisions and goals.

Predator control activities are coming under increased scrutiny from a public less tolerant of removing one species to benefit another. Public surveys have repeatedly shown that "animal suffering" and humaneness are important considerations in deciding what control techniques to use and when to apply them (Reiter et al. 1999).

Any agency that ignores the social implications of predator control stands to lose not only its credibility with the public, but also the ability to maintain management authority over predatory species. In 1971 the California Department of Fish and Game lost its ability to manage mountain lions as a result of a series of complex statutes that set a moratorium on mountain lion hunting (Mansfield 1994). Ironically, more problem mountain lions are now killed each year at public expense than were previously taken annually by hunters (Mansfield and Charlton 1998; J. Updike, personal communication, 2004). Several ballot initiatives in the 1990s proposed to prohibit or limit trapping in Arizona (1994), Colorado (1996), and California (1998); this trend is likely to continue.

Several years ago, the Alaska Department of Fish and Game highlighted research-based information indicating that predator control could increase moose and caribou populations, yet public outcry halted implementation plans. Ignoring public opinion of predators and their management and confining discussion to "just biology" is an ill-advised approach in today's social climate.

Conclusions

Effective short-term predator management has a place in cases of small, isolated populations that have been fragmented or have suffered habitat loss through human activities, such as pronghorn or bighorn sheep. Some local populations of deer may, for whatever reason, be considered important enough to spend a considerable amount of effort trying to protect them or help them recover more quickly. However, implementing predator control programs in the hope of increasing deer populations over large areas is not a feasible strategy for the long term. Localized and intensive predator removal may provide short-term help for an individual herd, but it will not make a difference from a statewide perspective, and it uses funds that could be better spent on long-lasting and biologically meaningful habitat enhancement projects.

Predator-prey relationships are complicated and must be evaluated by resource managers on a case-by-case basis; sweeping generalizations about the impacts of predation and predator control are guaranteed to be false in some areas at some times. Deer population dynamics are not simply a matter of one factor determining deer abundance, but instead are governed by extremely complex interrelationships among many environmental factors. Biologists and interested publics must look realistically at biological, logistical, and sociological issues surrounding predation and predator control, and consider for each case whether predator removal is an effective, feasible, or appropriate way to manage deer populations.

Diseases

Diseases can seriously affect a deer population. In other parts of the country disease episodes have sometimes resulted in serious losses to a deer population. Many serious disease episodes occur in populations that are nutritionally stressed, underlining the importance of malnutrition in population regulation (Prestwood et al. 1974). We know little about the extent of mortality caused by diseases, but recent interest in some diseases has heightened awareness of wildlife health issues.

There are four causative agents for disease in deer: virus, bacteria, protozoa, and proprion. Many of the infectious diseases caused by these agents occur at very low frequencies in a population or in localized areas and are not widespread (Smith 1976). Protozoan blood parasites are usually spread by ticks; none has been found to be significant in southwestern deer. Identification of disease epizootics affecting free-ranging deer herds is important to help us understand population dynamics, and to attempt to identify management actions that can lessen the severity of these losses.

Viral Infections

Viruses are very tiny segments of DNA (or RNA) surrounded by a protein coating, and some are then wrapped in a membrane (Dorsett 1985:753). They cannot move or reproduce on their own and appear more dead than alive. When a virus comes into contact with receptors on a healthy cell, it is engulfed by the cell and inserts its genes into the genetic material of the healthy cell. It then takes over the reproductive processes in that cell, replicating its own genetic material and producing thousands of copies of itself. When the host cell dies, these copies are released to infect other cells.

Hemorrhagic Disease

Hemorrhagic disease (HD) is caused by infection with either bluetongue (BT) or epizootic hemorrhagic disease (EHD). These two diseases are usually discussed together because they are very similar viruses that produce symptoms indistinguishable from each other. This disease has been the most significant infectious disease in North American deer (Trainer and Jachim 1969). It is a widespread problem primarily in the Southeast, Midwest, and Pacific coast. There are five types of BT (serotypes 2, 10, 11, 13, and 17) and two types of EHD (serotypes 1 and 2) active in the United States (Davidson and Nettles 1997). The viruses circulate together in the same pool of host species, composed of wild and domestic ungulates, and are spread by biting midges or "no-see-ums" of the genus *Culicoides*. Ungulate species vary in susceptibility to hemorrhagic disease, with white-tailed deer generally considered to be the most susceptible of the wild species. The distribution of *Culicoides* (especially *C. sonorensis*) is thought to determine the occurrence of hemorrhagic disease in North America. Because the disease is spread by tiny midges, it occurs in late summer months and disappears with the first significant frost. Seventeen species of *Culicoides* have been identified in Arizona alone (C. Olson, personal communication, 1999); nothing, however, is known about their role in the occurrence of HD. Studies to date indicate that these viruses are widespread in the Southwest. However, actual disease among deer is not often observed.

The HD virus is killed very quickly by decomposition, making confirmation of the disease very difficult with traditional laboratory techniques. The development of a genetic technique called polymerase chain reaction (PCR) allows for more effective identification of the virus. Even without a confirmation from virus isolation, trained observers can be reasonably sure in diagnosing HD by the clinical symptoms indicative of this ailment. In addition, analysis of blood samples for antibodies can indicate the history of a deer herd's exposure to HD.

Clinical indications of HD infection include loss of appetite, nervousness, weakness, and excessive salivation. Deer are usually found near water, possibly in an attempt to reduce their elevated body temperature. Hooves may peel and come off in layers, and sores sometimes develop on the tongue and the roof of the mouth. As the name implies, widespread hemorrhaging occurs throughout the body. The skin around the eyes, mouth (tongue), and anus becomes rosy or blue as blood vessels degrade. Internal organs become bloody, with most hemorrhaging occurring in the heart, liver, spleen, kidney, lung, and intestinal tract (Hoff and Trainer 1981).

Because this disease strikes in late summer, it can affect the final stages of antler growth. Deer in the captive facilities at Mississippi State University became infected by HD that caused an incomplete mineralization of the antler tips in some bucks (H. A. Jacobson, personal communication, 1999). Antler tips of those bucks were black and pithy rather than fully mineralized. In addition, a few captive bucks in Mississippi appeared to become sterile after surviving the infection and remained in velvet.

Effects on deer populations are variable. The fatality rate is reported to be near 90 percent in whitetails, but mule deer seem to be less susceptible to the disease (Hoff and Trainer 1981). Mortality rates are generally less than 25 percent of the population, with more severe die-offs in areas of nutritional stresses and overabundant deer populations. Deer that survive an infection carry antibodies that protect them from infection for a few years. The protection is extended only to the particular serotype of BT or EHD that infected the individual, but it is common for deer to be infected by more than one serotype. Because of this temporary immunity, HD infections occur in cycles, returning to a population only after the level of immunity has declined. Nothing can be done at present to treat infected animals in wild populations or to stop a disease epizootic from running its course. In a captive situation, preventive vaccinations may help, but vaccines for all BT and EHD serotypes would have to be administered, and vaccines for some serotypes have not yet been developed.

This disease is present in the Southwest, but does not appear to be a major source of outright mortality. Seven desert mule deer from extreme southeastern

California were all found to have evidence of exposure to bluetongue virus (Celentano and Garcia 1984). A later survey of infectious diseases throughout California revealed that mule deer in the southern part of the state ($n=47$) had the highest exposure to BT (44 percent) and EHD (48 percent) of any area of the state (Chomel et al. 1994). An analysis of forty-six mule deer in southern New Mexico found at least 28 percent were positive or suspected to be positive for exposure to EHD and BT (Couvillion et al. 1980). Pittman and Bone (1987) conducted an extensive analysis of several diseases in Texas mule deer in the mid-1980s. They detected an exposure to bluetongue virus in 76.7 percent of 150 deer in West Texas and 93 percent in 57 deer in the Panhandle.

In 1971–72, 134 mule deer from around Arizona were sampled for exposure to BT (Smith 1976). On the Kaibab Plateau, only one of the 104 deer had an elevated antibody level indicating past exposure. Interestingly, the exposure rate was much higher in central and southern Arizona; positive exposure was recorded in eight of nineteen (42 percent) on the Three Bar Wildlife Area and five of eleven (46 percent) in the Santa Rita Mountains. White-tailed deer were also surveyed in that sampling effort, and four of twelve were positive on Three Bar, while none of the thirty-seven samples from southern Arizona showed evidence of exposure. A more extensive effort to determine exposure to EHD and BT in southeastern Arizona was conducted in 1994 and 1995. The collection of 179 blood samples from mule deer showed that 79.9 percent had evidence of exposure to EHD virus, and 87.7 percent were exposed to BT virus (Heffelfinger and Olding 1996). This indicates that a very high proportion of the mule deer population was exposed to these diseases and survived.

Only recently has this disease been confirmed as a source of outright mortality in deer in the Southwest. In the late summer of 1993 a whitetail and mule deer were diagnosed with EHD in southeastern Arizona (Noon et al. 2002). A necropsy of the deer revealed sores, extensive hair loss, and hemorrhaging consistent with hemorrhagic disease. Blood samples contained elevated antibody titers to EHD virus, and the EHD type 2 virus was successfully isolated from one of the deer. Interestingly, data from a check station in the same area several months later indicated dramatically fewer yearling bucks in the harvest (Heffelfinger and Olding 1994). Because high fawn:doe ratios were recorded in this area the previous year (46–63 fawns:100 does), this indicates that the disease episode may have caused a disproportionately high mortality in one-year-old deer (Arizona Game and Fish Department 1994). In 1996 another deer was diagnosed with EHD in northwestern Arizona (Noon et al. 2002).

Yearlings would be expected to be the most vulnerable because they have not been around long enough to be exposed and to acquire antibodies to protect them. Fawns apparently acquire a detectable level of antibodies from their mother, which provides a measure of protection for at least four months (Gaydos et al. 2002). After that period the fawns' antibody levels drop, which could leave them unprotected and susceptible during the fall of their yearling year.

There is no established public health risk to humans handling deer infected with HD or even consuming venison from infected deer. Likewise, humans are not known to contract the disease by being bitten by infected midges. Of course, all wild game meat should always be fully cooked, and any deer that appears outwardly sick should not be consumed. Although constraints of time and money prevent long-term monitoring of disease exposures, HD in the Southwest appears to be limited to isolated, small outbreaks because of the mode of transmission and relatively low deer densities. It remains a relevant mortality factor, but is probably not of consequence in southwestern deer population fluctuations on the large scale.

Adenovirus Hemorrhagic Disease

Adenovirus hemorrhagic disease (AHD) was originally discovered in deer in California. In late 1993 an estimated one thousand mule and black-tailed deer in seventeen counties of northern California died between July and December from a previously unrecognized disease (Woods et al. 1997). Investigations revealed that rather than bluetongue as first suspected, the disease was caused by a different type of virus called an adenovirus. The symptoms of this disease are similar to those of bluetongue, but hemorrhaging may be less widespread (Woods et al. 1997). Microscopic adenovirus inclusions are usually present in the cells lining the blood vessels, especially the lungs. Infected deer may be observed with diarrhea, vomiting, seizures, mouth sores, and poor body condition; death follows quickly after the onset of symptoms.

This disease may affect deer populations through increased fawn and yearling mortality. Woods et al. (1996) reported that of thirteen deer they examined that died of AHD, eleven were fawns and another was a yearling. So far there is no evidence of this disease in southwestern deer, but it has been found in Oregon, Washington, Idaho, and has been suspected in southern California. Because little was known about AHD until recently, and because of its similarity to bluetongue, it will be necessary to monitor this disease in the future.

Skin Fibromas

Also called papillomas or fibropapillomas, black warts or tumors on the skin range in size from less than an inch to over 8 inches in diameter (Davidson and Nettles 1997) (Fig. 40). These tumors are caused by a virus and can proliferate in clusters covering large areas of the animal's body in the most severe cases. Robinette et al. (1977) reported that out of "thousands" of mule deer examined in Utah, they found only three fibromas larger than 3–4 inches in diameter. This disease is thought to be spread through contact and infection via biting insects. It does

Figure 40. Fibromas like this one on the eyelid are unsightly growths and usually cause concern when encountered in the field, but they pose no health hazard to those handling the deer. The presence of this fibroma may have indirectly resulted in the breakage and malformation of the antler. *Photo by author*

not seriously affect the deer's survival unless it significantly impairs normal feeding, hearing, or sight. The fibromas have been reported to disappear completely, and the animal may then be immune to further infection (Leopold et al. 1951:45). These unsightly growths usually cause concern when encountered on deer in the field, but they pose no health hazard to those handling the deer. The tumors are usually attached only to the skin and do not affect the underlying tissue or preclude the consumption of the meat.

Foot-and-Mouth Disease

Foot-and-mouth disease (FMD) is a highly contagious, viral-caused vesicular disease that is known to occur in domestic ungulates throughout the world. This ailment had a short but dramatic history as a deer disease in the Southwest. It first appeared in California in July 1924, and was eradicated by 1929. It has not since reappeared in the United States. However, during that one year over twenty-two thousand mule deer were destroyed in the Stanislaus National Forest to eradicate the disease (Leopold et al. 1951). It was reported that 10 percent of those shot had clinical signs of the disease. Mortality can be substantial in young ungulates of some species.

This viral disease is spread by direct contact with infected animals or their body fluids and through air-borne particles. The virus can also be maintained in the environment (soil and other inanimate objects) for months. Animals develop blisters on their tongue and in the lining of their mouth that rupture in about twenty-four hours, releasing virus-infected fluid. The feet also develop blisters between the toes and at the top of the hooves. Subsequent outbreaks occurred in Canada (1952) and Mexico (1946–54). A widespread outbreak of foot-and-mouth disease throughout Great Britain in the spring of 2001 resulted in authorities there killing over 2 million farm animals in an attempt to limit the disease. If this disease occurs again in the United States, it could have substantial effects on deer herds in local areas (Hibler 1981).

Vesicular Stomatitis

Vesicular stomatitis (VS) is known to occur in the warmer regions of North and South America and can be a serious disease in domestic livestock, particularly horses. Vesicular stomatitis in the Southwest normally appears in summer. Although VS is less severe and less contagious than foot-and-mouth disease, the clinical signs are indistinguishable (Davidson and Nettles 1997). In domestic livestock, blister-like lesions appear in the mouth and on the tongue, lips, nostrils, and teats. Affected animals may appear lame. Experimental infection of white-tailed deer with the New Jersey virus serotype resulted in deer showing no signs of infection (Comer et al. 1995). This disease is likely spread by an insect vector, given the seasonality of infection. VS seems to run its course in individual animals in a few weeks, while outbreaks can last through the summer.

An outbreak of this disease occurred in domestic livestock in central Arizona in May 1982 and spread to at least ten other states (including New Mexico, California, Utah, and Colorado) by the end of the summer (Jenney et al. 1984, Webb et al. 1987). In recent years, there have been several VS outbreaks in livestock, which have spread generally northward throughout the Southwest. There were no indications that deer populations were affected during these outbreaks. In the early 1990s Dr. John Maré and his students from University of Arizona tested nine whitetails and twenty-eight mule deer for exposure to VS, and all were negative. Vesicular stomatitis remains a disease to monitor because of its similarity to other, more serious diseases.

Miscellaneous Viruses

Several other viral diseases occur in domestic livestock that pose a potential threat to southwestern deer populations under the right conditions. Deer are exposed to these diseases and respond by developing elevated levels of antibodies that are detectable through analysis of blood samples.

Malignant catarrhal fever (MCF) is caused by a group of herpes viruses, the main ones of which are found in sheep and African wildebeest (*Connochaetes* spp.). It has been a problem in the deer-farming industry, where deer are held in proximity to exotic animals, especially wildebeest. Exotic gemsbok (*Oryx gazella*) inhabiting southern New Mexico are also carriers of this disease (Bender et al. 2003). Infected animals develop sores in the mouth, blindness, and bloody diarrhea. MCF has been documented in mule deer and white-tailed deer and is normally an irreversible, fatal disease, but it does not occur in humans (Pierson et al. 1974). The significance of this disease to wild deer herds may increase in the future if deer farming becomes more popular and widespread.

Infectious bovine rhinotracheitis (IBR) is an upper respiratory disease of cattle that has not been documented to cause deer mortality, but evidence of exposure to this disease has been detected in deer. In Arizona, Smith (1976) reported 36 of 104 (34.6 percent) mule deer from the Kaibab Plateau on the Arizona-Utah border were seropositive for IBR. In addition, eight of thirty (26.7 percent) mule deer south of the Grand Canyon were positive. None of the twelve whitetail samples on the ungrazed Three Bar Wildlife Area was positive; however, over one-third (12 out of 37) of those collected in southern Arizona showed that the deer had been previously exposed.

Bovine viral diarrhea (BVD) is another viral infection that affects livestock and also appears in serological surveys of deer. In domestic animals, this disease causes an inflammation of the digestive and upper respiratory tracts. Infected animals also show lack of fear, weakness, impaired vision and hearing, profuse diarrhea, dehydration, and emaciation. After documenting a deer dying of BVD, researchers in Wyoming found 60 percent of 124 mule deer in the area had elevated antibodies for this disease (Van Campen et al. 2001). A survey of deer in southern New Mexico revealed that 34 percent of 76 mule deer had been exposed to this virus and had elevated antibody levels in the blood (Couvillion et al. 1980). In Arizona only 14.1 percent (19 out of 134) of the mule deer and 20.4 percent (10 out of 49) of the whitetails sampled in 1971–72 showed exposure to BVD deer (Smith 1976).

Parainfluenza virus 3 (PI_3) is widespread in many wild mammals and is probably not a serious mortality source alone, but like other diseases, it has the potential to contribute to mortality in conjunction with malnutrition, parasites, or bacterial infections. None of the seven desert mule deer tested in extreme southeastern California was positive for exposure to this virus (Celentano and Garcia 1984). Smith (1976) found evidence of exposure to PI_3 in 22.1 percent (23 out of 104) of Arizona mule deer from the Kaibab Plateau, but in none of thirty deer from central and southern Arizona. None of the twelve whitetails collected on the ungrazed Three Bar had been exposed, but nine of thirty-seven (24.3 percent) from southern Arizona rangelands were positive.

Little is known of the effects of these miscellaneous viral pathogens on wild southwestern deer populations. It is interesting that ungrazed areas seem to have consistently lower exposure rates to these cattle-borne diseases (Smith 1976).

Bacterial Infections

Dermatitis

Scaly, crusty skin (dermatitis) can be caused by many things; however, a bacterial infection (*Dermatophilus congolensis*) causes one of the most common forms, called dermatophilosis. This condition affects wild and domestic animals, and is spread by animals physically touching each other, and possibly by biting insects. Infected animals usually have scaly scabs overlying sores on the face and neck. Skin appears gray and crusty, with patches of hair missing. Adult deer usually recover from infections, but fawns may die if the infection is severe (Davidson and Nettles 1997). As with any bacteria, humans can become infected, but the infection is easily treated with antibiotics. This is not known to be a widespread problem in arid climates, but sometimes causes concern when seen occasionally in southwestern deer.

Bovine Tuberculosis

Bovine tuberculosis (TB) is caused by bacteria (*Mycobacterium bovis*) and has been nearly eradicated in cattle in the United States. However, it persists as a problem in white-tailed deer in New York and Michigan. TB is characterized by the formation of small nodules containing a cheese-like yellow pus throughout the body. The nodules progressively enlarge and ultimately cause death. Transmission to other animals occurs by inhalation of airborne bacteria or through exposure to contaminated feed or water.

This disease was rare in deer before 1994, but has appeared since that time in a few counties in northern Michigan. In that area of Michigan, deer densities are extremely high (about 60 deer/mi^2) and winter feeding by the public is rampant, creating an ideal situation for the spread of the disease. Disease surveys of more than 67,000 white-tailed deer in that area from 1994 to 2001 have discovered at least 340 (0.5 percent) that tested positive (Schmitt 2001). To date, no livestock in that area have been infected. Because the spread of this disease relies on close contact and dense deer populations, it is not likely to become significant for deer in the desert Southwest. Humans can be infected with bovine TB, but it is rare in the United States since the advent of pasteurized milk. Field dressing a deer with TB lesions is a significant health hazard for humans.

After a 1993 outbreak of TB on a private elk ranch in Montana, disease surveys detected two wild mule deer (out of forty-one) with lesions, or nodules, indicative of TB. One of these deer was subsequently confirmed to have the disease (Rhyan et al. 1995).

Lumpy Jaw

Actinomycosis, or "lumpy jaw," is simply an infection by *Actinomyces* bacteria in the mouth of deer. This condition can occur if the lining of the mouth is punctured, allowing the bacteria to infect the wound. The infection sometimes moves to the jawbone and causes a swollen and deformed jaw. This condition occurs occasionally in the Southwest (Swank 1958:52, Day 1964:80) and is sometimes reported by the public, but is not a significant source of mortality.

Brain Abscess

A brain abscess can be caused by many different kinds of bacteria. It is usually associated with infections under the skin near the antlers of bucks. Mature bucks (more than three years old) are by far the most commonly infected animals, which is probably related to their more vigorous use of antlers during fighting and the practice of rubbing their forehead on trees and brush. Infected deer are disoriented and uncoordinated, and may walk in circles or run into objects. Animals are easily diagnosed by the presence of large pockets of pus in the brain cavity. Infected deer should not be consumed.

Brain abscesses in deer have been investigated more thoroughly in recent years. Researchers at the University of Georgia examined 683 deer from the southeastern United States and found that brain abscesses caused the death of 4 percent (Davidson and Nettles 1997). Illustrating that this condition occurs more in older bucks, they reported that brain abscesses killed 20 percent of the fifty-six mature bucks (older than three years) they examined.

In December of 1997 a five-year-old mule deer buck was observed walking in circles near a residence in Prescott, Arizona. By the time wildlife agency personnel arrived on the scene a mountain lion had already killed the deer. After chasing the lion away, they transported the deer to the University of Arizona Veterinary Diagnostic Lab for analysis. Besides the lion scratches and bite marks, the buck had a pus-filled abscess about 4 inches in diameter in his brain cavity (A. F. Fuller, personal communication, 1999).

Leptospirosis

Leptospirosis is a worldwide disease caused by the bacterium *Leptospira* spp. There are over 180 different types of *Leptospira* that affect many different species of animals. The disease is mainly spread through contact with infected urine. Infected white-tailed deer suffer from weight loss, fever, anemia, abortion, and death. Any widespread infection that affects reproduction through abortions could have the potential to seriously influence deer abundance. Infected humans experience fever, headaches, muscle ache, vomiting, and possible kidney failure.

Evidence of past exposure has been identified in wild deer, but leptospirosis has not been recognized as a serious source of mortality. Livestock producers are sometimes concerned that wild deer can serve as reservoirs for disease, but most serious livestock diseases are uncommon in deer. Researchers in Nebraska tested 478 whitetails for *Leptospira* and found that only 4 (0.8 percent) were positive. Of 2,245 Nebraska mule deer tested, 11 samples (0.5 percent) were positive, indicating that this pathogen is not a serious problem in deer (Bailey 1964). A survey of 792 mule deer from several populations throughout California found 25 (3 percent) that showed positive exposure to *Leptospira* (Behymer et al. 1989). Of all the populations sampled, a disproportionate number of positives came from two southern populations. Fifty-six Arizona mule and white-tailed deer and twenty-five southwestern Colorado mule deer were negative for *Leptospira*. (Day 1964, Mierau and Schmidt 1981).

Bovine Brucellosis

Bovine brucellosis, or Bang's disease, is caused by the bacterium *Brucella abortus*. The scientific name refers to the fact that it causes abortion, infertility, and reduced milk production in livestock. All these symptoms could be significant to deer population growth if similar symptoms occur in deer. The disease is spread by exposure to oral fluids, eyes, or wounds, or through contact during breeding, and has been essentially eradicated in cattle in the United States. It is a comparatively rare disease in deer in the United States, but has been the source of perpetual controversy in the management of elk, bison, and cattle in the greater Yellowstone National Park area. Deer have long been incorrectly blamed as a reservoir for this disease. Studies of about twenty thousand deer in twenty-four states showed that deer were not important in the occurrence of bovine brucellosis (Davidson and Nettles 1997). In Nebraska, 1,826 mule deer and 328 whitetails were tested for brucellosis, and only one of each species was classed as "suspect" (Bailey 1964). Tests in Utah ($n=89$), Arizona ($n=90$), southwestern Colorado ($n=98$), and Texas ($n=207$) failed to find a deer that showed past infection by the bacterium *Brucella* (Illige 1954, Day 1964, Mierau and Schmidt 1981, Merrell and Wright 1987, Pittman and Bone 1987). Out of 355 mule deer and 1,613 black-tailed deer tested in California, one of each subspecies was considered positive for past exposure (Drew et al. 1992).

Proprion

Chronic Wasting Disease (CWD)

Chronic wasting disease was first discovered in 1967 in a captive wildlife facility near Fort Collins, Colorado. Infected deer were listless, drank often, urinated frequently, drooled, and stood in their pens with drooping ears and heads held low (Williams et al. 2001). Researchers ran a barrage of tests but failed to identify the cause of the disease. In 1978 microscopic evaluation of brains of infected deer identified CWD as one of several diseases called transmissible spongiform

Figure 41. Animals with chronic wasting disease (CWD) first lose their fear of humans, hold their head lower, and generally appear sick. In later stages severe weight loss, droopy ears, and excessive salivation occur. *Photo by Tom Thorne, Wyoming Game and Fish Department*

encephalopathies (TSEs). Other TSEs include scrapie in sheep, bovine spongiform encephalopathy in cattle (dubbed "mad cow disease" by the media), and the rare Creutzfeldt-Jakob disease in humans.

For over a decade, deer died in the Colorado deer pens and also in a Wyoming facility that contained deer from the Colorado pens (Fig. 41). The disease claimed captive mule deer, whitetails, and elk before researchers discovered its cause. CWD is caused by altered proteins, called prions ("pree-ons"). These mutated proteins are apparently able to alter the shape of other normal proteins, thereby proliferating in the infected animal. Testing live deer is difficult because the pathogen accumulates primarily in the tissues of the brain, spinal cord, eyes, tonsils, spleen, and lymph nodes. The most reliable way to test for the disease is through microscopic evaluation of the base of the brain, but biopsies of tonsils and lymph nodes offers an alternative. Scientists are still not entirely sure how the disease is spread, but animals have contracted the disease while living in pens that were previously occupied by CWD-infected animals or infected carcasses. Symptoms have been documented in deer as young as 1.5 years, but most cases occur in deer 2–6 years old.

In 1981 the first case of CWD in the wild was documented in an elk from north-central Colorado; the first case in Wyoming appeared five years later (Williams et al. 2001). Over the next twenty years about a hundred clinical cases were documented in wild deer and elk. Intensified surveillance for CWD in recent years all over the United States has resulted in infected deer being discovered in many other states. Disease surveys in the focal area in Colorado and Wyoming found about 5 percent of the deer positive for CWD (Miller et al. 2000).

Before 2002 the disease did not seem to be spreading naturally in the wild, or if it was, it was doing so very slowly. Discovery of infected mule deer in an adjacent county in the Nebraska Panhandle was not a cause for alarm since they were less than ten miles from occurrences in Colorado and Wyoming. In the spring of 2001 the Canadian province of Saskatchewan discovered CWD in a few wild mule deer harvested near a game farm that had chronic problems with CWD. South Dakota and Wisconsin discovered CWD in 2002 in their wild whitetail herds, causing Wisconsin to implement drastic population reduction measures in an attempt to stop any further spread of the disease. In 2002 additional cases were also discovered in northeastern Utah and south-central New Mexico. No theory currently explains the source of infection in these two southwestern CWD areas. The spread from the focal area in the Rocky Mountains to several isolated locales caught the attention of deer biologists, administrators, legislators, and the public. The popular press helped fuel a near hysterical reaction.

Since 2002 there has been an increase in the number of cases reported, but this is probably directly related to the heightened awareness and substantially increased surveillance for this disease in the focal area. Several additional cases in *captive* deer and elk have been documented outside the focal area (Kansas, Minnesota, Montana, New York, Oklahoma, South Dakota, Wisconsin, Alberta, Saskatchewan). Many of these new focal areas can be traced back to the movement of animals from infected captive facilities.

This disease is being monitored closely because of its similarity to scrapie and bovine spongiform encephalopathy. There is no evidence that CWD can be transferred to humans. Despite thousands of deer consumed in the primary area of occurrence in Colorado and Wyoming, no human being has contracted the disease. In addition, there is no increased incidence of the human form, Creutzfeldt-Jakob disease, in the area of CWD prevalence. Research shows there are biological barriers to transmission of CWD to species other than deer and elk (Raymond et al. 2000). The closely related scrapie has been known to occur in domestic sheep for over 250 years with no known human infections. Although there is no connection with human illness, common sense dictates hunters should avoid consuming any animal that appears sick. Hunters should also avoid handling the eyes, spleen, lymph tissue, brain, or spinal tissues, and should wear disposable gloves when field dressing animals. Deer managers throughout the West are monitoring this disease closely because of the uncertainties in mode of transmission and because there is no known cure once an animal shows symptoms. This disease will remain at the forefront of western deer disease investigations in the coming years.

Parasites

Although both internal and external parasites are common in deer populations, they rarely cause widespread mortality. Like many diseases, parasites may be present in the population and only become a significant source of mortality when the deer population is under stress for another reason, such as nutritional deficiency during an extended drought. At such times, parasite loads may increase and kill the host animals; however, it is not to the parasite's advantage to kill its host. When deer die with heavy parasite infestations, it may be difficult to assign the real cause of death, since such animals may have been malnourished and predisposed to many other causes of mortality.

External Parasites

Nasal Bots

Nasal bots are the larval stage of the bot fly (*Cephenemyia* spp.). This is one of the most commonly encountered parasites in deer. Adult flies lay eggs on the wet skin around the nose or mouth of deer (Davidson and Nettles 1997:84). The eggs hatch, and the small larvae migrate into the moist nasal passages and molt into fully developed tan or yellowish larvae nearly 1 inch long (Fig. 42). The larvae drop out of the nose or mouth and complete their development into adult flies on the ground. The presence of nasal bots generally does not affect the survival of the deer, but can be a source of irritation if they are present in high numbers.

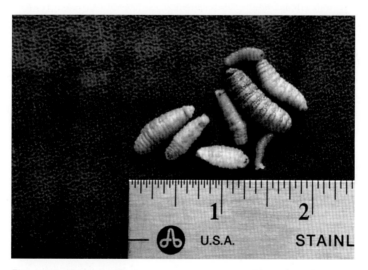

Figure 42. Nasal bot fly larvae are common and harmless to humans, but hunters notice them in the nasal passages and the throat (where they land after dropping out of the rear opening of the nasal passages). *Photo by author*

These nasal bots are harmless to humans, but frequently cause concern among hunters when they are discovered in the nasal passages or throat (having dropped out of the rear opening of the nasal passages) of harvested deer. They are very common throughout the Southwest and do not affect the quality of the meat in any way.

Mites

Although rarely reported in deer, several varieties of mites cause related but different pathogenic conditions. Some types of mites (*Psoroptes* spp.) infest the ears of deer and produce a waxy crust in the outer ear canal, sometimes called "scabies." Most deer do not appear sick, but a bacterial infection of the inner ear can accompany heavy infestations, causing body movements to be uncoordinated (Davidson and Nettles 1997). Heavy accumulations of crust in the ear canal may affect the animal's ability to hear predators or approaching vehicles.

Other mites cause mange by burrowing into the skin (*Sarcoptes* spp.) or hair follicles (*Demodex* spp.), causing intense irritation and excessive scratching. The scratching causes large areas of hair to fall out, and the underlying skin becomes thick and crusty (dermatitis). Hibler (1981) reported that two of sixteen deer in Bloody Basin, Arizona, had demodectic mange (*Demodex* spp.) on their heads and faces. Mange and ear mites may, in rare cases, reduce survival of individual deer, but do not play an important role in population fluctuations.

Although extensive mange is unsightly, there are no reports of health risks to humans. Each host species usually carries a different type of mite that is host-specific and seldom infests a different species. Since the mites reside in the skin, the underlying meat is unaffected.

Ticks

Ticks are recognized by most people because of their widespread distribution and the variety of animals they parasitize (Fig. 43). Eighteen species of ticks have been documented on white-tailed deer; the most common in the Southwest are winter or wood ticks (*Dermacentor* spp.) and the spinose ear tick (*Otobius megnini*) (Hibler 1981). Some ticks complete their life cycle with only one host (winter tick), while others may require three years and three different host animals (*Otobius* spp. and *Ixodes* spp.).

The winter tick develops from the larval stage to adulthood on the same host. Then the adult female mates, takes a large blood meal, drops off, lays her eggs, and dies. The larval ticks then lie dormant through the summer and find a host of their own in the fall (Davidson and Nettles 1997). Other species use small mammals or reptiles as hosts when they are developing through three life stages (larvae, nymph, adult).

Ticks usually attach themselves in or behind the ears; along the neck, back, or chest; and in the anal area. Infestations are usually light and rarely a significant

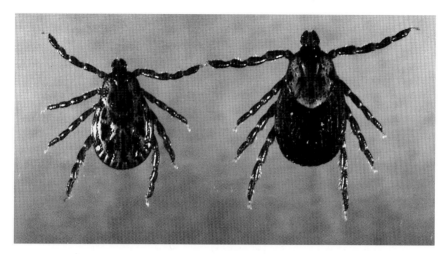

Figure 43. Ticks are widespread and recognizable external parasites of deer. They are usually present in low numbers and do not cause outright mortality under normal circumstances. *Photo by Roger Drummond*

cause of deer mortality. Severe tick infestations can reduce the survival of individual deer by causing weakness, anemia, paralysis, and further complications due to bacterial infection at tick attachment sites (Davidson and Nettles 1997). Heavier infestations are often found on deer in poor nutritional condition (Day 1964:78).

There is no human health hazard in consuming affected deer, but in some parts of the country tick bites can transmit diseases to humans, such as Lyme disease, tick paralysis, or Rocky Mountain spotted fever. Deer ticks or black-legged ticks (*Ixodes* spp.) are the carriers of Lyme disease and are normally limited to northern or coastal areas with higher moisture levels. One form of this tick (*I. pacificus*) is present along the West Coast from British Columbia to Baja California, Mexico, with eastern populations discovered in Utah, Nevada, and northwestern Arizona. Another (*I. scapularis*) ranges through Coahuila and Nuevo León in northeastern Mexico, southern and eastern Texas, and into the northern states. About 4 percent of the deer ticks collected in northwestern Arizona in 1991 showed evidence of containing bacteria that cause Lyme disease (Olson et al. 1992). Typically, a tick must be attached to the host (such as a human) for at least twenty-four hours before the transfer of pathogens occurs. This, coupled with the extremely limited distribution of deer ticks in the Southwest, makes transmission of Lyme disease much less likely.

Louse Flies

Louse flies (*Neolipoptena* spp. and *Lipoptena* spp.), sometimes called keds, are harmless to deer and humans. They are sometimes noticed by deer hunters and mistaken for ticks. Unlike ticks, which have eight legs, louse flies have only six legs.

Adult forms of louse fly hatch on the ground, and the winged hatchlings fly off in search of a host. When it locates a suitable host, the louse fly lands, and shortly thereafter its wings fall off. The louse fly lives and breeds on the host animal. A single ovum is produced at a time and matures within the female uterus; when fully grown, it is extruded from the mother, falls to the ground, and pupates. A single female can produce eight or more offspring during her lifetime (C. Olson, personal communication, 2000).

Fleas and Lice

Fleas (*Pulex* sp.) are reported on deer, but do not cause outright mortality. Two types of lice (*Bovicola* [=*Damalinia*] sp. and *Solenopotes* sp.) are most commonly reported on deer in the southwestern United States and Mexico (Illige 1954, Hanson and McCulloch 1955, Swank 1958, Mendez 1981). Day (1964) found large numbers of another type of louse (*Trichodectes* sp.) on white-tailed deer in the Chiricahua Mountains when deer populations were above carrying capacity and their physical condition was poor. These small parasites are normally merely a source of minor irritation to deer and not a factor in widespread mortality. An exception to this was reported on the Tejon Ranch in southern California in 1967. Thirty-four deer were found dead from an extreme infestation of the exotic African blue louse (*Linognathus africanus*), imported accidentally from Africa (Brunetti and Cribbs 1971). This louse occurs in other deer populations throughout California and has been found to cause other deer deaths. Its deadliness can be attributed to the fact that deer did not evolve with that particular parasite, making it all the more harmful when they do become infested with it. Like most parasites, the species of lice that occur on deer will not readily survive on humans.

Screwworm

Primary screwworm infestations are caused by the larval stage of the fly *Cochliomyia hominovorax*. The fly lays its eggs on the skin of live mammals near open wounds. The eggs hatch in twelve to twenty-four hours, and the tiny larvae begin feeding on the wound (USDA 1998). After five to seven days they are full grown (¼ inch to ¾ inch) and drop to the ground where they burrow into the soil. After seven to ten days, the flies hatch and work their way to the surface of the soil and fly away. At three to five days of age the flies are sexually mature and mate. Females mate only once during their lifetime, but males will mate several times. After the females are five to six days old, they deposit their eggs near an open wound, and the cycle continues.

This unique life cycle allowed screwworms to increase rapidly in numbers when conditions were favorable, but also allowed them to be controlled. Because the females mate only once in their life, the U.S. Department of Agriculture treated male screwworm flies with gamma radiation to make them sterile and then released them in large numbers throughout the Southwest. The large number of sterile males far outnumbered the existing fertile males, and as a result, most matings were with sterile males. After many years of this innovative program the screwworm was eliminated from the United States in 1966. Mexico was declared free of screwworm in 1991, but several outbreaks have occurred since that time.

Primary screwworms sometimes attacked the bloody velvet of bucks' antlers as it was being shed and also the navels of newborn fawns (USDA 1998). Primary screwworm infestations were a significant source of mortality for young fawns in the southeastern United States and Texas. The past impact on southwestern deer populations is not known; however, this parasite caused concern among the public in the 1950s and 1960s when heavily infected deer were seen near residences and water sources (Brown 2001).

Besides the "primary" screwworm, there is another species, the secondary screwworm (*C. macellaria*), that feeds mostly on dead tissue. This species has not been the subject of eradication attempts, probably because it does not cause extensive live tissue damage, it may breed more than once, and it is much more widespread than the primary screwworm.

Internal Parasites

Elaeophorosis

Elaeophorosis is sometimes called "sorehead" or "clear-eyed blindness." It first came to the attention of wildlife managers in the late 1960s, when blind elk with deformed ears and muzzles, apparent brain damage, and abnormal antler development were noticed in eastern Arizona and western New Mexico forests. Subsequent investigations revealed that a mule deer parasite was the culprit.

The disease is caused by *Elaeophora schneideri*, a small, white, round, blood worm (nematode) about 2–5 inches long that is usually found inside the carotid arteries of the neck near the angle of the jaw. Adult worms residing there produce microscopic offspring called microfilariae, which are carried in the blood to the small blood vessels at the surface of the skin of the forehead. When a horsefly bites an infected animal, it acquires some of the microfilariae, which then develop into larvae while in the fly (secondary host) in about two weeks. The larvae then find their way to the fly's mouth and, with the next fly bite, are transferred to another deer or other hoofed animal. Once in the deer, they find their way to the carotid arteries in three to four weeks and begin another life cycle (Hibler 1981).

Mule deer are the natural host and are not affected, but whitetails have been reported to suffer from blockage of the arteries and food impactions (Davidson and Nettles 1997:75). Day (1964) failed to recover any of these parasites from Arizona whitetails in the Chiricauhua Mountains. Like many diseases and parasites, *Elaeophora* infections seem only to kill or debilitate animals that are not the natural host, such as domestic sheep, bighorn sheep, elk, and some exotic deer species.

When these unnatural hosts are infected, the adult *Elaeophora* block the main arteries in the neck and head, causing the malformed ears, muzzle, and antlers, and sometimes blindness.

Elaeophora are common in mule deer throughout the Southwest (Hanson and McCulloch 1955, Swank 1958, Pence and Gray 1981, Pederson et al. 1985). However, they are most common in eastern Arizona and western New Mexico, where about 90 percent of the mule deer living above 5,000 feet elevation carry this parasite (Hibler 1981).

Foot Worm/Leg Worm

The foot worm (*Onchocerca cervipedis*) is a very common, but apparently harmless round worm (Hibler 1981). This parasite is long (2–10 inches) and thread-like. It resides mostly under the skin of the lower legs, but can be found in other locations, such as the brisket, neck, and shoulder. Hunters sometimes notice it coiled or extended in the connective tissue while skinning the deer. Like *Elaeophora*, foot worms produce tiny microfilariae that are found under the skin (especially the skin of the ears) and are spread by biting flies. Hibler (1981) examined over 200 mule deer in Arizona, Colorado, and New Mexico, and found that all carried these worms. Foot worms were also found in more than half of the 344 mule deer examined in California by Herman and Bischoff (1946). In rare cases, an especially heavy infection can render the venison unfit for human consumption, but this worm is not likely to affect deer abundance.

Gastrointestinal Nematodes (Stomach Worm)

Most deer are parasitized by at least one of several species of nematodes (such as *Haemonchus contortus*) in the digestive tract. Like many other parasites, these round worms usually do not cause sickness or death in the deer. High levels of these parasites rarely occur in healthy deer herds. Sickness due to a heavy infestation is most frequent in young deer (less than one year old) and is marked by a loss of body condition, rough hair coat, diarrhea, and anemia (Hibler 1981). The adult worms produce eggs in the stomach that are deposited on the ground when the deer defecates. The eggs hatch into larvae that molt twice more in one or two weeks and pass into a third larval stage before being accidentally eaten by a deer. After three to four weeks in the stomach, the larvae develop into adults, and the cycle begins again. This parasite is not a significant source of mortality except in conjunction with malnutrition in overpopulated deer ranges.

Abdominal Worm

The adult abdominal worm (*Setaria yehi*) is a relatively long (5–10 inches), white, thread-like worm that is sometimes noticed by hunters while dressing their deer. This worm differs from intestinal nematodes in that it is present on the *outside* surface of the internal organs. It is not attached to the organs and is the only round worm that occurs freely in the body cavity. The adult worms produce tiny microfilariae (larvae) that enter the bloodstream and circulate through the host's body (Davidson and Nettles 1997). The worms are spread to other animals when an insect bites the infected individual and ingests microfilariae along with a blood meal. Within the insect vector the microfilariae then develop into an infectious stage of the parasite and are injected into a new host when the insect feeds again.

This parasite is more common in younger animals, possibly because infected animals develop an immunity to further microfilariae infection. It is fairly common, but few deer ever have more than thirty adult worms. Deer in the Southwest are most likely to average one to five adult worms per individual (Hibler 1981). It presents no serious harm to the deer and poses no health hazard to humans handling or consuming the meat.

Lungworm

Lungworms (*Dictyocaulus* sp., *Protostrongylus* sp., *Parelaphostrongylus* sp.) are small (less than 2.5 inches), white or reddish brown round worms that infect the windpipe, bronchial tubes, and smaller air passages of the lungs. Heavy worm infestations can cause blocked airways and patchy pneumonia that appears as dark red or grayish firm areas in the lungs (Davidson and Nettles 1997). Adult worms produce eggs that hatch in the lungs; the larvae make their way up the windpipe and are swallowed and passed in the deer's feces. With sufficient moisture, they molt on the ground into an infectious stage and are then consumed accidentally by deer feeding on vegetation close to the ground. The larvae then migrate to the lungs and develop into adults in the new host animal. This parasite is found throughout the Southwest (Longhurst et al. 1952, Mierau and Schmidt 1981), but heavy infestations are not common because optimum temperature and moisture conditions are required to complete their life cycle. Hibler (1981) examined mule deer from throughout New Mexico and found only a small percentage of deer with lungworms, and even the infected animals carried fewer than ten adult worms.

Tapeworm

Western deer are intermediate hosts for several different types of larval (*Cysticercus* sp. and *Echinococcus* sp.) and adult (*Moneizia* sp. and *Thysanosoma* sp.) tapeworms. Deer accidentally ingest tapeworm eggs, which hatch in the small intestine and either develop into an adult tapeworm or remain larval and migrate through the intestine wall and into the body cavity. The tapeworm larvae appear as fluid-filled sacs in the liver or other internal organs (Fig. 44). These cysts do not develop into tapeworms in deer, but rather reside in the deer until eaten by a carnivore, such as a coyote. The cysts then develop into adult tapeworms in the carnivore's small intestine and may reach 13 feet in length. Larval and adult tapeworms are not harmful to their host, nor is it a human health hazard for those eating properly cooked venison. Hibler (1981:145) reported that northern Arizona was a

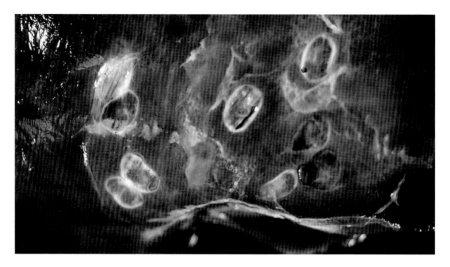

Figure 44. Tapeworm larvae visible on the surface of a deer liver. *Photo by Ted Noon, University of Arizona Veterinary Diagnostic Lab*

focal point for tapeworm larval infections, but this parasite is common in other southwestern deer herds (Longhurst et al. 1952, Swank 1958, Russo 1964:103, Stubblefield et al. 1987).

Unrecovered Deer

We know little about the frequency of deer legally shot but not recovered. This class of mortality is sometimes termed "crippling" loss. Crippling is an inaccurate term because it implies that the deer is left with a long-term disability, when extended suffering is probably rare (Nettles et al. 1976). Unrecovered deer are those that are shot and not recovered by the hunter; this includes deer that were killed and not found and also deer that recover from the injury (McCaffery 1985).

Although certainly desirable, it would be unrealistic to expect all shots to result in an instantaneous death. Predators certainly do not attain that ideal, and they also wound deer without bringing them down. Deer bones from Arizona archaeological sites showed that Native Americans were also not always able to recover the deer they shot. Olson (1990) reported deer remains from an archaeological dig that had healed with a flint arrowhead embedded in its sternum.

One of the biggest problems encountered when evaluating this type of information is that the results have been reported in many different ways. Some studies report the percentage of shots that resulted in an unrecovered deer; some tally the number of dead deer found in the field after the hunt and express it as a percentage of the total harvest; still others estimate the percentage of wounded deer in the population.

The studies that have attempted to estimate the extent of this type of mortality are highly variable. A review of research on unrecovered deer in eastern whitetail herds reported a range of "negligible" to 175 percent of the legal harvest (Nettles et al. 1976). Looking specifically at mule deer, Connolly (1981:309) reported that a combination of illegal harvest and unrecovered deer ranged from 8 percent to 92 percent of legal harvest. About 1 percent of the hunters responding to the annual AGFD post-hunt questionnaire report hitting a deer they could not recover (Amber Munig, personal communication, 2003). Clearly, the importance of this cause of mortality varies greatly with hunter density, thickness of vegetation, terrain, and hunt structures. Like many mortality factors, the number of deer left in the field after the hunting season would be very difficult to estimate accurately, and any estimate would not be applicable to wide areas or subsequent years.

Illegal Harvest

Illegal harvest includes the intentional shooting of deer out of season or taking of deer with illegal methods, such as with a spotlight at night. These activities are clearly "poaching" and cannot be tolerated. Unintentionally breaking the law—for example, forgetting to tag the deer, taking a deer in the wrong management unit, or accidentally shooting the wrong species—also represents a form of illegal harvest. Regulations are becoming more complicated (Heffelfinger and Olding 1998), and with this increasing complexity comes the potential for more unintentional violations. These are generally honest mistakes, but nevertheless elicit a citation.

Estimating the amount of deer lost to illegal harvest is difficult. In effect, it is an attempt to find out how many deer are killed that you do not find out about. Research in the eastern United States has estimated that poaching of whitetails is 40–80 percent of the recorded legal harvest (Wright 1980). Several wildlife agencies in the West have tried to estimate the illegal kill by employing undercover agents to shoot deer out of season to see how many times the agents were caught in the act. In Idaho a researcher shot 31 deer from January through June, none of which was reported to enforcement personnel. A similar study in New Mexico employed a person to shoot 19 deer and simulate 125 other instances by leaving deer parts in plain view when no deer season was open (Pursley 1977). Out of the 144 cases of poaching or simulated poaching, he was observed at least 43 times but reported to authorities only once.

In the mid-1980s Arizona biologists used the heads, hides, and entrails of deer collected for another research project to test the public's response to evidence of simulated poaching. Remains of deer were placed in twenty-four highly visible spots near the town of Patagonia. It was estimated that at least three hundred people observed the remains, but only five reports were received by the Arizona Game and Fish Department (K. Bahti, personal communication, 1998). It was evident that only a small fraction of the observed poachings are reported to law enforcement authorities. Poachers who make an effort to hide the evidence of their deeds would be observed less frequently and consequently reported at a lower rate.

In Mexico, a much different social and economic environment prevails. Illegal harvest probably affects Mexican deer populations more than any other factor near areas of human habitation. The lack of adequate resources for law enforcement, traditional attitudes toward wildlife, and an economy that requires rural people to live off the land are all causes for the widespread illegal deer kill. Deer are more abundant on large ranches that control access and do not tolerate illegal killing of deer. Protecting the deer from this form of mortality is one of the more pressing needs that will have to be addressed if Mexico is to move toward proper management and sustainable use of deer throughout the country (Ezcurra and Gallina 1981).

Accidents

Vehicle Collisions

An estimated 1.5 million deer/vehicle collisions occur each year in the United States, and over 90 percent of the deer involved die as a result of their wounds (Conover et al. 1995). This means well over 1 million deer die each year from vehicle collisions. Many of these collisions occur in states more densely populated with deer and people; eastern states with high whitetail populations report deer mortality rates of 18–33 percent due to vehicle collisions. Southwestern deer are also subjected to a certain amount of mortality due to collisions; however, the higher density deer populations are generally not located in the areas where most people live. The human population in much of the Southwest is concentrated, with large expanses that are sparsely populated, rather than being spread out evenly in small rural communities connected by a network of roads. Although this source of mortality varies drastically with road density and deer abundance, a few generalizations can be made. Most collisions occur during the early morning and evening hours when deer are more active and commuter traffic is highest (Allen and McCullough 1976). Also, during the rut, bucks range farther and are less attentive to potential dangers, resulting in a peak in road kills for bucks at that time of the year.

An average of seventy-seven deer per year (range of 41–142) were reported to the Arizona Game and Fish Department as killed by vehicles between 1988 and 1998 (Brent Vahle, personal communication, 1999). This probably represents no more than half the total annual kill, since research has estimated only about half of the deer hit by vehicles are ever reported (Allen and McCullough 1976). Of those reported killed in Arizona, 88.7 percent (821 out of 926) were mule deer. This is not surprising, given the lack of high-speed roads in most of whitetail habitat in the state. Although there are gaps in the data, Arizona records do not show an obvious peak in deer-vehicle collisions during the rut as expected. Instead, collisions seem to peak slightly during the summer monsoons. In desert mule deer, the availability of widespread water and forage after the rains allows for greater movement (Rautenstrauch and Krausman 1989), which may account for this peak.

Many devices have been used in an attempt to reduce the number of deer-vehicle collisions, but no good solution has been found. Experiments have been conducted with roadside lights, reflectors and mirrors, vehicle-mounted whistles, deer underpasses, and fencing. Roadside reflectors and similar devices are expensive to install and maintain, and their effectiveness at reducing collisions is inconsistent (Schafer and Penland 1985, Reeve and Anderson 1993). Vehicle-mounted whistles are ineffective (Romin and Dalton 1992); research indicates deer cannot hear them. Underpasses and deer crosswalks have had limited success, but temporary warning signs placed in migration corridors in the spring and fall can be effective (Sullivan et al. 2004) . Although very expensive, fencing will keep deer from crossing short stretches of highway and may be useful in redirecting deer to less dangerous locations to cross. The only consistent thing that seems to result in fewer deer-vehicle collisions on a large scale is fewer cars or fewer deer.

Considering deer population levels and the estimates of mortality due to collisions, it is apparent this form of mortality is not a significant drain on southwestern deer herds. This is not to say, however, it could not locally affect a deer population that was relatively isolated and experiencing low recruitment or high mortality from other sources as well.

Other Accidents

Intense fights between bucks during the breeding season are rare, but can be life-threatening. Bucks spar frequently without injury, but in unusual circumstances a serious brawl ensues. During these dominance fights a buck can sustain puncture wounds from the sharp antler tips of his opponent. These injuries usually occur on the face, eyes, neck, or rib cage. Losers of these fights may also be punctured in the hindquarters as they make a hasty retreat. Fight injuries are not numerous, but are probably more common that most observers realize. Geist (1981) used close personal observation and examined tanned deer hides to evaluate the frequency of punctures and cuts sustained by deer. He found that wounds due to antler punctures were very difficult to find even when actively bleeding. Cuts on the tanned skin showed that deer are injured and recover far more often than realized.

Fences also directly kill an untold number of deer each year by entanglement. Deer are usually caught while jumping a barbed wire fence when the lower part of their hind legs goes through the top two strands and the momentum of the jumping deer twists the wire onto the legs. Animals struggle until they become free or die. Wildlife biologists recommend fences be built with more space between the upper two strands, with a lower strand of smooth wire (no barbs) at least 16 inches above the ground. This facilitates free movement under the fence and reduces the number of deer that jump over.

Drowning may seem like an odd way for a southwestern deer to die, but this form of mortality can be a problem locally. Several buck carcasses have been retrieved from water tanks and irrigation canals where they fall or walk in and cannot extract themselves. Almost all of the drownings occur in the early summer months before the summer rains relieve the dry conditions. The Mohawk Canal, east of Yuma, Arizona, has been problematic since it was built. From 1968 to 1980, over two hundred mule deer drowned in the canal (Rautenstrauch and Krausman 1986). Intensive monitoring found that deer fell into a 9.5-mile section of the canal at least 279 times from June 1982 through September 1985. Twelve deer drowned during the study, but only three within the 9.5-mile study section.

Deer may die from an almost endless array of other accidental deaths. Deer mortality has been reported due to snake bites, lightning strikes, collisions with trains, falls in mine shafts or off cliffs, and broken necks (Bleich and Pierce 2001).

Chapter
9

Deer Management

Management Authority

THERE IS sometimes considerable confusion about the function and purpose of all the different natural resource agencies. The National Park Service (NPS), U.S. Forest Service (USFS), and Bureau of Land Management (BLM) are all federal agencies and are collectively called "land management agencies," because they are responsible for managing the land-based resources (timber, vegetation, and so forth) rather than wildlife populations. In addition, the U.S. Fish and Wildlife Service (USFWS) has both land management and wildlife management responsibilities. Adding to the confusion is the propensity of the media and others to mix and hybridize names of these agencies, creating fictional entities like the "U.S. Game and Fish Department" and the "New Mexico Forestry Service."

The U.S. Forest Service was established in 1905 to provide water and timber for the nation's benefit. The management goals of the USFS were later expanded to a multiple-use doctrine. Thus the USFS began to manage forested areas to provide fish and wildlife habitat, wood products, recreation, grazing, watershed protection, and preservation of historic or scientific values. The Southwest Region of the USFS is made of eleven national forests covering 20.6 million acres, most of which is the most productive deer habitat in the southwestern United States. Almost all of the Coues whitetail habitat in the United States is on national forest land.

The NPS was established in 1916 to manage the growing number of parks placed under the jurisdiction of the Department of the Interior. These parks were established to preserve representative and unique ecological characteristics and to provide for the public enjoyment without diminishing the value of the resources for future generations. Preservation was important at that time because game populations were at dangerously low levels and the recently enacted regulations protecting wildlife and wildlife habitats were only beginning to be formulated and implemented. Now that conservation by way of regulated use has shown itself to be so successful, the role of the NPS is primarily one of visitor management and interpretation of nature.

The Grazing Service and the General Land Office were merged in 1946 to form the BLM within the Department of the Interior. The BLM was given jurisdiction over all remaining federal land not already being managed by the other federal agencies—the USFS, NPS, Department of Defense, Bureau of Reclamation, or Bureau of Indian Affairs. The BLM dealt primarily with promoting mining and grazing during the early years, but that mission changed in 1964 when a multiple-use directive guided the BLM toward becoming the more comprehensive natural resource agency it is today. The 1964 mandate required the BLM to manage for wildlife, recreation, and soil and water resources, in addition to the traditional range, forestry, lands, and minerals. Today the BLM oversees more land than any other federal agency in the Southwest. Nearly all that land is home to mule deer.

The U.S. Fish and Wildlife Service is also a federal land management agency that administers the 93-million-acre national wildlife refuge system, which includes 530 refuges. In addition, it takes the lead on the management of migratory birds, threatened and endangered wildlife, and coastal fisheries. The USFWS is also the agency that collects the Federal Aid in Fish and Wildlife Restoration (Pittman-Robertson and Dingell-Johnson) funds from the sales of fishing- and hunting-related equipment. It then disburses this money to the state wildlife agencies (see chapter 2). This money has been the main source of funding for management of native wildlife in the United States.

All of these federal agencies employ wildlife biologists, and all have strong mandates for managing wildlife habitat. In contrast, land owned by the states does not always have strong environmental mandates. For example, the State Land Department (SLD) in Arizona administers 9.4 million acres (13 percent of Arizona) of land granted to the state by the federal government at the time of statehood in 1912. The primary responsibility of the SLD is to generate as much income as possible from the land, primarily by selling land and leasing grazing rights. The money generated from these activities goes into an account that is disbursed to beneficiaries such as schools, colleges, hospitals, and charitable institutions. Although the mission statement specifically directs that state land will be managed "consistent with sound stewardship, conservation and business management principles," wildlife needs and habitat quality are not considerations in land management decisions.

Other government agencies also control large areas of land throughout the Southwest. For example, the Department of Defense owns large tracts and cooperates with wildlife management agencies in the management of wildlife on those areas. Tribal lands, in contrast, are beyond the jurisdiction of state wildlife agencies. The wildlife management and environmental issues on these lands are the sole responsibility of the Bureau of Indian Affairs and individual tribes. The various tribal lands have natural resource departments in differing degrees of development. Some have highly skilled biologists running advanced game management programs, while others have no game programs at all.

The wildlife management authority in Mexico resides with the federal government, as outlined in the 1952 Mexican Game Law (Ley Federal de Caza). The federal agency SEMARNAT (Secretaría de Medio Ambiente Recursos Naturales) is responsible for administering regulations concerning deer and other big game through an office in the capital of each Mexican state (C. Alcalá-Galván, personal communication, 2003).

Each state has an agency responsible for managing resident wildlife in that state. There are almost as many names for these state wildlife agencies as there are states (Game and Fish; Fish and Game; Wildlife, Fisheries, and Parks; Division of Wildlife; or Department of Natural Resources). These agencies have the authority and responsibility to manage native wildlife throughout their respective states regardless of land ownership (except lands under Tribal Trust status). As noted in chapter 2, the authority of the state agencies to manage native wildlife such as deer evolved because wildlife belongs to the public and is held in public trust. Additionally, because of the complex mosaic of land ownership, it would make no sense for this authority to change as wildlife moved through parcels of land owned by different agencies and entities.

Management Data

Proper management of deer populations requires that decisions be based on solid information collected in a consistent way by professional staff. Although all information should be considered in management decisions, it is sometimes tempting to place too much emphasis on anecdotal information. Only through standard and scientifically designed surveys can deer managers collect useful information that has value as current-year estimates of population parameters and also for tracking of long-term trends.

The deer population dynamics throughout the Southwest differ in response to local environmental and habitat conditions. Management objectives for one area may be poorly suited for another area, so sound management requires units to be defined in order to tailor management to individual populations. Each state is divided into units of management generally called game management units (GMUs), big game management units, deer management units, or simply hunt areas. Deciding what to include in a "population" is difficult when deer are distributed contiguously across the state. In some areas delineating deer populations is much clearer because they inhabit isolated mountain ranges with limited dispersal among them. However, other populations have been designated somewhat arbitrarily, with substantial movement between "populations." This does not negate the value of these designations, since the nearby interchanging populations are probably being managed in a similar way.

Most states have a biologist responsible for the management of each unit or a group of units. This biologist becomes familiar with local deer distribution, movements, harvest patterns, and the history of the management actions in that area. This on-the-ground knowledge adds considerably to the quality and continuity of management.

Ideally all survey and harvest data are collected in each GMU to facilitate the management of deer in that area as an individual population. In reality, shrinking budgets or reallocation of funds has resulted in some states reducing the number or frequency of surveys. Sometimes a subset of units are selected to serve as representative sample areas from which information is extrapolated to unsurveyed units. Many states collect management data in a different way or at a different time because of varying factors like weather, terrain, budgets, or simply agency doctrine. Some states collect slightly different kinds of information as well, but most data used in management falls into one of two main categories: survey or harvest-related.

Survey Data

Sometimes deer surveys are erroneously thought of as a count of how many deer are in a game management unit. This is not possible from a technical or budgetary standpoint. Instead, biologists sample a representative portion of the population to obtain information that helps to estimate deer density, provide an index to deer abundance, and/or estimate the relative composition of sex and age groups (buck:doe and fawn:doe ratios).

Estimating actual density is expensive and time-consuming, and is usually not done annually or else is done on a small subset of GMUs. Indices to deer abundance are based on the number of deer seen or harvested per unit of effort. Trends in many forms of information can serve as an index to deer abundance, such as the number of deer observed per hour or mile of survey, total number seen on standard survey routes, deer fecal pellets counted on permanent plots, days of effort per deer harvested, and overall hunter success. Obtaining indices to deer abundance requires a less intensive effort than estimates of density, so they can be done more frequently and over a larger area.

Composition counts can also be done less intensively and more frequently over large areas. These surveys do not yield an index to deer abundance, but provide demographic data on reproduction (fawn:doe), sex ratios (buck:doe), and age structure (for example, percentage of bucks with more than two points). Composition counts must be conducted in fall or winter when the bucks still carry their antlers and fawns are not too large to be distinguished from older deer. Most states collect composition data while surveying to obtain an index to abundance.

Adhering to a standard survey period each year is important for measuring buck:doe and fawn:doe ratios consistently. Bucks are with doe groups during the rut, but usually occupy different areas at other times of the year. The best opportunity to obtain an accurate measure of the ratio of bucks to does is near the peak of the breeding season. Coincidentally, this time period is also after most of the annual hunting seasons, allowing biologists a look at the populations when buck:doe ratios are at their lowest.

Fawns are born in the summer and have a much higher mortality rate during their first year than do adults. Fawn:doe ratios measured in August will be much higher than those measured the following May because of this differential mortality rate. Collecting fawn:doe and buck:doe ratios at a consistent time each year allows the evaluation of annual trends while minimizing other variables.

Surveys are conducted primarily in the early morning and sometimes late afternoon hours when deer activity is relatively high and the low angle of the sun aids in detecting deer. The same area or survey route is surveyed each year so that multi-year trends can be evaluated (O'Brien 1984). Global positioning system (GPS) technology is being used now not only to maintain the same survey routes, but also to record the location of deer observed during the survey. An effort is usually made to survey only in areas hunters have access to; this ensures that survey data used to make management decisions reflect the conditions in areas where deer are hunted. Also, more accessible areas may have the lowest buck:doe ratios after the hunt, while less accessible areas will have higher ratios. Surveying hunted areas assures that the ratios obtained represent conservative estimates for that unit.

During the surveys, wildlife managers record the weather conditions at the time of the survey, the names of observers, the method of survey, and the time the survey was started and ended. When deer are seen, the observer records the time of observation, species, and the number of males, females, and fawns. The accurate classification of bucks, does, and fawns is important because these ratios form the basis for management. Deer that could not be observed well enough to classify accurately are marked as "unknown" so that the total number observed is accurate without jeopardizing the integrity of the ratios. Each group observed is entered in a separate line on the data sheet, and at the end of the survey all groups are totaled so that buck:doe ratios and fawn:doe ratios can be calculated. During the survey, the biologists also record some miscellaneous information, such as number of antler points of bucks, vegetation association where the deer were seen, range conditions, snow pack, condition of browse, and rutting activity.

Deer managers look not only at the present year's data, but also, and more important, at the trends in past surveys. Trend data from surveys commonly used to make management decisions include buck:doe ratios, fawn:doe ratios, distribution of antler classes, population estimates, and abundance indices.

Survey data can be gathered by several methods, each of which has many variations. Each method has its own advantages and disadvantages in relation to the others. No single method is the best in all situations, so deer managers pick and choose the best tool for the job depending on terrain, tree cover, deer densities, personnel available, and budget constraints.

Foot and Horseback Surveys

Surveys conducted on foot or horseback are much like hunting: the surveyor walks or rides through known deer habitat and records the deer he or she sees. Binoculars have always been used by surveyors and hunters alike, but not as intensively as they are now. Today, deer managers conducting surveys from the ground spend almost the entire survey sitting with a high-quality pair of binoculars. Mounting the binoculars on a tripod provides a stable base for a motionless field of view. Additionally, the tripod enables the observer to search an area systematically from top to bottom, side to side. This technique and other improvements (O'Brien 1984) have increased the number of deer seen on surveys and allow for more accurate classifications.

When conducting surveys on foot or horseback, the biologist spends days on the ground in deer habitat and has an opportunity to observe more than just deer. Time spent in deer habitat searching for deer allows the manager to assess the condition of the habitat, water availability, where the deer have been concentrated (tracks and pellets), and deer mortality, and also to observe rutting activity. Historically, most deer surveys were completed on foot or horseback, although vehicle routes became more common as four-wheel-drive trucks became available and the number and density of roads increased.

Ground Vehicle Surveys

Ground vehicle surveys consist of more than simply driving a route looking for deer through the windshield. The vehicle survey is similar to surveys conducted on foot or horseback, but the biologist is able to cover more deer habitat in less time. The disadvantage of this method is that the deer manager has less opportunity to see deer sign and observe current habitat conditions.

One type of vehicle survey uses the vehicle to drive between observation points or "glassing" locations. Surveyors drive a standard route, traveling from one location to another that provides a good vantage point for searching for deer. After spending time at one observation point, they drive the vehicle farther along the survey route until they reach the next glassing location. Certainly deer are also seen along the driving route, but in most cases the biologist spends a majority of the survey time stopped or outside the truck on a vehicle survey.

Another type of vehicle survey is the spotlight or line-transect survey through habitat that is representative of the unit or ranch being surveyed. This is usually done shortly after dark when deer are active and moving around. A driver navigates a vehicle along a permanently established route, while an observer (or two) shines a spotlight along the side of the route and records all deer seen. Also, observed deer are classified by sex and age (adult or juvenile). The number of deer seen per mile of route serves as an index to annual changes in deer abundance, and the sex/age composition provides trend information on population demographics.

In some cases, surveyors measure the distance to which deer can be observed from the road at regular intervals to calculate the area actually surveyed. Since the length of the route is known, they can multiply that by the average distance from the line that deer can realistically be seen and estimate the total area surveyed. The number of deer seen divided by that area yields a minimum estimate of deer density. Ideally the route is conducted several times and the results averaged to smooth out individual survey variability. Extrapolating the results of these surveys to overall deer density poses some problems. For example, if the road follows a drainage, or runs near agricultural fields or other attractive areas, the densities calculated along that route will not be representative of the larger area of interest. Also, deer may naturally avoid areas near roads because of disturbance or they may be attracted to the roadside because of better herbaceous growth.

Helicopter Surveys

Today state wildlife agencies are much more reliant on aerial observations; some states no longer conduct surveys from the ground. Because they are so maneuverable, helicopters offer a way to observe many deer in a short period of time (Fig. 45). Agencies started relying more heavily on helicopters for deer surveys in the early 1980s. The most obvious disadvantage of this method is the danger it poses to personnel. Helicopter and fixed-wing aircraft crashes are the leading cause of on-the-job deaths for wildlife biologists (Sasse 2003). Even though helicopter surveys are very dangerous to personnel and quite expensive ($560 or more per hour), they are a very cost-effective way to obtain survey data. In rough terrain, this method may be the only way to adequately sample a population.

In contrast to the effort required to travel foot routes, the same amount of time in a helicopter can yield ten times the number of deer observed. Besides cost and danger, another disadvantage is the increased difficulty of classifying deer accurately when they are scattering across a hillside or running in and out of cover. There is little opportunity to study animals closely when being flushed by a helicopter.

It is tempting to think all the deer in the area covered by the helicopter are seen and counted, but it is rare to see over 75 percent of the deer in the best of cases. Research consistently shows that 15–65 percent of the animals are not seen by observers (Bartmann et al. 1986, Pollock and Kendall 1987). Still, with consistent sur-

Figure 45. Helicopters are expensive and dangerous to use for surveys, but offer the opportunity to observe and classify a large sample of deer in rugged terrain. *Photo courtesy of AGFD*

vey methods the percentage of deer observed is assumed to be constant from year to year, allowing biologists to use the trends to track population parameters.

The number of deer observed from a helicopter per hour is a useful piece of trend data for indexing deer abundance. If helicopter surveys are flown over the same area each year, the number of deer per hour can be assumed to reflect deer density. This index is weakened when weather conditions differ or different areas are flown each year. Also, as helicopter time is limited due to budget constraints, only the highest density deer habitat is flown, which can inflate the number of deer observed per hour of flight.

Some states actually estimate the density of deer in some areas. This is too expensive and time-consuming to do on a statewide basis, but certain herds of high interest or those felt to be representative of much larger areas are sometimes singled out for intensive surveys. In these areas, the whole unit might be broken into blocks or quadrats and a random number of those blocks flown intensively (Kufeld et al. 1980). The number of deer seen in the blocks flown can be expanded to the whole area by assuming those not flown contained the same number of deer.

Another way to estimate actual abundance or density of deer is to conduct intensive research in representative areas to estimate the percentage of the population actually seen during the survey. This is usually done in conjunction with a large number of animals in the population being "marked" with radiocollars. By surveying intensively and recording the number of marked deer seen under different conditions compared to the number known to be there, biologists can estimate the percentage of the population they are seeing or, conversely, the percentage they are missing. These "sightability models" are designed to account for the fact that surveyors see a different percentage of the population depending on several factors. For example, with snow cover on the ground deer are much easier to see and a higher percentage of them are observed.

Fixed-wing Aircraft Surveys

Fixed-wing surveys are usually flown with a Cessna or Piper Cub and then only in open and flat terrain. As they do during helicopter surveys, fixed-wing pilots circle back when a group of deer is observed so surveyors can classify and count the deer. Because of the reduced maneuverability, it is more difficult for a plane to return to the group quickly enough to classify all animals accurately before losing them in the terrain or vegetative hiding cover. The main advantage of this method is the relatively lower cost ($100–$200 per hour) and the ability to cover large areas. Quadrat blocks and sightability models can still be used with this method, but blocks must be sufficiently large to accommodate long, parallel transects flown with the plane.

Pellet Group Survey

As described in chapter 6, this method involves clearing many plots to bare soil and returning after a specified time period to count the number of pellet groups on the plots. The number of pellet groups per acre can be estimated and converted to the number of deer by dividing by the number of times a deer defecates per day and the number of days the plots were exposed. For example, if you know a deer defecates 10 times per day and after 10 days you find 700 pellet groups per acre, you know seven deer were present (7 deer × 10 days × 10 pellet groups/day/deer).

Pellet group surveys originated in the 1930s and were popular in the early stages of deer management, but are no longer used systematically by wildlife agencies. There is good reason for this, as later evaluations have found that the defecation rate of deer is highly variable (Rodgers 1987). The estimates commonly used (12–21 pellet groups per deer per day) were calculated by watching deer in captivity being fed all the food they could eat. The real defecation rate of wild deer may vary with seasonal and annual changes in diet, animal health, age of animal, and available water (Neff 1968). Deer density estimates obtained with this method are very sensitive to small changes in the defecation rate. By using slightly different rates, biologists could arrive at substantially different density estimates.

Although density estimates obtained from pellet group surveys may not be accurate, this method has value as an inexpensive deer abundance index. If permanent pellet plots are randomly established throughout an area, to be cleared and "read" at the same time each year, many of the variables are controlled. The number of pellets found on the established plots then serves as an index to changes in deer abundance in that area.

Harvest Data

Along with information from deer surveys, harvest data are the second major source of deer management information. Harvest data consist of two general categories: hunter questionnaires and actual biological data collected from the harvested animals. All states question hunters in some way shortly after the hunt to obtain information that is useful in managing deer populations. In addition, most states collect biological data, such as weights, ages, antler classes, and antler measurements.

Hunter Surveys

Hunter questionnaires or phone surveys are a relatively inexpensive way to obtain information about the hunter effort and the harvest that occurred during the hunt. Information from hundreds of people afield at the same time is valuable for management purposes. While post-hunt questionnaires have been used to obtain information on sex/age composition of the deer population (Thompson and Bleich 1993), they are primarily used to gather harvest data. Questionnaires are usually mailed out to the deer-tag holders just before, or shortly after, the hunting season with a request to return them after the hunt. Phone surveys are conducted immediately after the season while the information is still fresh in the hunter's mind.

Hunters may initially be asked if they even went hunting, since not everyone holding a license or tag for a particular hunt actually goes hunting. Illness, workload, or other obligations may result in hunters not using their tags, so this question provides an estimate of the participation rate, actual number of hunters, and hunter density in each hunt.

"Did you harvest a deer?" is probably the most important question, as this allows the estimation of total deer harvest and hunter success (percentage of hunters who were successful) for that hunt. Regardless of whether they were successful, hunters are usually asked how many days they hunted and in what units they hunted. This provides the average number of days hunted per deer harvested (a measure of deer abundance) and total hunter-days (a measure of hunter pressure). For hunts that encompass more than one unit, the respondents might be asked to note how many days they spent in each unit.

Some surveys go further and ask hunters what wildlife they saw while hunting, if they shot a deer they were not able to recover, number of antler points on the deer they harvested, or what type of weapon they used. All this helps provide additional information on animal distribution, wounding loss, and weapon type preferences.

Figure 46. Properly managed, hunting is a beneficial use of a renewable natural resource that provides management data, funding, and support for wildlife conservation efforts. Pictured (*from left*) are three generations of Heffelfingers: Levi (with his first deer), Jim, and Bob. *Photo by author*

Biological Data

Data from harvested animals have long been a cornerstone of wildlife management (Fig. 46). Managers can learn a great deal from looking at a sample of harvested deer. Few western states have permanent check stations. The longest running and best known deer check station in the Southwest is on the Kaibab Plateau in northern Arizona, which has been mandatory for all successful firearms hunters since the first hunt in 1924 (Russo 1964:60). Temporary deer check stations are sometimes established as part of research projects or to gather more intensive information in a specific area. Check stations are not the only way to collect biological data from harvested animals; having conservation officers and biologists out in the field during the deer hunts presents an opportunity to collect some of the same information through contacts with successful hunters in camp and throughout the unit. This method eliminates the need for an elaborate check station set-up and might actually result in more deer inspected, depending on the road density and access to/from the area being hunted.

The data collected usually include weight, antler measurements, body measurements, an estimate of age, and a general health check of the animal. The age of the deer harvested is an important piece of information. It is one way managers can determine the appropriateness of the harvest level in that unit: a population that has few bucks in the older age class is being harvested more intensively than one with many mature bucks in the harvest (and thus in the population). More conservative harvest allows a larger proportion of young bucks to survive into the older age categories. Also, it is important to record and analyze weight and measurement data by age class since these parameters increase with age to a certain point (Bone 1987).

Body weights not only provide an index to deer health, but also give an indication of the quality of the habitat conditions during the previous year. When deer populations exceed carrying capacity, or during dry years, the body weights of harvested animals reflect the lack of food in the previous year (McCulloch and Smith 1987). The weight of yearlings is most important because they developed from the fawn stage to nearly adult size in the previous twelve months, so their body weight will reflect the habitat conditions during that period.

Likewise, antlers measured in the fall have undergone their entire growth in the preceding summer. Antlers are sensitive to nutritional deficiencies because energy intake must first go to body maintenance and secondarily to antler growth. Antler measurements, such as basal diameter or circumference and mainbeam length (by age class), correlate well with nutritional intake during the previous year (Ullery 1983). The body weight of yearlings and antler growth trends were found to be a useful predictor of population size in relation to carrying capacity for mule deer on the Kaibab Plateau (McCulloch and Smith 1991).

Other measurements of ears, legs, tails, and body length allow comparisons with deer from other areas or other species. If location of harvest is plotted on a map, it provides a picture of where most of the harvest is taking place, or may show shifts in the distribution of harvest over a series of years.

In addition to weight, age, and antler/body measurements, check stations and field checks also provide the opportunity to study the carcass for injuries, parasites, abnormalities, fat levels, and signs of disease. Samples can be taken for use in disease surveillance or genetic studies to answer subspecies questions, help with law enforcement forensics, or investigate the occurrence of hybridization.

Other Information

Forage Monitoring

Forage monitoring is sometimes used in conjunction with other data. This method is normally used on winter range to determine the appropriateness of the current deer population size in relation to the habitat available to sustain those animals through the winter. In some areas it provides important information for determining the management of deer populations. Erratic fawn recruitment in the

Southwest results in deer populations that generally stay below the long-term carrying capacity of the habitat and do not overuse shrubs. However, if some winter range is lost due to development or fire, or if deer populations increase with favorable conditions, important forage plants by show signs of overuse. When this happens, deer populations must be reduced to protect the important forage plants. Overuse (usually more than 50 percent) by ungulates can cause long-term damage to the range and reduce the future carrying capacity of that deer habitat.

When evaluating the extent of browsing on native shrubs, researchers cannot distinguish deer from cattle use with any degree of accuracy. Some studies have used different exclosure designs to exclude both cattle and deer, or only cattle, to determine the percentage of the browsing attributable to those species (Day 1964). Other small, fenced areas, such as exclosures around water catchments, provide a sample of shrubs that are exposed to higher-than-average deer browsing, but are not used by domestic livestock. It is instructive to examine deer forage protected from cattle inside these small plots and compare the plants with those same plants outside a fenced exclosure.

Computer Models

Some states use computer models to aid in the management of deer. Models generally take survey and harvest data, and calculate a population estimate for an area of interest or provide a simulation of population fluctuations through time. These models are only as good as the information they use to perform the calculations. In some cases these total population estimates are compared to a population goal (an overwinter goal, for example), and this helps deer managers determine if any antlerless tags should be recommended.

If the model or the data it uses are not robust enough for the manager to have confidence in the absolute population estimate it produces, then he or she may use it as one of many pieces of information tracking trends in deer abundance. Regardless of the level of confidence in the model, it is desirable to use computer population models as one of several factors guiding hunt recommendations. Throughout the history of deer management, nothing has been more of a lightning rod for public criticism than population estimates. If there is no corroborating information to support mathematically derived computer population estimates, agencies are sometimes accused of selling tags for "paper deer."

Models do more than estimate numbers of deer. Some states use models that allow the manager to calculate how many buck permits should be recommended to achieve a desired post-hunt buck:doe ratio. If the pre-hunt buck:doe ratio and population size can be estimated, then it is a simple calculation to determine how many bucks can be removed to achieve a target post-hunt buck:doe ratio. If the population needs to be reduced with doe harvests simultaneously, the calculation is a little more complicated, but still fairly straightforward.

Deer Management in the Southwest

Managing deer in the Southwest differs from other regions of the continent in several important ways (Heffelfinger et al. 2003). Terrain, topography, and vegetation require, and allow for, different ways to survey deer populations and collect harvest data. Surveys can usually be conducted from the air because of the open vegetation. Indeed, this method is preferred over much of the area consisting of rough, rugged terrain with few roads. With low human population densities and long hunting seasons, it is not feasible to operate statewide check stations to collect harvest information, as is the case in the eastern and midwestern states. Harvest information generally comes from post-hunt questionnaires and some focused check stations for specific purposes.

Although all state agencies in the Southwest base their management on some form of survey and harvest data, there are differences even within this region in how they actually collect those data and formulate hunt recommendations (Table 19). Some of these differences are justified by ecological variations from one area to another or budgetary constraints in some states. In other cases, methods simply developed along parallel paths separated by nothing but a state line. Deer management would be better served by working toward more standardization in the collection and use of survey and harvest data among states (Carpenter et al. 2003).

Arizona

The Arizona Game and Fish Department (AGFD) has established about eighty game management units throughout the state. Almost every GMU has a resident wildlife manager who is a law enforcement officer and holds a bachelor's degree in wildlife management. Firearm deer harvest is controlled in all GMUs by a limited-entry drawing.

Survey Data

Because desert mule deer rut a little earlier than Coues whitetails, the survey periods are slightly different. Wildlife managers conduct mule deer surveys from December 1 to February 15, while whitetails are surveyed between December 15 and February 15 (Arizona Game and Fish Department 2001b). Helicopters account for at least half of the deer observations, with fixed-wing aircraft, vehicle, foot, and horseback surveys making up the rest.

Wildlife managers survey areas of each GMU where deer are found in higher concentrations, in order to maximize the number of animals seen. The advantage of this approach is that ratios are more robust with larger sample sizes and thus surveyors are maximizing survey efficiency. In addition to surveying the higher density areas, each GMU has one or two survey blocks established that are

TABLE 19. COMPARISON OF SURVEY AND HARVEST DATA USED BY STATE WILDLIFE MANAGEMENT AGENCIES IN THE SOUTHWEST

Location	Survey data		Harvest data		Supporting information[c]
	Data collected[a]	Method of collection	Data collected[b]	Method of collection	
Arizona	Post-hunt B:D and F:D ratios, deer/hour, number observed, number of antler points	Helicopter, plane, vehicle, foot, horseback	Harvest, hunt success, days/harvest, number of hunters afield, age, weight, antler measurements	Post-hunt questionnaire, check stations	Sightability model, some browse surveys, some population modeling
New Mexico	Post-hunt B:D and F:D ratios, abundance indices, population estimates	Mostly helicopter	Harvest, hunt-days, number of hunters afield, use of private/public land, hunter success and satisfaction	Post-hunt questionnaire	Sightability model
West Texas	Post-hunt B:D and F:D ratios, density index	Plane, ground, spotlight counts, some helicopter in areas with bighorn sheep	Harvest, hunt-days, hunt success, deer harvest/hunter and area, hunters afield, hunter densities, age, weight, antler measurements	Post-hunt questionnaire and field check	Some browse surveys
Southern California	Post-hunt B:D and F:D ratios, number of deer seen	Helicopter, water hole cameras, hunter observation questionnaire	Harvest, hunt success, number of antler points, body condition, antler measurements, age	Post-hunt questionnaire, field checks	...
Nevada	Post-hunt B:D and F:D ratios, number of deer seen	Helicopter (winter and spring)	Harvest, hunt success (%), antler points, participation rate, area hunted	Post-hunt questionnaire	Computer model

(continued)

TABLE 19. (CONTINUED)

Location	Survey data		Harvest data		
	Data collected[a]	Method of collection	Data collected[b]	Method of collection	Supporting information[c]
Utah	Post-hunt B:D and F:D ratios, number of deer seen, over-winter mortality	Foot and vehicle (winter and spring)	Harvest, hunt success (%), hunter-days, number of hunters, days/harvest, age, sex	Post-hunt telephone survey, check stations	Statewide browse monitoring program, computer population model
Colorado	Post-hunt B:D and F:D ratios, density	Helicopter	Harvest	Post-hunt telephone survey	Radiocollared deer provide survival estimates used for population modeling
Northern Mexico	Post-hunt B:D and F:D ratios, density	Spotlight counts, pellet counts, drive counts	Age, weight, antler measurements	Field checks on some ranches	No standard system of surveys and harvest data collection by state

[a] Fawn:doe ratios (F:D), buck:doe ratios (B:D), deer observed per hour of helicopter flight, and total number observed.
[b] Number of deer harvested (harvest), percentage of hunters who were successful in harvesting a deer (hunt success), total hunter-days expended, average days hunted per deer harvested (days/harvest), number of hunters afield during the hunt, age of each deer, weight of each deer.
[c] Sightability model is a method of using multiple observers to estimate the percentage of observable deer being missed during surveys. This can then be use to estimate deer density in many cases.

surveyed each year for the same amount of time. This provides a few areas where managers can look at trends over time in a more controlled setting: a comparison of the same area flown at the same time of year for the same duration. There is an ongoing effort to standardize surveys by surveying the entire unit with wide transects to better sample the deer population in that GMU rather than only a few areas of concentration. By flying transects, biologists can also use a simultaneous double-count technique that allows them to estimate the percentage of available deer they are actually seeing.

Harvest Data

Immediately after the hunts, harvest questionnaires are mailed to permit holders to obtain more harvest information. This program has remained basically unchanged since the 1960s, giving AGFD a long-term data set of harvest-related information gathered consistently. The information collected includes estimates of harvest, number of hunters, hunter-days, participation rate, days per deer harvested, unrecovered deer, weapon-type preference, and hunt success. The questionnaire card is sent to all hunters in each hunt up to a maximum of 800 questionnaires (Alexander 1997). These questionnaires are postcards that can be completed in about thirty seconds and returned postage paid.

From 45 percent to 60 percent of the hunters return the questionnaires to AGFD; those responses are then expanded to account for all hunters. There is a tendency for successful hunters to return questionnaires at a higher rate than do unsuccessful hunters. This results in harvest being overestimated by approximately 10 percent, but the bias is thought to be consistent from year to year and thus not to affect the validity of the trends. This overestimate of harvest may counterbalance the loss of unrecovered deer to some degree.

The AGFD has operated a check station in Coues whitetail and desert mule deer habitat south of Tucson since 1993 (Arizona Game and Fish Department 2000a). This check station collects data from deer harvested in most of three GMUs. Harvest data collected include weight, age, antler measurements, body measurements, parasites present, DNA samples, and body condition. As mentioned above, the check station on the Kaibab Plateau collects much of the same information and has aided the management of that herd for over seventy-five years. This check station has produced the longest running set of biological harvest data in the West and probably the entire country.

In addition to operating the established check stations, most wildlife managers collect harvest data as they patrol their GMUs during the hunt. Data collected during contacts with hunters are not as extensive as the information gathered at check stations, because the wildlife manager has law enforcement responsibilities during those patrols. However, ages of the harvested bucks and dressed weights of at least yearlings provide useful information about the harvest and the condition of the deer herd.

Other Data

Forage monitoring is an important part of deer management on the Kaibab Plateau. The management plan for this herd states that antlerless hunts will be considered when use of cliffrose plants on winter range by deer has exceeded 50 percent before mid-March (Arizona Game and Fish Department 2000b). This use is determined by estimating the percentage of twigs that have been bitten off. Generally, several permanently marked transects are surveyed, and 25 cliffrose plants per transect are evaluated. On each of the 25 plants, 10 stems at browsing height, representing the current year's growth, are checked for evidence of use. The percentage of the 250 (10 twigs $\times$ 25 plants) twigs on each transect gives the average browse use for the transect. In addition, photos are taken of the first cliffrose plant on each transect (B. Lemons, personal communication, 2001). Transects and photo-points are also established inside and outside 4-acre livestock exclosures.

With a computer population model, AGFD simulates deer population trends through time to aid in management (Heffelfinger and Piest 1996). The computer model uses historic survey and harvest data stored in database files to run through millions of calculations to simulate deer population fluctuations over a predetermined span of years. These models were not designed to estimate populations in individual GMUs, but rather to give the wildlife manager an indication of whether the deer population is increasing, decreasing, or stable. The information from the population modeling is used as supplemental information and does not directly drive permit levels.

The Hunt Recommendation Process

AGFD hosts public hunt meetings throughout the state to obtain public input regarding game management. This is one of many opportunities for the public to offer suggestions for new seasons, altered seasons, primitive weapons opportunities, or permit adjustments.

After the survey period, all survey and harvest data are entered into a database and summarized. For each species in each GMU, a form called the Management Summary Form is printed by extracting information from all the associated database files for the last five years. The data reported on this multi-page form are then compared with management guidelines to help formulate a permit recommendation for each GMU.

Permit adjustments are made using many different kinds of information (Arizona Game and Fish Department 1993). However, a set of Hunt Guidelines prescribe management ranges for three of the most important variables:

1. Buck:doe ratios
   Mule deer: 15–25 bucks per 100 does surveyed
   White-tailed deer: 20–30 bucks per 100 does surveyed

2. Fawn:doe ratios
   Mule deer: 40–50 fawns per 100 does surveyed
   White-tailed deer: 35–45 fawns per 100 does surveyed
3. Hunt success: 15–20 percent of hunters successful in harvesting a deer.

If the data from a unit fall below these ranges for all or most of the three variables, then a reduction in permits may be appropriate. If survey and harvest data are above guidelines, a permit increase may be warranted. The reasons are straightforward; if the buck:doe ratio is high, there are more bucks available to harvest. If the fawn:doe ratio is above guidelines, there will be more yearling bucks available the following year (the year for which the recommendations are being developed). If the hunt success was high the previous year, hunters found it relatively easier to harvest bucks, and since hunter skill does not vary appreciably from one year to the next, this variable provides a useful index to the relative abundance of bucks.

Those guidelines serve as the basis for the recommendation, but many other factors, biological and sociological, are also taken into account. The recommendation is primarily based on trends in the most useful data, such as hunt success, fawns:100 does, bucks:100 does, deer/hour observed from a helicopter, days/deer harvested, total harvest, forage monitoring, and age structure of harvested deer. Supplemental data, such as access, hunter densities, precipitation amounts and patterns, and population models, are also used to support the hunt recommendation.

The wildlife managers submit their recommendations to the regional office in late February. The regional office staff sometimes make small adjustments before the regional package is submitted to the state office the first week in March. After considering public input, the big game supervisor and game branch chief assemble all six regional packages into a statewide hunt recommendation. This is reviewed by the director and executive staff before being formally presented to the five-member commission in a public meeting in mid-April. After hearing and considering input from the public, the commission approves the recommendations, with or without alterations, and the seasons become final.

New Mexico

Deer management in New Mexico is based on fifty-eight big game management units, some of which are divided into subunits for specific management purposes. All field personnel (area game managers, sergeants, and conservation officers) have at least a bachelor's degree in wildlife or fisheries management or biology. The field personnel, along with survey help from the state headquarters office, are responsible for collecting management information from the units and making the field-based management recommendations.

Survey Data

Like many states, New Mexico conducted pellet group surveys in the 1950s and 1960s, but has replaced that method with direct observations of deer. Current surveys focus on collecting information to monitor abundance and demographics of deer populations. Some data are collected from ground surveys in some areas, but the vast majority comes from helicopter surveys. Helicopters are used in December and January to obtain post-hunt information on population composition (buck:doe and fawn:doe ratios). Not every unit is surveyed every year; a subset of units (about 20–25 percent statewide) are flown, and biologists assume that these represent other similar units that are not flown. Survey effort is rotated to different units each year with the intent of surveying most units periodically (B. Hale, personal communication, 2004). Some units receive annual surveys or more intensive surveys because of higher public interest, while other units (generally private land, tribal lands, and military reservations) are not surveyed at all.

Helicopter surveys were traditionally nonrandom surveys flown to obtain population composition data. In recent years, New Mexico has used a sightability model for deer and elk. This allows biologists to estimate the proportion of deer missed on each survey and develop a population estimate for each area surveyed.

Harvest Data

All general license holders are mailed a post-hunt questionnaire to obtain harvest data. In addition, those hunters lucky enough to draw a tag for the "deer entry units" find a mail-in card attached to their license when it is issued. Generally, questionnaire response rates range between 20 percent and 35 percent annually, with deer entry license holders responding at a higher rate compared to that of general license holders (B. Hale, personal communication, 2004). Like the questionnaires used in other states, the New Mexico form asks questions that allow the estimation of harvest, number of hunters, number of hunter-days, hunt success, and harvest rates by weapon type. Beginning in 2003, hunters were asked additional questions that include whether they hunted and harvested on private land and their level of satisfaction with their deer-hunting experience.

There are no hunter check stations in use for collecting biological data directly from harvested deer, except for one in south-central New Mexico to obtain samples for testing for chronic wasting disease (CWD) near the original outbreak area (White Sands Missile Range headquarters). Conservation officers sometimes age harvested deer, but widespread field checks are not conducted.

Other Data

No browse surveys are conducted annually for management purposes, but some individual projects have provided information about deer use of browse. A computer population model (DEERMODEL) was developed and used directly for making management decisions in the 1990s (Green-Hammond 1996). DEERMODEL estimates the information (survival rate, birth rate, natural mortality, and so forth) needed for evaluating the potential effect of regulations before they are put into practice (New Mexico Department of Game and Fish 1997). This model is no longer used directly for estimating populations because it was heavily dependant on harvest information, which may not have been accurate enough to drive the population model.

Hunt Recommendation Process

Field staff develop the first stage of the hunt recommendations for each unit and submit them to the area game manager in their area (northwest, northeast, southwest, southeast). The area game managers and sergeants discuss the recommendations with area administrators and forward them to the Wildlife Management Division at the state headquarters office in Santa Fe (B. Hale, personal communication, 2004). The Wildlife Management Division discusses the statewide package with the area and department administrators and then forwards it to the seven-member State Game Commission for review and approval. The public has the opportunity to comment at public meetings, on the department Web site, at individual group meetings, and at commission meetings.

Changes are made periodically to adjust harvest rates when needed. These adjustments can take the form of changing season length or license numbers, opening the hunt on weekdays, allowing harvest only during weekdays, or closing areas to hunting altogether. Not all these actions are popular with the public, but they can reduce buck harvest when that is necessary or desirable.

Complete closure of areas such the Zuni, Peloncillo, and Manzano Mountains to restore deer abundance was not successful (Haussamen 1995). Even before buck harvest in those areas was curtailed, nearly all does were being impregnated by the bucks still present. Stopping the removal of bucks from a population will increase the number of bucks in the population, but will not result in more fawns on the ground and hence, will not increase a population or keep it from declining.

West Texas

With over 97 percent of Texas land in private ownership, the state has some differences in how deer are managed as compared to states with large amounts of public land. Traditionally deer management has been done by county because of the historic control each county had over wildlife management. Management data are still collected and summarized by county, but a more refined approach for mule deer now includes thirty-two Mule Deer Management Compartments (Bone 2003). Each county has a game warden responsible for law enforcement, but regulatory biologists may have responsibility for up to five counties.

Survey Data

In the Trans-Pecos and Panhandle areas of Texas, fixed-wing aircraft and ground surveys are used to collect survey data from August through October. For aerial surveys, long survey lines are flown with two observers tallying deer seen on both sides of the plane within 100 yards of the flight line (Bone 2003). Using the length of the survey line and the width of observation regulatory biologists can calculate the area surveyed and thus the number of deer per square mile by county and deer management compartment. In areas with bighorn sheep populations, some deer herd composition data (sex and age information) are collected for all deer seen during helicopter flights in sheep habitat.

Ground surveys are conducted to estimate density and collect herd composition data. Standard spotlight counts have been established and are repeated annually. Biologists establish spotlight routes up to fifteen miles long in areas that appear to represent most of the habitat in that management compartment (Bone 2003). The actual counts begin one hour after sunset with two observers using hand-held spotlights from seats mounted in the back of a truck. As in the fixed-wing surveys, the length of the transect is known, so the surveyors must estimate the width of the area they are surveying to calculate the area covered. Periodically along the transect, biologists measure or estimate the distance from the line of travel to which a deer could be seen. The average observation distance multiplied by the length of the transect gives the area surveyed and assists in estimating deer density in that area. Ideally, each route is repeated two or three times and the observations are averaged to smooth out some of the variation due to weather, disturbance, or other factors.

As mentioned previously, there are some difficulties in calculating deer density with spotlight counts along specific routes. Routes cannot be selected at random because they are dependent on the location of existing roads. Densities estimated on unrepresentative routes and expanded to an entire county or compartment are inaccurate. Biologists realize this and commonly use "density" estimates more as an index to abundance than as a figure of exactly how many deer are in the county. Since the surveys are conducted in a consistent manner, mostly by the same people, this is a valuable way to track annual changes in deer abundance.

Harvest Data

A statewide big game harvest survey is mailed out after the hunts to a random selection of the nearly 1 million licensed hunters in Texas. About 30–40 percent of the roughly 25,000 questionnaire recipients actually return the information to the department (Texas Parks and Wildlife Department 2003). This survey allows biologists to estimate the harvest-related parameters that aid in evaluating the status of the deer population and the appropriateness of current hunt structures. The information collected by county and deer management compartment includes number of hunters, gender of hunters, harvest (buck and doe), hunter-days, percentage of hunters who were successful, average number of deer harvested per hunter, hunter densities, and number of deer harvested per thousand acres.

In addition to the harvest data provided by hunters, biological data are collected by regulatory biologists in the field, at meat lockers, and in hunting camps in each management compartment (C. Brewer, personal communication, 2004). Data such as age, number of antler points, inside antler spread, antler base circumference, ranch name, and county of harvest are recorded from all deer handled. This allows for the evaluation of age structure to ascertain how intensive the harvest is. The percentage of the harvest that was made up of yearlings, the proportion of deer older than three years, or average age are all measures of the relative maturity of the deer population. Various antler measurements are an index to relative changes in the health and well-being of the individual deer in the population. By tracking antler growth and body weight annually, biologists not only monitor herd health, but also are able to communicate trends in antler size to an interested constituency.

Other Data

Although no vegetative surveys are formalized on a statewide or regional basis, habitat and forage conditions are evaluated informally during ground and aerial surveys. In addition, browse surveys are conducted by many of the biologists in an effort to support harvest recommendations.

Hunt Recommendation Process

Deer hunting in Texas is not limited by a lottery-style draw. The harvest structures and seasons are set fairly broadly by Texas Parks and Wildlife Department, and are not altered much on an annual basis. With the predominance of private land in Texas, the landowners can control access and permission to use their property to be more restrictive than what is allowed by the state. Many landowners have specific goals and objectives, and by working with the local biologist, they develop a strategy to accomplish those goals. This is a higher level of resolution in management than in a state where the wildlife agency is managing deer on large tracts of easily accessible public land.

Regulatory biologists formulate their recommendations by county using the trends in abundance, herd composition, harvest, hunt success, and biological harvest data. The recommendations are then submitted to the state office in Austin where they are reviewed by staff, assimilated into a statewide recommendation, and forwarded to the nine-member commission for approval.

If populations are determined to be above the carrying capacity of the habitat, antlerless tags may be issued to the landowner to allow him or her to reduce the population. Antlerless tags are not often issued for mule deer, but are more common on the eastern edge of the Trans-Pecos region.

Southern California

California Department of Fish and Game (CDFG) manages deer in forty-five management units called deer zones, but only the southern portion of the state falls into the scope of this book. Originally each county in the state had its own biologist. Because of retirements and cuts in funding, one biologist now may be responsible for more than one county, but each deer zone still has a unit biologist in charge. Management strategies and survey methods differ between southern and northern California because deer inhabit very different environments in those two areas (Kucera and Mayer 1999).

Survey Data

Aerial surveys are conducted in southwestern California between late November and mid-December to index the population and collect herd composition data. A helicopter is used to fly transects about 500 yards apart within well-defined survey areas (R. Botta, personal communication, 2004). An attempt is made to classify at least two hundred deer in each herd to assure meaningful buck:doe and fawn:doe ratios. In the desert areas of southeastern California, low deer densities make aerial survey inefficient; too few deer are seen to justify the expense of surveying with this method.

In the dry deserts of southeastern California, deer are closely associated with the limited water sources remaining by mid-June of each year. CDFG biologists use this to their advantage and set up time-lapse cameras after ten to fifteen consecutive days of temperatures exceeding 100 degrees (G. Mulcahy, personal communication, 2004). The cameras are set to take pictures every twenty minutes once they are triggered by watering animals breaking an infrared beam. Film is changed every one to three days depending on the number of deer using the water. Three to six rolls of film are usually expended at each water source. Experience has shown that bucks return about every third day, while does may water daily depending on whether or not they are pregnant. Longer periods of camera operation recorded repeated visits by the same deer. After reviewing the pictures, biologists can evaluate the minimum number of bucks using the waters, the fawn:doe ratio, and buck:doe ratio. When the summer monsoon rains arrive, this method ceases to work because the deer disperse and visits to the water nearly stop.

In some cases biologists have obtained deer population data by asking hunters about their observations. An opening-day postcard questionnaire is mailed to deer hunters before the hunt for deer hunt zones D-12 and D-17 so the hunters know what information is being asked of them before they go afield. This postcard survey asks hunters if they hunted opening day, how many deer they saw, how many bucks they saw, number of hours they hunted, and how many burros (*Equus asinus*) they saw. Hunters are asked to mail in this postcard survey even if they did not hunt on opening day. With a return rate of 24–40 percent, this one-day survey offers a "snapshot" of potential deer numbers. Thompson and Bleich (1993) found that in the absence of robust deer population survey data, hunters could provide accurate data on sex and age composition of the deer herd.

Harvest Data

Under California law, all successful deer hunters must return the report card portion of their tags to CDFG with harvest information. This allows the department to estimate the harvest, hunter success, and number of antler points (G. Mulcahy, personal communication, 2004). Tag returns probably underestimate the actual buck harvest by 15 percent statewide because not all successful hunters comply. Still, this bias is probably consistent from year to year, providing valuable trend information.

During the hunts, the unit biologist and the game wardens collect harvest data during hunter contacts in the field. This yields general location of harvest, general body condition, and the number of bucks, does, and fawns seen. Personnel also collect teeth for age analysis and take a series of antler measurements from harvested bucks. Data are also collected on general body condition. Research projects to look at specific deer issues periodically gather other useful data on genetics, age, health, survival rates, fat reserves, recruitment rates, and habitat use.

Other Information

Some interstate deer herds in northern California are managed in part with computer simulation models. Because the southern deer herds have lower densities, they generally do not yield survey and harvest data with sufficient sample sizes and statistical reliability to model accurately. Lower density deer populations require innovative and creative means to acquire data for management decisions.

Hunt Recommendation Process

Hunt recommendations are developed by deer zone every two years and remain unchanged in the intervening year. Provisions built into the process allow for emergency changes to a deer zone if warranted. The biologists recommend whether permits should be increased or decreased in each zone based on trends in harvest, hunter success, and other data that were collected. The recommendations are submitted to the headquarters office in Sacramento and then forwarded to the State Fish and Game Commission for approval and adoption.

Nevada

Nevada has established 115 big game management units that Nevada Department of Wildlife (NDOW) uses for managing the deer harvest statewide. Several of these units are managed together in what are called "unit-groups," with a field biologist responsible for each unit-group. About thirty unit-groups form the basis of Nevada's deer management program.

Survey Data

In the past, surveys were conducted in the winter (November–January) and again in spring (March–April). Most winter surveys were discontinued in 2001, but three or four unit-groups are still surveyed at that time. These remaining units were selected because they were thought to be representative of other units in the state with similar vegetation. Data from these representative unit-groups are then applied to other, similar units. Winter surveys may be conducted every five years in some of the other units to provide comparative survey information. Spring surveys in March and early April are still conducted every year on winter range in about twenty-four of the thirty unit-groups (M. Cox, personal communiction, 2004).

Surveys are conducted almost entirely by helicopter in areas where field biologists know that deer are concentrated. Some observations may come from ground surveys, but generally only if weather prevents a timely completion with a helicopter. Biologists strive to observe at least five hundred deer per unit-group to assure that ratios reflect the areas surveyed. Since bucks do not have antlers during the spring surveys, a buck:doe ratio is not possible and the measure of reproductive recruitment is the fawn:adult ratio. Besides the traditional survey data obtained, field biologists also can observe deer distribution, trespass cattle, water levels, feral horse distribution, effects of fires and vegetation condition.

Harvest Data

Since 1976 all deer hunts in Nevada have been by limited quota (Hess 1983). Each deer tag is issued with a "return card" questionnaire with harvest-related questions. Hunters can fill out the card and mail it back or log onto the Internet and respond electronically. By 2004 over 80 percent of the hunters were responding via the Internet. The questionnaire asks hunters if they hunted, how many days, in what unit, if they harvested a deer, and how many antler points their buck had. Returning the information has been mandatory since 1979 and hunters who do not respond are not eligible to receive a deer tag the following year. However, paying a $50 penalty will get the hunter's name back on the eligible list. This mandatory system results in a 95 percent return rate of important harvest information. There are no permanent check stations or field check systems institutionalized, but some field biologists age and otherwise evaluate hunter-harvested deer they encounter in the field.

Other Data

Like many other states, Nevada used to conduct forage monitoring throughout deer habitat, but this is no longer an integral part of the hunt recommendation process. There was heavy use of browse in the late 1980s, but deer populations have declined since then. A computer population model built on a spreadsheet is used to estimate parameters necessary to generate harvest recommendations.

Hunt Recommendation Process

After survey and harvest information is assembled and summarized, the data are entered into a spreadsheet. This allows the biologist to calculate estimates for population size, bucks available to harvest, and pre-hunt buck:doe ratio using previously gathered data and the current year's fawn:adult ratio obtained from surveys. Goals are established for the desired post-hunt buck:doe ratio, so prescribing the number of buck tags to authorize is a matter of calculating how many bucks can be removed to arrive at the goal (M. Cox, personal communication, 2004). These tag quotas for management units change every year due to these calculations, but the season structure is evaluated every two years.

When formulated, the tag quota recommendations are sent to the NDOW commission and the public in late April. They are then forwarded to county advisory boards appointed by the county commissions. The advisory board members review the recommendations and sometimes make changes based on public input. The county advisory boards then send the recommendations to the NDOW commission for final approval. Passing the recommendations through the advisory boards gives the public direct input to the quota-setting process. This public involvement is a necessary part of the process, but sometimes local anecdotal information is used to overturn recommendations generated by trained biologists. It is not uncommon for 25 percent of the biologists' recommendations to be changed at the advisory board level.

Utah

Utah deer management is based on thirty wildlife management units (WMUs) spanning five regions of the state. Each region has a regional wildlife manager supervising three or four wildlife biologists. Each wildlife biologist has management responsibility for one or more WMUs.

Survey Data

In November and December each year, wildlife biologists and conservation officers survey herds in their units of responsibility on foot and by vehicle (S. Cranney, personal communication, 2004). No helicopter or fixed-wing aircraft are used for gathering deer survey data in Utah. These fall/winter surveys occur after the hunt and, as such, represent the minimum buck:doe ratios of the annual cycle. Another round of surveys is conducted the same way in spring (March–April). By conducting surveys twice per year, biologists can estimate overwinter mortality and relative loss of fawns. In colder climates, the percentage of fawns surviving the winter can vary with weather conditions and habitat quality. Surveying again in the spring allows biologists to measure fawn recruitment after the winter, one of the most important pieces of information for understanding deer population fluctuations.

Surveyors attempt to observe and classify at least two hundred adult does in each unit, but actual target sample sizes in the units vary with the overall size of the population. A representative area in each unit is surveyed during the early morning and late evening hours. During these surveys deer are classified as bucks, does, and fawns, information to be used in evaluating trends in population composition (UDWR 2001). To track the relative age structure of the population, biologists also summarize the percentage of bucks with more than two antler points on a side.

Harvest Data

Harvest-related information is obtained from hunters after the hunting season with a telephone survey of a random sample of hunters statewide. Data gathered this way by WMU allows for the estimation of harvest of bucks and does by weapon type, number of hunters afield, hunter-days, hunt success rate, and days per deer harvested. Telephone surveys were considered an improvement over the mail-out survey conducted in the past.

During the October firearms deer hunts, biologists from each of the five regions operate between three and six check stations. The locations of these check stations generally remain unchanged to make trends in the survey data more useful. Hunter compliance is voluntary, and age and sex of harvested animals are recorded to provide the biological data to track the age structure of the male segment of the population. In some cases a tooth is collected from the deer so it can be aged by cementum annuli.

Other Data

While many other states have discontinued extensive browse surveys, Utah has developed a statewide monitoring project to determine areas of use and relative degree of utilization. The objective of these established range trend transects is to monitor, evaluate, and record the trend in range conditions on important areas and allow deer biologists, land management agencies, and private landowners to detect changes in the quality of the habitat. Transects occur mostly on winter range in every WMU and are funded primarily with hunters' dollars. The beginning point of each transect is permanently marked so it can be relocated every five years when the analysis is repeated. From the starting point, a 500-foot transect is established to serve as the basis for the vegetative measurements. Every 100 feet along the transect, another transect is established perpendicular to the original one (UDWR 2001). A small metal frame is placed on the ground at intervals to measure what plants are growing along the transect and how much of the ground is exposed (percent ground cover). If shrubs are present along the transect, they are also described as to percent canopy coverage, relative age, and overall health. The overall trend in range condition is determined by biologists based on the density, health, age structure, and species composition of the vegetation along the transects.

The survey and harvest information is used in a computer population model to indicate whether the population is increasing, decreasing, or stable. This model simulates or re-creates deer population fluctuations and demographics using the buck age structure information from biological check stations, reproductive rate (fawn:doe ratio), estimates of overwinter fawn survival, and the harvest estimate compiled from the telephone questionnaire.

Hunt Recommendation Process

Regional wildlife managers and wildlife biologists in each region evaluate all data available to construct a hunt recommendation for each WMU. The fawn:doe ratio, buck:doe ratio, hunt success, days per harvested deer, trends in total number of deer surveyed, modeled population simulation, range trend analysis, and other data are evaluated in concert to determine if the current permit level in each WMU is appropriate for the number of available bucks. Utah is mandated to manage populations for a minimum of 15 bucks per 100 does on general-season units and 35 bucks per 100 does on premium, limited-entry units.

Wildlife managers submit recommendations for permits to a Regional Advisory Council (RAC). The RAC is composed of twelve to fifteen people who represent various interest groups, ranging from sportsmen to nonconsumptive users to agricultural interests. Upon receipt of a recommendation, the RAC considers public input and votes on the recommendation. After the last RAC meeting each year, the Utah Wildlife Board makes the final decisions on hunt recommendations. The Utah Wildlife Board consists of seven members appointed by the governor.

Until 1994, deer tags in Utah could be purchased over the counter with no limit by lottery-style drawing and Utah had in excess of 175,000 deer hunters (L. Cornicelli, personal communication, 2004). The winter of 1992–93 resulted in fawn losses approaching 100 percent and significant mortality in other age classes. Subsequently, a group of hunters made an organized effort to limit the number of hunters afield and thus the harvest rate. They were successful, and the legislature set a cap on the number of deer tags at 97,000, which was less than half the number of tags sold statewide under the unlimited system. Accordingly, harvest declined from more than 56,000 in 1992 to less than 30,000 in 1994 (UDWR 2001). This drastic reduction in buck harvest allowed more bucks to survive to the older age classes and also allowed the buck:doe ratio to increase.

Colorado

Colorado Division of Wildlife (CDOW) manages deer separately in fifty-four deer analysis units (DAU) statewide. Eighteen terrestrial wildlife biologists in the state are responsible for collecting survey and other management data.

Survey Data

Two types of deer surveys are conducted by helicopter: composition counts for sex and age ratios, and quadrat surveys to estimate density. The composition counts are completed before mid-January to assure that the count occurs before antlers are shed. This yields buck:doe ratios and fawn:doe ratios. Whitetails are tallied separately when they are observed, but most of the survey effort occurs in the western half of the state, which supports mainly mule deer. These surveys are conducted annually in fifteen to twenty DAUs in the state that are important or representative herds (J. Ellenberger, personal communication, 2004). Data collected in these units are assumed to represent other similar units. In addition, compositional counts are conducted in most other units every two to three years.

Density is estimated by surveying randomly located quadrats on winter range in nine DAUs (Kufeld et al. 1980). The same type of rotating system is used for this survey so that each unit is surveyed every two to three years. During the intervening years, the results in other nearby units are used to guide management decisions.

To gather more detailed information about important deer herds, four DAUs undergo intense scrutiny. Terrestrial wildlife biologists not only conduct annual composition and quadrat counts, but also obtain estimates of survival from radiocollared deer. The goal is to estimate the survival of adult does and fawns every year in those units, because this information is very important to accurately modeling a deer population (White and Bartmann 1997).

Harvest Data

An annual phone questionnaire is conducted to gather harvest data from hunters after the season (White and Bartmann 1997). The goal is to be able to estimate the actual harvest within 10 percent. Deer permits have been issued by limited-entry draw since 1999, so contact information is available for all hunters each year. However, it is not necessary to survey all hunters because statistical analysis can find the point at which additional questionnaires will not change the results substantially. That results in about 20–25 percent of the buck hunters and up to 50 percent of the doe hunters being surveyed (J. Ellenberger, personal communication, 2004).

There are no formal check stations operated, but some information is obtained from the deer captures conducted in the intensive units, such as age, body condition, presence of parasites, and physical measurements.

Other Data

Except for specific research projects, standard forage analyses or browse surveys are no longer conducted to monitor forage base. These types of data were collected by many agencies in the past when deer populations were higher and forage was experiencing the impact. Computer models are used to estimate and project population estimates and establish harvest objectives.

Hunt Recommendation Process

In 1999 Colorado changed from having most licenses sold over-the-counter to a limited-permit system. This statewide restriction on the number of hunters cut hunter pressure by about 40 percent (nearly 90 percent in some specific areas). Even though this dramatically limited hunter opportunity, most hunters seemed supportive when they saw positive changes in the demographics of the deer herds. Buck:doe ratios increased, and hunt success exceeded 40 percent in many areas.

Season structures are set every five years to maintain some consistency, but permit levels can change every year. The terrestrial wildlife biologists for each area are responsible for assembling the survey and harvest data, and formulating a hunt recommendation. The data are entered into a computer population model built in spreadsheet format, and the resulting population estimate is compared to the population goal for that DAU. If the population is within 90 percent of the goal and rising, or already over the goal, antlerless permits may be authorized to maintain the population within appropriate levels.

If antlerless permits are not needed, buck-only permits are authorized to meet a post-hunt buck:doe ratio target. This target or desired ratio is generally 15–20 bucks per 100 does (Colorado Division of Wildlife 2004). If the two-year average for any unit is below 15 bucks:100 does, biologists will take actions to improve that ratio. Some units may have different overall goals, depending on whether they are being managed as "premier" units. Units designated as premier are managed for lower hunter densities and strive for a ratio of 30 or more bucks per 100 does after the season. About 75 percent of the units in the eastern half of Colorado are premier units because of the predominance of private land. In western Colorado only about 25 percent are premier units.

After hunt recommendations are formulated by the terrestrial wildlife biologists, they are submitted to the regulations manager in the state office. The state office staff review the technical aspects of the recommendations and forward them to the commission for review and approval. The commission has nine regular members and two unofficial members (the directors of the Department of Natural Resources and Department of Agriculture).

Northern Mexico

Mexico's history of wildlife conservation is much different than that of the rest of the Southwest (see chapter 2). The states in Mexico are not divided into deer management units where hunters purchase hunting permits from a government agency and hunt on public land: there is no public land in Mexico. The administration of wildlife programs rests with the federal government rather than the states and is much more centralized than in the United States. The agencies responsible for various parts of natural resource management have frequently been, and continue to be, reorganized, and this creates a discontinuous and inconsistent administrative base for long-term programs. Traditionally, wildlife biologists in Mexico came from other fields of study, such as animal science, veterinary science, zoology, and general biology (R. Carrera, personal communication, 2003).

Wildlife conservation and habitat protection in Mexico is not based on money generated from the consumptive use of a few game species. Lacking a solid funding base, law enforcement is woefully inadequate to enforce the laws throughout much of the country. Instead, deer are sometimes conserved best when large private ranches protect deer against illegal harvest and manage them as a renewable resource with appropriate harvest levels. Unfortunately this is more the exception than the rule.

The government began to require the creation of management units, called Unidad de Manejo para la Conservacion de la Vida Silvestre (UMA), to provide a specific unit for the management of wildlife. The General Wildlife Law of 2000 (Ley General de Vida Silvestre del 2000) required the use of UMAs for the harvest of all living things (plants included). With the UMA come specific requirements for a management plan written by a wildlife professional to be authorized and monitored by the government (R. Carrera, personal communication, 2003).

Survey Data

Some private ranches that offer hunting opportunities have the incentive and funds to conduct surveys, but no government agency routinely surveys deer populations. Funds are normally not available for aerial surveys except on a few areas with sufficient financial resources. Methods used on most areas include spotlight counts, direct vehicle counts, pellet group surveys, and, in some cases, drive counts (Alcalá-Galván and Carrillo 1999). These surveys yield information that can be used to estimate deer density/abundance, buck:doe ratios, and fawn:doe ratios.

Surveys are an important part of the management plans submitted to fulfill the requirements of the UMA. Surveys conducted on UMAs are extremely variable in scientific rigor. Some deer survey data are collected by conducting extensive and well-designed surveys using the best available methodology and consistency, while other survey information is "collected" without leaving the ranch house. Since surveys of the deer population are the cornerstone of a deer management plan, it is critical that no shortcuts be taken during this phase.

Harvest Data

Before the year 2000, the annual hunting regulations (Calendario Cinegético Annual), which complemented the old Wildlife Law of 1952 (Ley Federal de Caza 1952), required every hunter to complete and mail a questionnaire providing harvest information, but these data were never compiled and used in management. Since the passage of the General Wildlife Law of 2000, no state or federal agency in Mexico organizes or sponsors a hunter questionnaire or field check stations to collect harvest data. Ranches with good management plans collect hunter harvest information from all hunters on the ranch and biological data from the deer harvested. Harvest data that represent the foundation of a good long-term data set include the number of hunters, number of deer harvested, days hunted, hunt success, body measurements, antler measurements, age, date of harvest, body condition, and number of deer seen (Alcalá-Galván and Carrillo 1999).

Hunt Recommendation Process

Deer can only be hunted on UMAs designated for that purpose. Deer permits for each UMA must be obtained from SEMARNAT (Secretariat of Environment and Natural Resources). Before passage of the General Wildlife Law of 2000, the Mexican government published annual hunting regulations (Calendario Cinegético Annual), but these did not change dramatically each year. These regulations contained the season dates and general rules for each species, but the lack of a strong law enforcement presence weakened their usefulness. After 2000 the Mexican government shifted responsibility to the landowners who are now required to provide the survey and harvest information, along with a management plan, to receive hunting permits. Some management plans are better than others, and there is no established system for ensuring that data are collected rigorously or that experienced people are writing the management plans. For those who are serious about collecting the necessary information, deer herds in many parts of Mexico can be managed to be more profitable than cattle grazing in the long term.

Antlerless Hunts

Harvesting bucks will not limit or reduce deer populations. The breeding strategy of deer assures that all does in estrus are located and impregnated during the breeding season. A single buck is physically capable of successfully breeding many does, so even after the removal of some bucks during the hunting season, nearly all does will still become pregnant. This is why closing a unit to buck harvest for a few years fails to increase the deer population. If does are not removed, then managers are not controlling deer abundance, and deer density attains a level that is governed by factors other than hunting.

Deer population management actually boils down to doe management, yet when populations need to be reduced, there is still sometimes opposition to harvesting antlerless deer (does). The most vocal opposition to doe hunts comes from hunters themselves. Some of the opposition is a holdover from the early days of wildlife management, because the protection of does was a cornerstone in the development of modern game management. During the course of Anglo settlement of America, game species were being killed much faster than they were reproducing naturally. Conservationists (mostly those hunting deer at the time) rallied for some level of protection, and some of the first game laws were restrictions on shooting does.

As deer populations rebounded throughout the country, deer managers were suddenly faced with a problem they had not anticipated—too many deer. Biologists in the field saw, and research confirmed, that deer forage plants were being damaged by overbrowsing, and they quickly recognized that deer populations had to be reduced. The problem was that they had done such a great job in selling the public on the protection of does, it was then difficult to convince hunters they now needed to kill does in great numbers. Regulated hunting of does in the Southwest for the purpose of population reduction began on the Kaibab Plateau in Arizona in 1929. The famous overpopulation of deer on the Kaibab overshadowed an identical situation occurring in the Black Canyon area in western New Mexico at about the same time (Warren 1997:121).

As states initiated doe hunts to control deer populations, they encountered strong public opposition. In some cases biologists recognized the obvious, but did not have the public support to remedy the situation. Some opposition to antlerless hunts seemed to stem from an innate feeling in the minds of some hunters that it is unsportsmanlike or a breach of masculinity to shoot female deer. This transspecies chivalry was poignantly illustrated in 1931 when members of the Border Game Protective Association of Deming, New Mexico, posted large red posters in nearby Black Canyon that read "Sportsmen Don't Kill Does" (Warren 1997:120).

Concern has also been raised that shooting female deer in the fall leads to a higher mortality of their fawns. Giuliano et al. (1999) concluded that shooting female deer in early November negatively affected white-tailed deer fawns in South Texas because orphaned fawns had smaller home ranges than did those that were not orphans. However, the survival rates and movements of orphaned versus unorphaned fawns were not significantly different, so it is doubtful there was a population-level effect. Other researchers have found that orphaning during fall hunts had no negative effect on the survival or development of fawns (Woodson et al. 1980, Demarais et al. 1988, Hölzenbein and Marchington 1992).

As deer populations became better managed, doe harvest was used not only for deer population irruptions, but also as a baseline management tool to maintain healthy deer populations. The reduction of deer populations through doe harvest in other parts of the country was found to result in more food available for the remaining deer in the population, which then had higher survival and a better reproductive rate than before the harvest. This has been demonstrated amply in the eastern whitetail populations, where deer managers were harvesting large numbers of does and the deer herds responded with increased reproduction.

It was assumed that all deer herds would respond in the same way, but management experiences showed just how different southwestern deer herds were. Between 1958 and 1968, 55,929 mule deer does and 10,501 whitetail does were harvested in Arizona (Arizona Game and Fish Department 2001a). New Mexico harvested an estimated 181,150 does between 1953 and 1982 (New Mexico Department of Game and Fish 1996). Deer populations in much of the Southwest declined during the 1960s and early 1970s, and there was widespread concern that the high doe harvests contributed to this decline, despite assurances to the contrary from deer biologists (Tallon 1965).

Research subsequently concluded that even a moderate increase in the female mortality rate could cause declines in southwestern deer populations where recruitment is chronically low. Short (1979) applied actual Arizona and New Mexico deer survey and harvest data to a computer model to ascertain the effects of various levels of doe harvest on deer populations. His simulations showed that increasing the harvest of does by just 5–10 percent would be enough to reduce a deer population that has limited productivity, such as those in desert or chaparral areas. This is supported by management data collected on the Three Bar Wildlife Area in Arizona, where the deer population declined dramatically with a small (less than 10 percent) increase in doe mortality (Smith et al. 1969). Considering the level of fawn recruitment that occurred during the time of the study, Smith et al. calculated that the population could only withstand an additional 4–7 percent mortality of does.

In the arid regions of the Southwest, the number of deer the habitat can support (carrying capacity) is extremely variable and heavily dependent on the amount and timing of rainfall. Because dry years and wet years occur erratically, deer are frequently over or under the carrying capacity. If the deer population is under carrying capacity (for example, in a wet year), deer are not limited by food resources, and a further reduction in the population with doe harvests will not result in the remaining deer being better nourished.

Although past doe harvests probably contributed to a general deer decline or, at the least, delayed recovery, they did not have a long-lasting effect on deer densities in the Southwest. With sufficient habitat and precipitation, deer populations have the capacity to increase rapidly.

Southwestern deer herds with relatively high and consistent fawn recruitment are those that are more likely to need periodic removal of does to keep the deer population at levels that are appropriate for the carrying capacity of the habitat. Likewise, deer herds that consistently migrate to lower elevations in the winter and subsist on a finite amount of browse on a restricted winter range are also candidates for antlerless hunts. In this situation the deer population must be maintained at a level commensurate with the amount of winter browse available. Wet weather patterns in the mid-1980s resulted in a population build-up in many areas, and antlerless hunts were used to manage the deer population at an appropriate level.

Sometimes antlerless hunts are needed not because the deer population increased, but because the habitat was reduced. An example of this occurred when a fire burned nearly half of the winter range on the west side of Arizona's Kaibab Plateau in 1996. The Arizona Game and Fish Department issued antlerless permits to reduce the deer population—a somewhat controversial move at the time because the Kaibab deer population was not at historically high levels. Wildlife managers, however, had an obligation to properly manage deer populations at an appropriate level to preserve the future health of their habitat. This situation may become more common as winter range disappears under the bulldozers of human encroachment.

Antlerless deer harvest is neither all good nor all bad. It is simply a tool available to the deer manager to influence the abundance of deer in an area. This tool is not needed very often in the Southwest because erratic precipitation keeps deer populations below carrying capacity most of the time. Unfortunately, when antlerless hunts are necessary, deep-seated and often unfounded opposition sometimes interferes with the timely reduction of a deer population. The decision-makers and the public must understand that antlerless harvest is a periodic part of responsible deer management.

Epilogue

EVERYONE enjoys seeing deer and deer sign when enjoying the outdoors. A camping, hiking, or hunting trip is much more enjoyable when people can tell others how many deer they saw. For many people, deer are the embodiment of nature itself. This popularity among people engaging in all forms of outdoor recreation creates a broad base of public interest in the status of deer populations.

A period of deer abundance in the Southwest throughout the mid-1980s followed several consecutive years of above-average precipitation, just as a deer decline in the 1990s followed a series of dry years. The causes of an even earlier period of abundance in the late 1950s are not as clear. We still have a lot to learn about what makes deer populations fluctuate. These dramatic changes in deer abundance are typical of the Southwest, but the extreme lows cause considerable consternation nonetheless. Years of aggressive fire suppression have allowed forests and chaparral to become thicker, shading out the growth of herbaceous plants and allowing woody browse plants to become decadent and unproductive. Invasion and introduction of exotic grasses, excessive timber harvest, overgrazing by livestock and wild ungulates, and weather patterns have changed the structure and plant composition of deer habitat.

Many factors contribute to deer population increase and decrease. In the short term, deer fluctuations are driven by the amount and timing of precipitation. When rainfall is at or above average, pregnant does have sufficient nutrition to produce healthy fawns born amid adequate hiding cover, and predators have plenty of other food items to keep them satiated. Below-average rainfall patterns result in weak and underweight fawns born on bare ground among hungry coyotes.

Long-term periods of deer abundance depend on the quantity and quality of deer habitat. Loss or degradation of habitat may result in deer population "highs" that are not as high and "lows" that are lower than historical fluctuations. There have certainly been dramatic habitat changes in the last century, some of which may have benefited deer (shrub encroachment and water distribution) and some that have not (lack of fire and human encroachment). It is mostly human activities that precipitated these changes, and so it is our responsibility to monitor and manage these and further habitat changes for the benefit of deer and all native species.

More could be done to rejuvenate the carrying capacity of deer habitat with prescribed burns, water development, and the manipulation of browse communities. Overgrazing, especially in low-elevation desert areas, results in heavy competition for browse and inadequate herbaceous growth for deer nutrition and fawn hiding cover (Fig. 47). Grazing allotments in the desert based primarily on browse (rather than grass) have the potential to seriously affect the ability of deer herds to withstand dry years and to recover from population lows. These issues are not easy to solve, but they must be addressed if we are serious about helping deer. Approaches like isolated predator control and deer transplants only divert money and energy away from the source of the real problem—habitat quality and quantity on a landscape scale. Focusing intensively on "deer" habitat is not selfish, single-species management. Conserving habitat on a large scale for deer benefits untold other, less charismatic species reliant on the health of the same vegetative communities.

Who will fight against housing developments, excessive off-road use, or inappropriate grazing in important deer habitat? It will be crucial for those interested in deer to join organizations that have the best interests of deer in mind. These groups serve as watchdogs and stand ready to spring into action whenever greed, politics, policies, or process gets in the way of proper management of deer and deer habitat. The most effective groups are those that focus on habitat acquisition, protection, and improvement. Southwestern deer herds in the future will need a unified and well-respected cadre of supporters willing to roll up their sleeves and get to work. As history has shown, this support base will be made up primarily of hunters. Deer conservation and hunting are intertwined because the funding base that supports wildlife management comes from hunting and also because of the active and passionate group of constituents that hunting fosters. However, support for deer conservation should not come only from hunters, for everyone has a stake in productive deer herds and healthy landscapes. More cooperation is

Figure 47. The future of deer conservation in the Southwest lies squarely in how well the habitat is managed to provide for the nutritional and cover requirements of the deer. *Photo by author*

needed between hunting organizations and preservation-oriented groups. They will never fully agree on all wildlife issues, but they have more in common than either side is willing to admit. By working together, we can mobilize the strengths of all organizations for the greater good of deer and many other species.

Habitat management plans directed at improving conditions for deer have become less common. As a result, everyone interested in deer will have to become more involved in land management decisions. Allotment management plans, forest plans, resource area management plans, and many other planning documents are continually being written and revised by all land management agencies. The public needs to become involved in the planning process to assure that the needs of deer are being addressed. If no one comes forward with concerns about the effect of land use decisions on deer, these issues will receive scant attention in land management plans.

Often land management agencies place too much emphasis on habitat for threatened and endangered species, while other important species are neglected. The squeaky species (or its advocates) gets the grease. This will require that the public learn about policies and procedures of government agencies. These policies and regulations are rarely appealing to anyone, but a working knowledge of these documents will be crucial to the future health of our deer herds. There are several excellent sources of information that provide ideas for improving deer habitat in the Southwest (Neff 1979, Kucera and Mayer 1999, Heffelfinger et al. 2006). Building a water catchment affecting hundreds of acres may seem more productive and beneficial to deer, but forcing a forest plan or allotment management plan to consider the needs of deer will benefit thousands or hundreds of thousands of acres for many years. No one said conservation was easy: nothing this important and worthwhile ever is.

Southwestern deer populations will continue to fluctuate with habitat quality and environmental conditions; our challenge is to wisely manage deer populations and to preserve and enhance their remaining habitat for future generations who will share our admiration for deer and will have an appreciation of those who came before them.

Appendix

Common and Scientific Names of Important Plants Eaten by Deer in the Southwest

Plant names in the following list are based on Lehr (1978).
Scientific names are given in parentheses following the common name.

Acacia, Catclaw (*Acacia greggii*)
Acacia, White Thorn (*Acacia constricta*)
Agave (*Agave* spp.)
Algerita (*Berberis haematocarpa*)
Anemone (*Anemone* sp.)
Antelope Bitterbrush (*Purshia tridentata*)
Apache Plume (*Fallugia paradoxa*)
Arrowweed (*Pluchea sericea*)
Aspen, Trembling (*Populus tremuloides*)
Aster (*Aster* spp.)
Ayenia (*Ayenia filiformis*)
Baccharis (*Baccharis* spp.)
Bahia (*Bahia pedata*)
Bearberry (*Arctostaphylos uva-ursi*)
Bigtooth Maple (*Acer grandidentatum*)
Birch, Water (*Betula occidentalis*)
Birdsbill Dayflower (*Commelina dianthifolia*)
Bitterbrush (*Purshia tridentata*)
Black Willow (*Salix nigra*)
Bladderpods (*Lesquerella* spp.)
Bluegrass (*Poa* spp.)
Borage (Boraginaceae family)
Brickellia (*Brickellia californica*)
Brittlebush (*Encelia farinosa*)
Brome, Red (*Bromus rubens*)
Broom Snakeweed (*Gutierrezia sarothrae*)
Buckbrush (*Ceanothus* spp.)
Buckwheat (*Eriogonum* spp.)
Buffalo Gourd (*Cucurbita foetidissima*)

Buffel Grass (*Pennisetum ciliare*)
Burro-weed (*Ambrosia dumosa*)
Cactus, Barrel (*Ferocactus* spp.)
Cactus, Prickly Pear (*Opuntia engelmannii*)
Careless Weed (*Amaranthus palmeri*)
Ceanothus (*Ceanothus* spp.)
Ceanothus, Desert (*Ceanothus greggii*)
Ceanothus, Fendler's (*Ceanothus fendleri*)
Cenizo (*Leucophyllum frutescens*)
Chamise (*Adenostoma fasciculatum*)
Chokecherry (*Prunus virginiana*)
Cholla (*Opuntia* spp.)
Cholla, Jumping (*Opuntia bigelovii*)
Cliffrose (*Cowania* [=*Purshia*] *mexicana*)
Clover (*Trifolium* spp.)
Condalia (*Condalia hookeri*)
Copperleaf (*Acalypha pringlei*)
Copperleaf, California (*Acalypha aspera*)
Cottonwood (*Populus* spp.)
Crown-beard (*Verbesina rothrockii*)
Daisy (*Erigeron* spp.)
Dalea (*Dalea* spp.)
Deer Weed (*Porophyllum gracile*)
Desert Hibiscus (*Hibiscus coulteri*)
Desert Vine (*Janusia gracilis*)
Desert Yaupon (*Schaefferia cuneifolia*)
Ditaxis (*Ditaxis neomexicana*)
Dogweed, Common (*Thymophylla* [=*Dyssodia*] *pentachaeta*)

Ebony, Texas (*Ebenopsis ebano*)
Elderberry (*Sambucus* spp.)
Elephant Tree (*Pachycormus discolor*)
Ephedra (*Ephedra aspera*)
Eriastrum (*Eriastrum* sp.)
Fairy Duster (*Calliandra eriophylla*)
False Indigo (*Amorpha californica*)
Fendlera or Fendlerbush (*Fendlera rupicola*)
Fescue (*Festuca* spp.)
Filaree (*Erodium cicutarium*)
Fir, Douglas (*Pseudotsuga menziesii*)
Fir, White (*Abies concolor*)
Flax (*Linum* sp.)
Fleabane (*Erigeron* spp.)
Galleta (*Pleuraphis rigida*)
Globemallow (*Sphaeralcea* spp.)
Goldeneye (*Viguiera* spp.)
Goldeneye, Heart-leaf (*Viguiera cordifolia*)
Goldeneye, Skeletonleaf (*Viguiera stenoloba*)
Gourd (*Cucurbita* spp.)
Grama Grass (*Bouteloua* spp.)
Grape or Canyon Grape (*Vitis arizonica*)
Grass, Brome (*Bromus* spp.)
Grass, Orchard (*Dactylis glomerata*)
Grass, Squirreltail (*Elymus elymoides*)
Grass Nuts (*Dichelostemma pulchellum*)
Green Sprangletop (*Leptochloa dubia*)
Ground Cherry, Fendler (*Physalis fendleri*)
Ground Cherry, Yellow (*Physalis viscosa*)
Guajillo (*Acacia beriandieri*)
Guayacan (*Porlieria angustifolia*)
Gumhead (*Gymnosperma glutinosum*)
Hackberry, Desert (*Celtis pallida*)
Hackberry, Mountain (*Celtis reticulata*)
Half-shrub Sundrop (*Calylophus serrulatus*)
Holly-leaf Buckthorn (*Rhamnus crocea*)
Horsemint or Giant Hyssop (*Agastache* spp.)
Huisache, Mexican (*Painteria leptophyllum*)
Indian Mallow (*Abutilon* sp.)
Indian Paint Brush (*Castilleja* spp.)
Ironwood (*Olneya tesota*)
James Bundleflower (*Desmanthus cooleyi*)
Javelina Bush (*Condalia ericoides*)
Jojoba (*Simmondsia chinensis*)
Juniper (*Juniperus* spp.)
Juniper, Alligator (*Juniperus deppeana*)
Juniper, One-seed (*Juniperus monosperma*)
Juniper, Utah (*Juniperus osteosperma*)
Kidney Wood (*Eysenhardtia polystachya*)

Lamb's Quarters (*Chenopodium fremontii*)
Larkspur, Tall (*Delphinium andesicola*)
Leatherweed Croton (*Croton pottsii*)
Lechuguilla (*Agave lechuguilla*)
Lehmann's Lovegrass (*Eragrostis lehmanniana*)
Lemonade Berry (*Rhus integrifolia*)
Lentisco (*Rhus lentii*)
Lewisia (*Lewisia* spp.)
Longstalk Greenthread (*Thelesperma longipes*)
Lupine (*Lupinus* spp.)
Madrone (*Arbutus arizonicus, A. glandulosa*)
Mallow (Malvaceae family)
Manzanita (*Arctostaphylos pungens*)
Manzanitia, Mission (*Arcostaphylos bicolor*)
Meadow Rue (*Thalictrum fendleri*)
Menodora (*Menodora* spp.)
Mesquite (*Prosopis glandulosa*)
Metastelma (*Metastelma arizonicum*)
Milkvetch or Locoweed (*Astragalus* spp.)
Milkvetch, Slender (*Astragalus recurvus*)
Milkwort (*Polygala* spp.)
Mimosa (*Mimosa* spp.)
Mimosa, Velvet-pod (*Mimosa dysocarpa*)
Mistletoe (*Phoradendron* spp.)
Monkey Flower (*Mimulus* spp.)
Morning Glory (*Ipomea* spp.)
Mountain Dandelion (*Agoseris glauca*)
Mountain Mahogany (*Cercocarpus* spp.)
Mountain Mahogany, Birchleaf (*Cercocarpus betuloides*)
Mutton Grass (*Poa fendleriana*)
Needleleaf Bluets (*Hedyotis acerosa*)
Nightshade (*Solanum* sp.)
Oak (*Quercus* spp.)
Oak, Arizona White (*Quercus arizonica*)
Oak, Emory (*Quercus emoryi*)
Oak, Gambel (*Quercus gambelii*)
Oak, Gray (*Quercus grisea*)
Oak, Mohr Shrub (*Quercus mohriana*)
Oak, Turbinella (*Quercus turbinella*)
Oak, Wavyleaf (*Quercus undulata*)
Ocotillo (*Fouquieria splendens*)
Oregon Grape (*Berberis repens*)
Palo Verde (*Cercidium* spp.)
Paperflower (*Psilotrophe* spp.)
Penstemon (*Penstemon* spp.)
Pigweed (*Amaranthus almeri*)

Pine, Pinyon (*Pinus edulis*)
Pine, Ponderosa (*Pinus ponderosa*)
Plantain (*Plantago insularis*)
Poppy (*Eschscholtzia* spp.)
Prickly Poppy (*Argemone* spp.)
Purslane (*Portulaca oleracea*)
Rabbitbrush (*Chrysothamnus* spp.)
Rabbitbrush, Rubber (*Chrysothamnus nauseosus*)
Ratany (*Krameria erecta*)
Redberry (*Rhamnus* sp.)
Rose (*Rosa* spp.)
Sage, Black (*Salvia mellifera*)
Sage, White (*Salvia apiana*)
Sagebrush, Big (*Artemisia tridentata*)
Sagebrush, Coastal (*Artemisia californica*)
Sagebrush, Sand (*Artemisia filifolia*)
Sagebrush, White (*Artemisia ludoviciana*)
Sagewort, Fringed (*Artemisia frigida*)
Saltbush (*Atriplex* spp.)
Saltbush, Four-wing (*Atriplex canescens*)
Sedge (*Carex* spp.)
Serviceberry (*Amelanchier oreophila*)
Silktassel (*Garrya wrightii*)
Skunkbush or Three-leaf Sumac (*Rhus trilobata*)
Snowberry (*Symphoricarpos* spp.)
Snowberry, Mountain (*Symphoricarpos oreophilus*)
Sotol (*Dasylirion leiophyllum*)
Sowthisle (*Sonchus oleraceus*)
Spikebent (*Agrostis exarata*)
Spurge (*Euphorbia* spp.)
Sumac, Evergreen (*Rhus virens*)
Sumac, Laurel (*Malosma laurina*)
Sumac, Littleleaf (*Rhus microphylla*)
Sumac, Mearns (*Rhus choriophylla*)
Sumac, Sugar (*Rhus ovata*)
Sumac, Threeleaf or Skunkbush (*Rhus trilobata*)
Sweetclover, White (*Melilotus albus*)
Talinum (*Talinum* spp.)
Tansymustard (*Descurainia* spp.)
Tidestromia (*Tidestromia* spp.)
Vervain (*Verbena bipinnatifida*)
Vetch, Deer (*Lotus* spp.)
Wheatgrass (*Agropyron* spp.)
Wild Rose (*Rosa* spp.)
Wire lettuce (*Stephanomeria pauciflora*)
Woodrush, Small-flowered (*Luzula parviflora*)
Yucca (*Yucca* spp.)

Literature Cited

Alcalá-Galván, C. H., and E. Enriquez-Carrillo. 1999. Manejo y aprovechamiento de venados. Folleto Tecnico 3. Campo Experimental Carbo, Instituto Nacional de Investigaciones Forestales, Agrícolas y Pecuarias. 24 pp.

Alessio-Robles, A. 1959. Memorial y proyectos de leyes sobre la fauna silvestre y el ejercicio de la caza en Mexico. Mexico City.

Alexander, A. 1997. I'm important! Send me back! Arizona Wildlife Views. Arizona Game and Fish Department, Phoenix. December.

Alexy, K. J., J. W. Gassett, D. A. Osborn, and K. V. Miller. 2001. Remote monitoring of scraping behaviors of a wild population of white-tailed deer. Wildlife Society Bulletin 29:873–78.

Allen, J. A. 1893. List of mammals and birds collected in northeastern Sonora and northwestern Chihuahua, Mexico, on the Lumholtz archaeological expedition, 1890–92. Bulletin of the American Museum of Natural History 5:27–42.

Allen, R. E., and D. R. McCullough. 1976. Deer-car accidents in southern Michigan. Journal of Wildlife Management 40:317–25.

Allred, K. W. 1996. Vegetative changes in New Mexico rangelands. New Mexico Journal of Science 36:168–231.

Alsheimer, C. J. 1999. Hunting whitetails by the moon. Krause Publications, Iola, Wisconsin. 251 pp.

Alvarez-Cárdenas, S., S. Gallina, P. Galina-Tessaro, and R. Dominguez-Cadena. 1999. Habitat availability for the mule deer (Cervidae) population in the relictual oak-pine forest from Baja California Sur, Mexico. Tropical Zoology 12:67–78.

Andelt, W. F., T. M. Pojar, and L. W. Johnson. 2004. Long-term trends in mule deer pregnancy and fetal rates in Colorado. Journal of Wildlife Management 68:542–49.

Anderson, A. E. 1981. Morphological and physiological characteristics. Pp. 27–98 *in* O. C. Wallmo, ed., Mule and black-tailed deer of North America. Wildlife Management Institute and University of Nebraska Press, Lincoln. 605 pp.

Anderson, A. E., L. G. Frary, and R. H. Stewart. 1964. A comparison of three morphological attributes of mule deer from the Guadalupe and Sacramento Mountains, New Mexico. Journal of Mammalogy 45:48–53.

Anderson, A. E., D. E. Medin, and D. C. Bowden. 1974. Growth and morphometry of the carcass, selected bones, organs, and glands of mule deer. Wildlife Monograph No. 39. Wildlife Society. 122 pp.

Anderson, A. E., W. A. Snyder, and G. W. Brown. 1965. Stomach content analyses related to condition in mule deer, Guadalupe Mountains, New Mexico. Journal of Wildlife Management 29:352–66.

Anderson, A. E., W. A. Snyder, and G. W. Brown. 1970. Indices of reproduction and survival in female mule deer, Guadalupe Mountains, New Mexico. Southwestern Naturalist 15:29–36.

Anderson, A. E., and O. C. Wallmo. 1984. *Odocoileus hemionus*. Mammalian Species Special Publication No. 219. American Society of Mammalogists. 9 pp.

Anthony, R. G. 1976. Influence of drought on diets and numbers of desert deer. Journal of Wildlife Management 40:140–44.

Anthony, R. G., and N. S. Smith. 1974. Comparison of rumen and fecal analysis to describe deer diets. Journal of Wildlife Management. 38:535–40.

Anthony, R. G., and N. S. Smith. 1977. Ecological relationships between mule deer and white-tailed deer in southeastern Arizona. Ecological Monographs 47:255–77.

Arizona Game and Fish Department. 1993. Species management guidelines. Arizona Game and Fish Department, Phoenix.

Arizona Game and Fish Department. 1994. Game survey and harvest data summary. Arizona Game and Fish Department, Phoenix.

Arizona Game and Fish Department. 1997. Wildlife water developments in Arizona: A technical review. Arizona Game and Fish Department, Phoenix. Briefing document. 74 pp.

Arizona Game and Fish Department. 2000a. Big game annual performance report, Federal Aid Project W-53-M. Arizona Game and Fish Department, Phoenix.

Arizona Game and Fish Department. 2000b. Alternative mule deer management plan. Arizona Game and Fish Department, Phoenix.

Arizona Game and Fish Department. 2001a. Hunt Arizona. Arizona Game and Fish Department, Phoenix.

Arizona Game and Fish Department. 2001b. Species management guidelines. Arizona Game and Fish Department, Phoenix.

Atkeson, T. D., and R. L. Marchington. 1982. Forehead glands in white-tailed deer. Journal of Mammalogy 63:613–17.

Atkeson, T. D., R. L. Marchington, and K. V. Miller. 1988. Vocalizations of white-tailed deer. American Midland Naturalist 120:194–200.

Atkeson, T. D., V. F. Nettles, R. L. Marchington, and W. V. Branan. 1988. Nasal glands in the Cervidae. Journal of Mammalogy 69:153–56.

Audubon, J. J. 1989. Quadrupeds of North America. Wellfleet Press, Secaucus, New Jersey. 440 pp.

Baber, D. W. 1987. Gross antler anomaly in a California mule deer: The "cactus" buck. Southwestern Naturalist 32:404–406.

Bahre, C. J. 1991. A legacy of change: Historic human impact on vegetation in the Arizona borderlands. University of Arizona Press, Tucson. 231 pp.

Bahti, T. 1973. Southwestern Indian tribes. KC Publications, Las Vegas, Nevada. 72 pp.

Bailey, B. 1964. Serological surveys of brucellosis and leptospirosis in Nebraska deer and antelope. Unpublished report. 16 pp.

Bailey, V. 1931. Mammals of New Mexico. North American Fauna No. 53. U.S. Department of Agriculture, Bureau of Biological Survey, Washington, D.C. 412 pp.

Baird, S. F. 1859. Mammals of the Mexican boundary: Part II. Pp. 1–55 *in* W. H. Emory, Report on the U.S. and Mexican boundary survey. Washington, D.C. 764 pp.

Banks, W. J. 1966. Antlerogenesis and osteoporosis. Master's thesis, Colorado State University, Fort Collins. 208 pp.

Banks, W. J. 1974. The ossification process of the developing antler in the white-tailed deer (*Odocoileus virginianus*). Calcium Tissue Research 14:257–74.

Barrows, P., and J. Holmes. 1990. Colorado's wildlife story. Colorado Division of Wildlife, Denver. 450 pp.

Barsch, B. K. 1977. Distribution of the Coues deer in pinyon stands after a wildfire. Master's thesis, University of Arizona, Tucson. 52 pp.

Bartel, R. A., and M. W. Brunson. 2003. Effect of Utah's coyote bounty program on harvester behavior. Wildlife Society Bulletin 31:736–43.

Bartmann, R. M., L. H. Carpenter, R. A. Garrott, and D. C. Bowden. 1986. Accuracy of helicopter counts of mule deer in pinyon-juniper woodland. Wildlife Society Bulletin 14:356–63.

Bartmann, R. M., G. C. White, and L. H. Carpenter. 1992. Compensatory mortality in a Colorado mule deer population. Wildlife Monograph No. 121. Wildlife Society. 39 pp.

Batman, R. 1984. James Pattie's West: The dream and the reality. University of Oklahoma Press, Norman. 378 pp.

Beale, D. M., and N. W. Darby. 1991. Diet composition of mule deer in mountain brush habitat of southwestern Utah. Federal Aid Final Report, W-65-R, Job B-9, Publication No. 91–14. Utah Division of Wildlife Resources. 70 pp.

Beasom, S. L., and E. P. Wiggers. 1984. A critical assessment of white-tailed and mule deer productivity. Pp. 68–79 *in* P. R. Krausman and N. S. Smith, eds., Deer in the southwest: A workshop. Arizona Cooperative Wildlife Research Unit and University of Arizona, Tucson. 131 pp.

Behymer, D., D. Jessup, K. Jones, C. E. Franti, H. Riemann, and A. Bahr. 1989. Antibodies to nine infectious disease agents in deer from California. Journal of Zoo and Wildlife Medicine 20:297–306.

Beier, P., and D. R. McCullough. 1990. Factors influencing white-tailed deer activity patterns and habitat use. Wildlife Monograph No. 109. Wildlife Society. 51 pp.

Bellantoni, E. S., P. R. Krausman, and W. W. Shaw. 1993. Desert mule deer use of an urban environment. Transactions of the North American Wildlife and Natural Resource Conference 58:92–101.

Beltran, E. 1966. La administración de la fauna silvestre. Pp. 225–59 *in* Instituto Mexicano de Recursos Naturales Renovables, Mesas redondas sobre problemas de caza y pesca deportivas en México. Ediciones del Instituto Mexicano de Recursos Naturales Renovables, Mexico City.

Bender, L. C. 2003. Identification of factors limiting mule deer populations and development of corrective management strategies along the Upper Santa Fe Trail, New Mexico. Annual Report for Santa Fe Trail Mule Deer Adaptive Management Partnership. U.S. Geological Survey, New Mexico Cooperative Fish and Wildlife Research Unit, Las Cruces, New Mexico. 54 pp.

Bender, L. C., H. Li, B. C. Thompson, P. C. Morrow, and R. Valdez. 2003. Infectious disease survey of gemsbok in New Mexico. Journal of Wildlife Diseases 39:772–78.

Bischoff, A. I. 1958. Productivity in some California deer herds. California Fish and Game 44:253–59.

Bleich, V. C., and B. M. Pierce. 2001. Accidental mass mortality of migrating mule deer. Western North American Naturalist 61:124–25.

Boeker, E. L., V. E. Scott, H. G. Reynolds, and B. A. Donaldson. 1972. Seasonal habits of mule deer in southwestern New Mexico. Journal of Wildlife Management 36:56–63.

Bone, T. L. 1987. Trends in mule deer age and antler characteristics in the Trans-Pecos. Texas Parks and Wildlife Department, Austin. Unpublished report. 7 pp.

Bone, T. L. 2003. Mule deer harvest recommendations. Federal Aid Report W-127-R-11, Project 3. Texas Parks and Wildlife Department, Austin.

Bone, T. L. 2004. Mule deer harvest recommendations. Federal Aid Report W-127-R-12, Project 3. Texas Parks and Wildlife Department, Austin. 22 pp.

Bowyer, R. T. 1986a. Habitat selection by southern mule deer. California Fish and Game 72:153–69.

Bowyer, R. T. 1986b. Antler characteristics as related to social status of male southern mule deer. Southwestern Naturalist 31:289–98.

Bowyer, R. T. 1991. Timing of parturition and lactation in southern mule deer. Journal of Mammalogy 72:138–45.

Bowyer, R. T., and V. C. Bleich. 1984. Distribution and taxonomic affinities of mule deer, *Odocoileus hemionus,* from Anza-Borrego Desert State Park, California. California Fish and Game 70:53–57.

Bristow, K. L. 1992. Effects of simulated hunting during the rut on reproduction and movement of Coues white-tailed deer. Master's thesis, University of Arizona, Tucson. 39 pp.

Brown, D. E. 1984. In search of the bura deer. Pp. 42–44 *in* P.R. Krausman and N.S. Smith, eds., Deer in the southwest: A workshop. Arizona Cooperative Wildlife Research Unit and University of Arizona, Tucson. 131 pp.

Brown, D. E. 2001. Turning on the worm that turns. Arizona Wildlife Views Magazine. May–June.

Brown, D. E., ed. 2002. The wolf in the Southwest: The making of an endangered species. High-Lonesome Books, Silver City, New Mexico. 195 pp.

Brown, D. E., and R. S. Henry. 1981. On relict occurrences of white-tailed deer within the Sonoran desert in Arizona. Southwestern Naturalist 26:147–52.

Brown, G. W. 1961. Investigations of big game and ranges—deer food habits study. Federal Aid Final Report W-75-R-8, Work Plan 14, Job 5. New Mexico Department of Game and Fish, Santa Fe. 48 pp.

Brown, M. T. 1984. Habitat selection by Coues white-tailed deer in relation to grazing intensity. Pp. 1–6 *in* P. R. Krausman and N. S. Smith, eds., Deer in the southwest: A workshop. Arizona Cooperative Wildlife Research Unit and University of Arizona, Tucson. 131 pp.

Brownlee, S. 1971. Conception rates and breeding potential of desert mule deer. Federal Aid Project Number W-48-D-22. Texas Parks and Wildlife, Austin. 9 pp.

Brownlee, S. 1979. Water development for desert mule deer. Texas Parks and Wildlife Department, Austin.

Brownlee, S. 1981. Mule deer herd productivity. Final report, Federal Aid Project Number W-109-R-3. Texas Parks and Wildlife, Austin. 37 pp.

Broyles, B. 1995. Desert wildlife water developments: Questioning use in the Southwest. Wildlife Society Bulletin 23:663–75.

Broyles, B. 1997. Wildlife water-developments in southwestern Arizona. Journal of Arizona-Nevada Academy of Sciences 30:30–42.

Brunetti, O., and H. Cribbs. 1971. California deer deaths due to massive infestation by the louse (*Linognathus africanus*). California Fish and Game 57:138–53.

Bubenik, A. B. 1983. Behavioral aspects of antlerogenesis. Pp. 389–450 *in* R. D. Brown, ed., Antler development in Cervidae. Caesar Kleberg Wildlife Research Institute, Kingsville, Texas.

Bubenik, A. B. 1990. Epigenetical, morphological, physiological, and behavioural aspects of evolution of horns, pronghorns, and antlers. Pp. 3–113 *in* G. A. Bubenik and A. B. Bubenik, eds., Horns, pronghorns, and antlers: Evolution, morphology, physiology, and social significance. Springer-Verlag, New York. 562 pp.

Bubenik, G. A. 1983. Shift of seasonal cycle in white-tailed deer by oral administration of melatonin. Journal of Experimental Zoology 225:155–56.

Bubenik, G. A. 1990a. Neuroendocrine regulation of the antler cycle. Pp. 265–97 *in* G. A. Bubenik and A. B. Bubenik, eds., Horns, pronghorns, and antlers: Evolution, morphology, physiology, and social significance. Springer-Verlag, New York. 562 pp.

Bubenik, G. A. 1990b. The role of the nervous system in the growth of antlers. Pp. 339–58 *in* G. A. Bubenik and A. B. Bubenik, eds., Horns, pronghorns, and antlers: Evolution, morphology, physiology, and social significance. Springer-Verlag, New York. 562 pp.

Bubenik, G. A. 1993. Morphological differences in the antler velvet of Cervidae. Pp. 56–64 *in* N. Ohtaishi and H. I. Sheng, eds., Deer of China. Elsevier Science Publishers, Amsterdam. 164 pp.

Bubenik, G. A. 1994. A coat of many functions. Pp. 41–49 *in* D. Gerlach, S. Atwater, and J. Schnell, eds., Deer. Stackpole Books, Mechanicsburg, Pennsylvania. 384 pp.

Bubenik, G. A. 1996. Morphological investigations of the winter coat in white-tailed deer: Differences in skin, glands, and hair structure of various body regions. Acta Theriologica 41:73–82.

Bubenik, G. A., J. M. Morris, D. Schams, and C. Klaus. 1982. Photoperiodicity and circannual levels of LH, FSH, and testosterone in normal and castrated male, white-tailed deer. Journal of Physiological Pharmacology 60:788–93.

Bubenik, G. A., and P. S. Smith. 1987. Circadian and circannual rhythm of melatonin in adult male white-tailed deer: The effect of oral administration of melatonin. Journal of Experimental Zoology 241:81–89.

Bubenik, G. A., P. S. Smith, and D. Schams. 1986. The effect of orally administered melatonin on the seasonality of deer pelage exchange, antler development, LH, FSH, prolactin, testosterone, T_3, T_4, cortisol, and alkaline phosphatase. Journal of Pineal Research 3:331–49.

Burt, W. H. 1938. Faunal relationships and geographic distribution of mammals in Sonora, Mexico. Miscellaneous Publications No. 39. Museum of Zoology, University of Michigan, Ann Arbor.

Buss, I. O., and F. H. Harbert. 1950. Relation of moon phases to the occurrence of mule deer at a Washington salt lick. Journal of Mammalogy 31:426–29.

Cantu, R., and C. Richardson. 1997. Mule deer management in Texas. Texas Parks and Wildlife Department, Austin. 22 pp.

Carmony, N. B. 1985. Elliot Coues and the white-tailed deer of the Southwest. Arizona Wildlife. Special edition of Arizona Wildlife Views. Arizona Game and Fish Department, Phoenix.

Carmony, N. B. 2002. The Grand Canyon deer drive of 1924: The accounts of Will C. Barnes and Mark E. Musgrave. Journal of Arizona History 43:41–64.

Carpenter, L. H., D. Lutz, and D. Weybright. 2003. Mule deer data types, uses, analysis, and summaries. Pp. 163–75 *in* J. C. deVos Jr., M. R. Conover, and N. E. Headrick, eds., Mule deer conservation: Issues and management strategies. Berryman Institute Press, Utah State University, Logan. 240 pp.

Carr, S. M., S. W. Ballinger, J. N. Derr, L. H. Blankenship, and J. W. Bickham. 1986. Mitochondrial DNA analysis of hybridization between sympatric white-tailed deer and mule deer in west Texas. Proceedings of the National Academy of Science 83:9576–80.

Carr, S. M., and G. A. Hughes. 1993. Direction of introgressive hybridization between species of North American deer (*Odocoileus*) as inferred from mitochondrial cytochrome-b sequences. Journal of Mammalogy 74:331–42.

Carrel, W. K., R. A. Ockenfels, R. E. Schweinsburg. 1999. An evaluation of annual migration patterns of the Paunsaugunt mule deer herd between Utah and Arizona. Technical Report No. 29. Arizona Game and Fish Department, Phoenix.

Cathey, J. C., J. W. Bickham, and J. C. Patton. 1998. Introgressive hybridization and nonconcordant evolutionary history of maternal and paternal lineages in North American deer. Evolution 52:1224–29.

Celentano, R. R., and J. R. Garcia. 1984. The burro deer herd management plan. California Department of Fish and Game, Sacramento. 90 pp.

Chomel, B. B., M. L. Carniciu, R. W. Kasten, P. M. Castelli, T. M. Work, and D. A. Jessup. 1994. Antibody prevalence of eight ruminant infectious diseases in California mule and black-tailed deer (*Odocoileus hemionus*). Journal of Wildlife Diseases 30:51–59.

Clark, F. D. 1953. A study of the behavior and movements of the Tucson Mountain mule deer. Master's thesis, University of Arizona, Tucson. 111 pp.

Clutton-Brock, T. H. 1982. The functions of antlers. Behaviour 79:108–25.

Collins, W. B., and P. J. Urness. 1981. Habitat preferences of mule deer as rated by pellet-group distributions. Journal of Wildlife Management 45:969–75.

Colorado Division of Wildlife. 2004. Five-year big game season structure (BGSS) 2005–2009 semi-final policy recommendations, May 11, 2004. Colorado Division of Wildlife, Fort Collins. 9 pp.

Comer, J. A., D. E. Stallknecht, and V. F. Nettles. 1995. Incompetence of white-tailed deer as amplifying host of vesicular stomatitis virus for *Lutzomyia shannoni* (Diptera: Psychodidae). Journal of Medical Entomolgy 32:738–40.

Connolly, G. E. 1981. Assessing populations. Pp. 287–345 *in* O. C. Wallmo, ed., Mule and black-tailed deer of North America. Wildlife Management Institute and University of Nebraska Press, Lincoln. 605 pp.

Connolly, G. E., and W. M. Longhurst. 1975. The effects of control on coyote populations: A simulation model. Cooperative Extension Service Bulletin No. 1872. University of California, Berkeley. 37 pp.

Conover, M. R., W. C. Pitt, K. K. Kessler, T. J. DuBow, and W. A. Sanborn. 1995. Review of human injuries, illness, and economic losses caused by wildlife in the United States. Wildlife Society Bulletin 23:407–14.

Cooper, C. F. 1960. Changes in vegetation, structure, and growth of southwestern pine forests since white settlement. Ecological Monographs 30:129–64.

Cordell, L. 1997. Archaeology of the Southwest. Academic Press, San Diego, California. 522 pp.

Couvillion, C. E., E. W. Jenney, J. E. Pearson, and M. E. Coker. 1980. Survey for antibodies to viruses of bovine virus diarrhea, bluetongue, and epizootic hemorrhagic disease in hunter-killed mule deer in New Mexico. Journal of the American Veterinary Medical Association 177:790–91.

Cowan, I. M. 1933. The mule deer of southern California and northern lower California as a recognizable race. Journal of Mammalogy 14:326–27.

Cowan, I. M. 1936. Distribution and variation in deer (Genus *Odocoileus*) of the Pacific Coastal Region of North America. California Fish and Game 22:155–246.

Cowan, I. M. 1961. What and where are the mule and black-tailed deer? Pp. 334–59 *in* W.P. Taylor, ed., The deer of North America. Stackpole, Harrisburg, Pennsylvania. 668 pp.

Cronin, M. A. 1991a. Mitochondrial-DNA phylogeny of deer (Cervidae). Journal of Mammalogy 72:533–66.

Cronin, M. A. 1991b. Mitochondrial and nuclear genetic relationships of deer (*Odocoileus* spp.) in western North America. Canadian Journal of Zoology 69:1270–79.

Cronin, M. A. 1997. Systematics, taxonomy, and the endangered species act: The example of the California gnatcatcher. Wildlife Society Bulletin 25:661–66.

Cronin, M. A., and V. C. Bleich. 1995. Mitochondrial DNA variation among populations and subspecies of mule deer in California. California Fish and Game 81:45–54.

Cronin, M. A., E. R. Vyse, and D. G. Cameron. 1988. Genetic relationships between mule deer and white-tailed deer in Montana. Journal of Wildlife Management 52:320–28.

Cunningham, S. C., L. A. Haynes, C. Gustavson, and D. D. Haywood. 1995. Evaluation of the interaction between lions and cattle in the Aravaipa-Klondyke area of southeast Arizona. Technical Report No. 17. Arizona Game and Fish Department, Phoenix. 64 pp.

Cutting, H., T. G. Pearson, T. W. Tomlinson, and J. B. Burnham. 1924. Report of Kaibab deer investigating committee. U.S. Forest Service Files. 32 pp.

D'Antonio, C. M., and P. M. Vitousek. 1992 Biological invasions by exotic grasses, the grass/fire cycle, and global change. Annual Review of Ecological Systems 23:63–87.

Davidson, W. R., and V. F. Nettles. 1997. Field manual of wildlife diseases in the southeastern United States. Southeastern Cooperative Wildlife Disease Study, Athens, Georgia. 417 pp.

Davis, G. P., Jr. 2001. Man and wildlife in Arizona: The American exploration period, 1824–65. Edited by N. B. Carmony and D. E. Brown. Arizona Game and Fish Department, Phoenix. 225 pp.

Day, G. I. 1960. Carrying capacity of various vegetation types for white-tailed deer. Federal Aid Project W-78-R-4, WP5, J6. Arizona Game and Fish Department, Phoenix. 20 pp.

Day, G. I. 1964. An investigation of white-tailed deer (*Odocoileus virginianus couesi*) forage relationships in the Chiricahua Mountains. Master's thesis, University of Arizona, Tucson. 101 pp.

Day, G. I. 1980. Characteristics and measurements of captive hybrid deer in Arizona. Southwestern Naturalist 25:434–38.

Demarais, S., R. E. Zaiglin, and D. A. Barnett. 1988. Physical development of orphaned white-tailed deer fawns in southern Texas. Journal of Range Management 41:340–42.

Derr, J. N. 1990. Genetic interactions between two species of North American deer, *Odocoileus virginianus* and *Odocoileus hemionus*. Ph.D. dissertation, Texas A&M University, College Station. 111 pp.

Desai, J. H. 1962. A study of the reproductive pattern in the desert mule deer, *Odocoileus hemionus crooki* (Mearns), related to range conditions in the Trans-Pecos region of Texas. Master's thesis, Texas A&M University, College Station. 61 pp.

DeStefano, S., S. L. Schmidt, and J. C. deVos Jr. 2000. Observations of predator activity at wildlife water developments in southern Arizona. Journal of Range Management 53:255–58.

DeYoung, C. A. 1989. Aging live white-tailed deer on southern ranges. Journal of Wildlife Management 53:519–23.

DeYoung, D. W., P. R. Krausman, L. E. Weiland, and R. C. Etchburger. 1993. Baseline ABRs in mountain sheep and desert mule deer. International Congress on Noise as a Public Health Problem 6:251–53.

Dickinson, T. G. 1978. Seasonal movements, home ranges and home range use of desert mule deer in Pecos County, Texas. Master's thesis, Sul Ross State University, Alpine, Texas. 156 pp.

Dickinson, T. G., and G. W. Garner. 1979. Home range and movements of desert mule deer in southwestern Texas. Proceedings of the Annual Conference of Southeastern Fish and Wildlife Agencies 33:267–78.

Diem, K. L. 1958. Fertile antlered mule deer doe. Journal of Wildlife Management 22:449.

Dixon, J. S. 1934. A study of the life history and food habits of the mule deer in California. California Fish and Game 20:1–142.

Donaldson, J. C., and J. K. Doutt. 1965. Antlers in female white-tailed deer: A four-year study. Journal of Wildlife Management 29:699–705.

Dorsett, P. H. 1985. Fundamentals of animal virology. Pp. 718–57 *in* B. A. Freeman, ed., Textbook of microbiology. W. B. Sanders, Philadelphia, Pennsylvania. 1038 pp.

Drew, M. L., D. A. Jessup, A. A. Burr, and C. E. Franti. 1992. Serologic survey for brucellosis in feral swine, wild ruminants, and black bear of California, 1977 to 1989. Journal of Wildlife Diseases 28:355–63.

Dutton, B. P. 1975. Navajo and Apache: The Athabascan people. Prentice-Hall, Englewood Cliffs, New Jersey. 97 pp.

Eberhardt, L. E., and H. C. Pickens. 1979. Homing in mule deer. Southwestern Naturalist 24:705–706.

Eisenberg, J. F. 1987. The evolutionary history of the Cervidae with special reference to the South American radiation. Pp. 61–64 *in* C. M. Wemmer, ed., Biology and management of Cervidae. Smithsonian Institution Press, Washington, D.C. 577 pp.

Elder, J. B. 1954. Notes on summer water consumption by desert mule deer. Journal of Wildlife Management 18:540–41.

Elder, J. B. 1956. Watering patterns of some desert game animals. Journal of Wildlife Management 20:368–78.

Elton, C. 1930. Animal ecology and evolution. Clarendon Press, London. 96 pp.

Etheredge, O. F. 1949. Trapping and transplanting of mule deer. Final report, Federal Aid in Wildlife Restoration Act. Texas Game, Fish and Oyster Commission, Austin. 10 pp.

Ezcurra, E., and S. Gallina. 1981. Biology and population dynamics of white-tailed deer in northwestern Mexico. Pp. 79–107 *in* P. F. Ffolliot and S. Gallina, eds., Deer biology, habitat requirements, and management in western North America. Institute de Ecologio Publication No. 9. Hermosillo, Sonora, Mexico. 238 pp.

Ezcurra, E., S. Gallina, and P. F. Ffolliott. 1980. Manejo combinado del venado y el ganado en el Norte de Mexico. Rangelands 2:208–209.

Ferg, A., and W. B. Kessel. 1987. Subsistence. Pp. 49–86 *in* A. Ferg, ed., Western Apache material culture: The Goodwin and Guenther collections. University of Arizona Press, Tucson. 205 pp.

Findley, J. S., A. H. Harris, D. E. Wilson, and C. Jones. 1975. Mammals of New Mexico. University of New Mexico Press, Albuquerque. 360 pp.

Flader, S. L. 1974. Thinking like a mountain. University of Nebraska Press, Lincoln. 284 pp.

Fowler, M. E. 1983. Plant poisoning in free-living wild animals: A review. Journal of Wildlife Diseases 19:34–43.

Fox, K. B., and P. R. Krausman. 1994. Fawning habitat of desert mule deer. Southwestern Naturalist 39:269–75.

Frels, Jr., D. B., E. Fuchs, and W. E. Armstrong. 2002. Genetic and environmental interaction in white-tailed deer. Federal Aid Project W-127-R-9, Job 96. Final Report. Texas Parks and Wildlife Department, Austin.

Frick, C. 1937. Horned ruminants of North America. Bulletin of the American Museum of Natural History 69:1–669.

Frison, G. C. 1986. Prehistoric, Plains-Mountain, large-mammal, communal hunting strategies. Pp. 177–223 *in* M. H. Nitecki and D. V. Nitecki, eds., The evolution of human hunting. Plenum Press, New York. 464 pp.

Galindo-Leal, C. 1992. Overestimation of deer densities in Michilia Biosphere Reserve, Durango, Mexico. Southwestern Naturalist 37:209–12.

Galindo-Leal, C., A. Morales, and M. Weber. 1994. Utilizacion de habitat, abundancia y dispersión del venado de Coues: Un experimento seminatural. Pp. 315–32 *in* C. Vaughan and M. A. Rodriguez, eds., Ecologia y manejo del venado cola blanca en México y Costa Rica. Editorial de la Universidad Nacional, Mexico City. 462 pp.

Galindo-Leal, C., and M. Weber. 1998. El venado de la Sierra Madre Occidental. Ediciones Culturales, S.A. de C.V., Mexico City. 272 pp.

Gallina, S. 1991. El venado cola blanca en la Reserva La Michilia. Vida Silvestre 1(1):18–22.

Gallina, S. 1994a. Dinamica poblacional y manejo de la población del venado cola blanca en la reserva de la biosfera la Michilia, Durango, Mexico. Pp. 207–34 *in* C. Vaughan and M. A. Rodriguez, eds., Ecologia y manejo del venado cola blanca en México y Costa Rica. Editorial de la Universidad Nacional, Mexico City. 462 pp.

Gallina, S. 1994b. Uso del habitat por el venado cola blanca en la reserva de la biosfera la Michilia, Mexico. Pp. 299–314 *in* C. Vaughan and M. A. Rodriguez, eds., Ecologia y manejo del venado cola blanca en México y Costa Rica. Editorial de la Universidad Nacional, Mexico City. 462 pp.

Gallina, S., P. Galina-Tessaro, and Sergio Alvarez-Cárdenas. 1992. Habitat y dinamica poblacional del venado bura. Pp. 297–327 *in* A. Ortega, ed., Uso y manejo de los recursos naturales en la Sierra de La Laguna, Baja California Sur. Publication No. 5. Centro de Investigaciones Biológicas de Baja California Sur, A.C.

Gallina, S., S. Mandujuano, J. Bello, and C. Delfin. 1998. Home range size of white-tailed deer on northeastern Mexico. Pp. 47–50 *in* J. C. deVos Jr., ed., Proceedings of the 1997 Deer/Elk Workshop, Rio Rico, Arizona. Arizona Game and Fish Department, Phoenix. 224 pp.

Gallina, S., E. Maury, and V. Serrano. 1978. Hábitos alimenticios del venado cola blanca (*Odocoileus virginanus* Rafinesque) en la reserva La Michilía, estado de Durango. Pp. 59–108 *in* G. Halffter, ed., Reservas de la Biosfera en el estado de Durango. Publication No. 4. Instituto de Ecologia, A.C., Mexico City.

Gallizioli, S. 1956. Field observations of deer in the Prescott study area. Federal Aid Project W-71-R-3, WP3, Job 2. Arizona Game and Fish Department, Phoenix. 29 pp.

Gallizioli, S. 1977. Statement of Steve Gallizioli, Chief of Research, Arizona Game and Fish Department. Pp. 90–96 *in* R. J. Smith and J. E. Townsend, eds., Improving fish and wildlife benefits in range management. U.S. Fish and Wildlife Service, U.S. Department of the Interior, FWS/OBS-77/1. 118 pp.

Gassett, J. W., D. P. Wiesler, A. G. Baker, D. A. Osborn, K. V. Miller, R. L. Marchington, and M. Novotny. 1996. Volatile compounds from interdigital gland of male white-tailed deer (*Odocoileus virginianus*). Journal of Chemical Ecology 22:1689–96.

Gastil, G., J. Minch, and R. Phillips. 1983. The geology and ages of the islands. Pp. 13–25 *in* T. J. Case and M. L. Cody, eds., Island biogeography in the Sea of Cortez. University of California Press, Berkeley. 495 pp.

Gaydos, J. K., D. E. Stallknecht, D. Kavanaugh, R. J. Olson, and E. R. Fuchs. 2002. Dynamics of maternal antibodies to hemorrhagic disease viruses (Reoviridae: Orbivirus) in white-tailed deer. Journal of Wildlife Diseases 38:253–57.

Geist, V. 1966. The evolution of horn-like organs. Behaviour 27:175–214.

Geist, V. 1981. Behavior: Adaptive strategies in mule deer. Pp. 157–224 *in* O.C. Wallmo, ed., Mule and black-tailed deer of North America. Wildlife Management Institute and University of Nebraska Press, Lincoln. 605 pp.

Geist, V. 1986. Super antlers and pre–World War II European research. Wildlife Society Bulletin 14:91–94.

Geist, V. 1992. Endangered species and the law. Nature 357:274–76.

Geist, V. 1994. Origin of the species. Pp. 2–16 *in* D. Gerlach, S. Atwater, and J. Schnell, eds., Deer. Stackpole Books, Mechanicsburg, Pennsylvania. 384 pp.

Geist, V. 1998. Deer of the world: Their evolution, behaviour and ecology. Stackpole Books, Mechanicsburg, Pennsylvania. 421 pp.

Germany, J. C. 1969. Mule deer habitat preference on the pinyon-juniper ranges of Fort Stanton, New Mexico. Master's thesis, New Mexico State University, Las Cruces. 25 pp.

Gingerich, P. D. 1993. Plio-Pleistocene rates of evolution. Pp. 84–106 *in* R. A. Martin and A. D. Barnosky, eds., Morphological change in Quaternary mammals of North America. Cambridge University Press, New York. 415 pp.

Giuliano, W. M., S. Demarais, R. E. Zaiglin, and M. L. Sumner. 1999. Survival and movements of orphaned white-tailed deer fawns in Texas. Journal of Wildlife Management 63:570–74.

Goldman, E. A. 1939. A new mule deer from Sonora. Journal of Mammalogy 20:496–97.
Goodman, G. 1987. The social divisions and economic life of the western Apache. Pp. 41–47 *in* A. Ferg, ed., Western Apache material culture: The Goodwin and Guenther collections. University of Arizona Press, Tucson. 205 pp.
Goss, R. J. 1983. Deer antlers: Regeneration, function, and evolution. Academic Press, New York. 316 pp.
Gotch, A. F. 1995. Latin names explained. Cassell Publishing, London. 714 pp.
Gould, S. J. 1974. The origin and function of "bizarre" structures: Antler size and skull size in the "Irish elk," *Megaloceros giganteus*. Evolution 28:191–220.
Gray, A. P. 1972. Mammalian hybrids: A check-list with bibliography. Commonwealth Agricultural Bureaux, Farnham Royal, England. 262 pp.
Green-Hammond, K. A. 1994. Assessment of impacts to populations and human harvests of deer and elk caused by the reintroduction of Mexican wolves. Special Report No. 20181–4-0201. U.S. Fish and Wildlife Service Endangered Species Office. 25 pp.
Green-Hammond, K. A. 1996. New Mexico Department of Game and Fish deer model users manual. New Mexico Department of Game and Fish, Santa Fe. December 1992 version with March 1996 revisions. 161 pp.
Guthrie, R. D. 1984. Mosaics, allelochemics and nutrients: An ecological theory of late Pleistocene megafaunal extinctions. Pp. 259–98 *in* P. S. Martin and R. G. Klein, eds., Quaternary extinctions. University Arizona Press, Tucson. 892 pp.
Guynn, D. C., Jr., J. R. Sweeney, R. J. Hamilton, and R. L. Marchington. 1988. A case study in quality deer management. South Carolina White-tailed Deer Management Workshop 2:72–79.
Hakonson, T. E., and F. W. Whicker. 1971. The contribution of various tissues and organs to total body mass in the mule deer. Journal of Mammalogy 52:628–30.
Hall, E. R. 1925. Kaibab deer. Unpublished manuscript. 82 pp.
Hamlin, K. L. 1997. Survival and home range fidelity of coyotes in Montana: Implications for control. Intermountain Journal of Sciences 3:62–72.
Hancock, N. V. 1981. Mule deer management in Utah—past and present. Pp. 2–27 *in* Utah mule deer workshop. Utah Cooperative Wildlife Research Unit.
Hansen, R. M., and L. D. Reid. 1975. Diet overlap of deer, elk, and cattle in southern Colorado. Journal of Range Management 28:43–47.
Hanson, W. R. 1955. Field observations of deer in the Prescott study area. Federal Aid Project W-71-R, WP3, Job 2. Arizona Game and Fish Department, Phoenix. 33 pp.
Hanson, W. R., and C. Y. McCulloch. 1955. Factors influencing mule deer on Arizona brushlands. Transactions of the North American Wildlife Conference 20:568–88.
Harmel, D. E., J. D. Williams, and W. E. Armstrong. 1989. Effects of genetics and nutrition on antler development and body size of white-tailed deer. Federal Aid Report Series No. 26. Texas Parks and Wildlife Department, Austin. 55 pp.
Haury, E. 1950. The stratigraphy and archaeology of Ventana Cave, Arizona. University of New Mexico Press, Albuquerque. 599 pp.
Haussamen, W. 1995. Mule deer management in New Mexico: Understanding the problem. New Mexico Department of Game and Fish, Santa Fe. 5 pp.
Hayes, C. L., and P. R. Krausman. 1993. Nocturnal activity of female desert mule deer. Journal of Wildlife Management 57:897–904.
Haywood, D. D., R. L. Brown, R. H. Smith, and C. Y. McCulloch. 1987a. Migration patterns and habitat utilization by Kaibab mule deer. Federal Aid Project W-78-R, WP2, Job 18. Arizona Game and Fish Department, Phoenix. 29 pp.

Haywood, D. D., R. Miller, and G. I. Day. 1987b. Effects of hunt design and related factors on productivity in mule deer and white-tailed deer in Arizona: A problem analysis report. Federal Aid Project W-78-R, WP2, Job 30. Arizona Game and Fish Department, Phoenix. 22 pp.

Hazam, J. E., and P. R. Krausman. 1988. Measuring water consumption of desert mule deer. Journal of Wildlife Management 52:528–34.

Heffelfinger, J. R. 1997. Age criteria for Arizona game animals. Special Report No. 19. Arizona Game and Fish Department, Phoenix. 40 pp.

Heffelfinger, J. R. 1999. Hybridization in large mammals. Pp. 27–37 in P. R. Krausman and S. Demarais, eds., Conservation and management of large mammals in North America. Stackpole Books, Harrisburg, Pennsylvania.

Heffelfinger, J. R. 2000. Status of the name *Odocoileus hemionus crooki (Mammalia: Cervidae)*. Proceedings of the Washington Biological Society 113:319–33.

Heffelfinger, J. R., C. Brewer, C. H. Alcalá-Galván, B. Hale, D. Weybright, B. F. Wakeling, and L. Carpenter. 2006. Mule deer habitat guidelines: Southwest deserts. Western Association of Fish and Wildlife Agencies, Mule Deer Working Group.

Heffelfinger, J. R., L. H. Carpenter, L. C. Bender, G. L. Erickson, M. D. Kirchhoff, E. R. Loft, and W. M. Glasgow. 2003. Ecoregional differences in population dynamics. Pp. 63–91 in J. C. deVos, M. R. Conover, and N. E. Headrick, eds., Mule deer conservation: Issues and management strategies. Western Association of Fish and Wildlife Agencies and Jack H. Berryman Institute, Logan, Utah. 240 pp.

Heffelfinger, J. R., and R. J. Olding. 1994. Mule deer performance report. Federal Aid Project W-53-M. Arizona Game and Fish Department, Tucson. 13 pp.

Heffelfinger, J. R., and R. J. Olding. 1996. Occurrence and distribution of epizootic hemorrhagic disease in mule deer. Federal Aid Project W-53-M, WP6, Job 9. Arizona Game and Fish Department, Phoenix. 22 pp.

Heffelfinger, J. R., and R. J. Olding. 1998. The increasing complexity of deer management: Is more better? Pp. 51–61 in J. C. deVos Jr., ed., Proceedings of the Western States Deer and Elk Workshop, Rio Rico, Arizona. 224 pp.

Heffelfinger, J. R., and L. M. Piest. 1996. Big game population modelling: A user's guide. Special Report. Arizona Game and Fish Department, Tucson. 23 pp.

Henry, R. S., and L. K. Sowls. 1980. White-tailed deer of the Organ Pipe Cactus National Monument, Arizona. Technical Report No. 6. National Park Service and University of Arizona, Tucson. 85 pp.

Herman, C. M., and A. I. Bischoff. 1946. The footworm parasite of deer. California Fish and Game 32:182–90.

Hervert, J. J., and P. R. Krausman. 1986. Desert mule deer use of water developments in Arizona. Journal of Wildlife Management 50:670–76.

Hess, M. 1983. Nevada's use of change-in-ratio estimates to establish deer hunting quotas. Pp. 67–81 in S. L. Beasom and S. F. Roberson, eds., Game harvest management. Caesar Kleberg Wildlife Research Institute, Texas A&M University—Kingsville. 374 pp.

Hibler, C. P. 1981. Diseases. Pp. 129–56 in O. C. Wallmo, ed., Mule and black-tailed deer of North America. Wildlife Management Institute and University of Nebraska Press, Lincoln. 605 pp.

Hillman, J. R., R. W. Davis, Y. Z. Abdelbaki. 1973. Cyclic bone remodeling in deer. Calcium Tissue Research 12:323–30.

Hilton-Taylor, C., comp. 2000. *2000 IUCN Red List of Threatened Species*. International Union for Conservation of Nature and Natural Resources, Gland, Switzerland, and Cambridge, U.K. 61 pp.

Hirth, D. H. 1977. Social behavior of white-tailed deer in relation to habitat. Wildlife Management Monograph 53. Wildlife Society. 55 pp.

Hobson, M. D. 1990. Effects of predator control on desert mule deer numbers. Federal Aid Project W-125-R-1, Job 50. Texas Parks and Wildlife Department, Austin. 60 pp.

Hoff, G. L., and D. O. Trainer. 1981. Hemorrhagic diseases of wild ruminants. Pp. 45–53 in J. W. Davis, L. H. Karstad, and D. O. Trainer, eds., Infectious diseases of wild mammals. Iowa State University Press, Ames. 446 pp.

Hoffmeister, D. F. 1986. Mammals of Arizona. University of Arizona Press, Tucson, and Arizona Game and Fish Department, Tucson. 602 pp.

Holechek, J. L., and D. Galt. 2000. Grazing intensity guidelines. Rangelands 22:11–14.

Hölzenbein, S., and R. L. Marchington. 1992. Emigration and mortality in orphaned male white-tailed deer. Journal of Wildlife Management 56:147–53.

Horejsi, R. G. 1982. Mule deer fawn survival on cattle-grazed and ungrazed desert ranges. Federal Aid Project W-78-R, WP2, Job 17. Arizona Game and Fish Department, Phoenix. 43 pp.

Horejsi, R. G., D. D. Haywood, and R. H. Smith. 1988. The effects of hunting on a desert mule deer population. Federal Aid Project W-78-R, WP2, Job 27. Arizona Game and Fish Department, Phoenix. 25 pp.

Horejsi, R. G., and R. H. Smith. 1983. Impact of a high density deer population on desert vegetation. Federal Aid Project W-78-R, WP4, Job 20. Arizona Game and Fish Department, Phoenix. 15 pp.

Hornocker, M. G. 1970. An analysis of mountain lion predation upon mule deer and elk in the Idaho Primitive Area. Wildlife Monograph No. 21. Wildlife Society. 39 pp.

Hosley, N. W. 1961. Management of the white-tailed deer in its environment. Pp. 187–260 in W. P. Taylor, ed., The deer of North America. Stackpole, Harrisburg, Pennsylvania. 668 pp.

Howard, V. W., Jr. 1966. Mule deer density in relation to habitat on the Fort Stanton range. Master's thesis, New Mexico State University, Las Cruces, New Mexico. 30 pp.

Howard, V. W., Jr., and T. A. Eicher. 1984. Hunting pressure, morphological characteristics and ages of mule deer bucks in the Sacramento Mountains, New Mexico. Pp. 26–34 in P. R. Krausman and N. S. Smith, eds., Deer in the Southwest: A workshop. Arizona Cooperative Wildlife Research Unit and University of Arizona, Tucson. 131 pp.

Huckell, B. B. 1996. The archaic prehistory of the North American Southwest. Journal of World Prehistory 10:305–73.

Huey, L. M. The mammals of Baja California, Mexico. Transactions of the San Diego Society of Natural History 13:85–168.

Humphreys, D., and A. Elenowitz. 1988. New Mexico's mule deer population/environment/hunt computer model. Final report, Federal Aid in Wildlife Restoration Project W-124-R-11. New Mexico Department of Game and Fish, Santa Fe.

Hungerford, C. R. 1970. Response of Kaibab mule deer to management of summer range. Journal of Wildlife Management 34:852–62.

Hunt, D. L., Jr. 1978. Diet and habitat utilization of tame mule deer in a pinyon-juniper woodland. Master's thesis, New Mexico State University, Las Cruces. 82 pp.

Hutchings, S. S. 1946. Drive the water to the sheep. National Wool Grower 36:10–11, 48.

Illige, D. J. 1954. The analyzation of deer specimens from various Arizona deer herds. Federal Aid Project W-71-R-1, WP6, Job 2. Arizona Game and Fish Department, Phoenix. 43 pp.

Illige, D. J. 1956. Deer and cattle diet components. Federal Aid Project W-75-R-3, WP8, Job 3. New Mexico Department of Game and Fish, Santa Fe. 22 pp.

Ingersoll, E. 1906. The life of animals: The mammals. Macmillan, New York. 555 pp.

Jacobs, G. H., J. F. Deegan II, J. Neitz, B. P. Murphy, K. V. Miller, and R. L. Marchington. 1994. Electrophysiological measurements of spectral mechanisms in the retinas of two cervids:

White-tailed deer (*Odocoileus virginianus*) and fallow deer (*Dama dama*). Journal of Comparative Physiology A 174:551–57.

Jacobson, H. A. 1992. Deer condition response to changing harvest strategy, Davis Island, Mississippi. Pp. 48–55 *in* R. D. Brown, ed., The biology of deer. Springer-Verlag, New York. 596 pp.

Jacobson, H. A. 1995. Age and quality relationships. Pp. 103–11 *in* K. V. Miller and R. L. Marchington, eds. Quality whitetails. Stackpole Books, Mechanicsburg, Pennsylvania. 322 pp.

Jacobson, H. A., D. C. Guynn Jr., R. N. Griffin, and D. Lewis. 1979. Fecundity of white-tailed deer in Mississippi and periodicity of corpora lutea and lactation. Proceedings of the Annual Conference of the Southeastern Association of Fish and Wildlife Agencies 33:30–35.

Jacobson, H. A., and S. D. Lukefahr. 1998. Case study: Genetics research on captive white-tailed deer at Mississippi State University. Pp. 47–60 *in* The role of genetics in white-tailed deer management. Symposium proceedings. Texas A&M University, College Station.

Jaczewski, Z. 1990. Experimental induction of antler growth. Pp. 371–95 *in* G. A. Bubenik and A. B. Bubenik, eds., Horns, pronghorns, and antlers: Evolution, morphology, physiology, and social significance. Springer-Verlag, New York. 562 pp.

Jenney, E. W., G. A. Erickson, and M. L. Snyder. 1984. Vesicular stomatitis outbreaks and surveillance in the United States, January 1980 through May 1984. Proceedings of the United States Animal Health Association 88:337–45.

Johnson, A., and W. G. Swank. 1957. Population dynamics of Arizona's important deer herds. Federal Aid Project W-78-R-1, WP4, Job 1. Arizona Game and Fish Department, Phoenix. 22 pp.

Julander, O. 1937. Utilization of browse by wildlife. Transactions of the North American Wildlife Conference 2:276–87.

Kamler, J. F., W. B. Ballard, and D. A. Swepston. 2001. Range expansion of mule deer in Texas. Southwestern Naturalist 46:378–79.

Keller, G. L. 1975. Seasonal food habits of desert mule deer (*Odocoileus hemionus crookii* [sic]) on a specific mule deer–cattle range in Pecos County, Texas. Master's thesis, Sul Ross State University, Alpine, Texas. 80 pp.

Knipe, T. 1977. The Arizona whitetail deer. Special Report No. 6. Arizona Game and Fish Department, Phoenix. 108 pp.

Knox, K. L., J. G. Nagy, and R. D. Brown. 1969. Water turnover in mule deer. Journal of Wildlife Management 33:389–93.

Koenen, K. K. 1999. Seasonal densities and habitat use of desert mule deer in a semidesert grassland. Master's thesis, University of Arizona, Tucson. 70 pp.

Koenen, K. K., S. DeStefano, P. R. Krausman. 2002. Using distance sampling to estimate seasonal densities of desert mule deer in a semidesert grassland. Wildlife Society Bulletin 30:53–63.

Koenen, K. K., and P. R. Krausman. 2002. Habitat use by desert mule deer in a semidesert grassland. Southwestern Naturalist 47:353–62.

Koerth, B. H. 1981. Habitat use, herd ecology, and seasonal movements of mule deer in the Texas Panhandle. Master's thesis, Texas Tech University, Lubbock. 61 pp.

Koerth, B. H., and F. C. Bryant. 1982. Home ranges of mule deer bucks in the Texas Panhandle. Prairie Naturalist 14:122–24.

Koerth, B. H., B. F. Sowell, F. C. Bryant, and E. P. Wiggers. 1985. Habitat relations of mule deer in the Texas panhandle. Southwestern Naturalist 30:579–87.

Krausman, P. R. 1978a. Forage relationship between two deer species in Big Bend National Park, Texas. Journal of Wildlife Management 42:101–107.

Krausman, P. R. 1978b. Dental anomalies of Carmen Mountains white-tailed deer. Journal of Mammalogy 59:863–64.

Krausman, P. R. 1985. Impacts of the Central Arizona Project on desert mule deer and desert bighorn sheep. Final report, U.S. Bureau of Reclamation, Contract 9–07–30-X069, Tucson, Arizona.

Krausman, P. R., and E. D. Ables. 1981. Ecology of the Carmen Mountains white-tailed deer. Scientific Monograph Series No. 15. U.S. Department of the Interior, National Park Service, Washington, D.C. 114 pp.

Krausman, P. R., and R. C. Etchberger. 1993. Effectiveness of mitigation features for desert ungulates along the Central Arizona Project. Final report, U.S. Bureau of Reclamation, Contract 9-CS-32–00350, Tucson, Arizona.

Krausman, P. R., and R. C. Etchberger. 1995. Response of desert ungulates to a water project in Arizona. Journal of Wildlife Management 59:292–300.

Krausman, P. R., A. J. Kuenzi, R. C. Etchberger, K. R. Rautenstrauch, L. L. Ordway, and J. J. Hervert. 1997. Diets of desert mule deer. Journal of Range Management 50:513–22.

Krausman, P. R., B. D. Leopold, R. F. Seegmiller, and S. G. Torres. 1989. Relationships between desert bighorn sheep and habitat in western Arizona. Wildlife Monograph 102. Wildlife Society. 66 pp.

Krausman, P. R., B. D. Leopold, and N. S. Smith. 1984. Morphological characteristics of desert mule deer in Texas and Arizona. Pp. 35–41 *in* P. R. Krausman and N. S. Smith, eds., Deer in the Southwest: A workshop. Arizona Cooperative Wildlife Research Unit and University of Arizona, Tucson. 131 pp.

Krausman, P. R., K. R. Rautenstrauch, and B. D. Leopold. 1985. Xeroriparian systems used by desert mule deer in Texas and Arizona. Pp. 144–49 *in* R. Johnson et al., eds., Riparian ecosystems and their management: Reconciling conflicting uses. General Technical Report RM-120. U.S. Forest Service.

Krausman, P. R., D. J. Schmidly, and E. D. Ables. 1978. Comments on the taxonomic status, distribution, and habitat of the Carmen Mountains white-tailed deer (*Odocoileus virginianus carminis*) in Trans-Pecos Texas. Southwestern Naturalist 23:577–90.

Krueger, W. C., W. A. Laycock, and D. A. Price. 1974. Relationships of taste, smell, sight, and touch to forage selection. Journal of Range Management 27:258–62.

Krysl, L. J. 1979. Food habits of mule deer and elk, and their impact on vegetation in Guadalupe Mountains National Park. Master's thesis, Texas Tech University, Lubbock. 131 pp.

Kucera, T. E. 1978. Social behavior and breeding system of the desert mule deer. Journal of Mammalogy 59:463–76.

Kucera, T. E., and K. E. Mayer. 1999. A sportsman's guide to improving deer habitat in California. California Department of Fish and Game, Sacramento. 95 pp.

Kufeld, R. C., D. C. Bowden, and D. L. Schrupp. 1988. Habitat selection and activity patterns of female mule deer in the Front Range, Colorado. Journal of Range Management 41:515–22.

Kufeld, R. C., D. C. Bowden, and D. L. Schrupp. 1988. Influence of hunting on movements of female mule deer. Journal of Range Management 41:70–72.

Kufeld, R. C., J. H. Olterman, and D. C. Bowden. 1980. A helicopter quadrat census for mule deer on Uncompahgre Plateau, Colorado. Journal of Wildlife Management 44:632–39.

Kufeld, R. C., O. C. Wallmo, and C. Feddema. 1973. Foods of the Rocky Mountain mule deer. Research Paper RM-111. U.S. Forest Service. 31 pp.

Kurtén, B. 1971. The age of mammals. Weidenfeld and Nicolson, London. 250 pp.

Kurtén, B. 1988. Before the Indians. Columbia University Press, New York. 156 pp.

Laing, S. P., and F. G. Lindzey. 1993. Patterns of replacement of resident cougars in southern Utah. Journal of Mammalogy 74:1056–58.

Lang, E. M. 1957. Deer of New Mexico. New Mexico Department of Game and Fish, Santa Fe. 41 pp.

Lang, R. W., and A. H. Harris. 1979. The faunal remains from Arroyo Hondo Pueblo, New Mexico. Arroyo Hondo Archaeological Series Vol. 5. School of American Research Press, Santa Fe, New Mexico. 316 pp.

Lawrence, R. K., S. Demarais, R. A. Reylea, S. P. Haskell, W. B. Ballard, and T. L. Clark. 2004. Desert mule deer survival in Southwest Texas. Journal of Wildlife Management 68:561–69.

Lawrence, R. K., R. A. Relyea, S. Demarais, and R. S. Lutz. 1994. Population dynamics and habitat preferences of desert mule deer in the Trans-Pecos. Final report, Federal Aid Project W-127-R, Job 94. Texas Parks and Wildlife Department, Austin. 178 pp.

LeCount, A., 1977. Causes of fawn mortality. Federal Aid Project W-78-R, WP2, Job 11. Arizona Game and Fish Department, Phoenix. 19 pp.

Lehr, H. J. 1978. A catalogue of the flora of Arizona. Northland Press, Flagstaff, Arizona. 203 pp.

Leon, F. G., III, C. A. DeYoung, and S. L. Beasom. 1987. Bias in age and sex composition of white-tailed deer observed from helicopters. Wildlife Society Bulletin 15:426–29.

Leopold, A. S. 1954. Dichotomous forking in the antlers of white-tailed deer. Journal of Mammalogy 35:599–600.

Leopold, A. S. 1959. Wildlife of Mexico. University of California Press, Berkeley. 568 pp.

Leopold, A. S., T. Riney, R. McCain, and L. Tevis Jr. 1951. The jawbone deer herd. Game Bulletin No. 4. California Department of Natural Resources, Berkeley. 139 pp.

Leopold, B. D., and P. R. Krausman. 1986. Diets of three predators in Big Bend National Park, Texas. Journal of Wildlife Management 50:290–95.

Leopold, B. D., and P. R. Krausman. 1987. Diurnal activity patterns of desert mule deer in relation to temperature. Texas Journal of Science 39:49–53.

Leopold, R. A. 1933. Game management. Charles Scribner's Sons, New York. 481 pp.

Leopold, R. A. 1966. A Sand County almanac. Oxford University Press and Random House, New York. 295 pp.

Ligon, J. S. 1927. Wildlife of New Mexico: Its conservation and management. New Mexico Department of Game and Fish, Albuquerque. 207 pp.

Lindzey, F. G., B. B. Ackerman, D. Barnhurst, and T. P. Hemker. 1988. Survival rates of mountain lions in southern Utah. Journal of Wildlife Management 52:664–67.

Lindzey, F. G., W. G. Hepworth, T. A. Mattson, and A. F. Reeve. 1997. Potential for competitive interactions between mule deer and elk in the western United States and Canada: A review. Report to the Rocky Mountain Elk Foundation, Wyoming Cooperative Fish and Wildlife Research Unit, Laramie. 112 pp.

Lindzey, F. G., W. D. Van Sickle, B. B. Ackerman, D. Barnhurst, T. P. Hemker, and S. P. Laing. 1994. Cougar population dynamics in southern Utah. Journal of Wildlife Management 58:619–24.

Lingle, S. 1989. Limb coordination and body configuration in the fast gaits of white-tailed deer, mule deer, and their hybrids: Adaptive significance and management implications. Master's thesis, University of Calgary, Alberta. 287 pp.

Linsdale, J. M., and P. Q. Tomich. 1953. A herd of mule deer. University of California Press, Berkeley. 567 pp.

Lister, R. H., and F. C. Lister 1983. Those who came before. University of Arizona Press, Tucson. 184 pp.

Loft, E. R., J. W. Menke, J. G. Kie, and R. C. Bertram. 1987. Influence of cattle stocking rate on the structural profile of deer hiding cover. Journal of Wildlife Management 51:655–64.

Logan, K. A., and L. L. Sweanor. 2001. Desert puma: Evolutionary ecology and conservation of an enduring carnivore. Island Press, Washington, D.C. 463 pp.

Logan, K. A., L. L. Sweanor, T. K. Ruth, and M. G. Hornocker. 1996. Cougars of the San Andres Mountains, New Mexico. Federal Aid in Wildlife Restoration Project W-128-R. New Mexico Department of Game and Fish, Santa Fe. 280 pp.

Longhurst, W. M., A. S. Leopold, and R. F. Dasman. 1952. A survey of California deer herds, their ranges and management problems. Game Bulletin No. 6. California Department of Fish and Game, Sacramento. 136 pp.

Lukefahr, S. D. 1997. Genetic and environmental parameters for antler development traits in white-tailed deer using an animal model. Final Report, contract number 386-0692. Texas Parks and Wildlife Department, Austin.

Lukefahr, S. D., and H. A. Jacobson. 1998. Variance component analysis and heritability of antler traits in white-tailed deer. Journal of Wildlife Management 62:262–68.

Lydekker, R. 1898. The deer of all lands: A history of the family cervidae living and extinct. Rowland Ward, Ltd., London. 329 pp.

Maghini, M. T. 1990. Water use and diurnal ranges of Coues white-tailed deer. Master's thesis, University of Arizona, Tucson. 86 pp.

Mahgoub, E. F. 1984. Seasonal food habits of mule deer in the foothills of the Sacramento Mountains, New Mexico. Ph.D. dissertation, New Mexico State University, Las Cruces. 89 pp.

Mahgoub, E. F., R. D. Pieper, J. L. Holecheck, J. D. Wright, and V. W. Howard Jr. 1987. Botanical content of mule deer diets in south-central New Mexico. New Mexico Journal of Science 27:21–27.

Mann, W. G. 1941. The Kaibab deer: A brief history and the present plan of management. U.S. Forest Service files. Unpublished report. 55 pp.

Mann, W. G., and S. B. Locke. 1931. The Kaibab deer: A brief history and recent developments. U.S. Forest Service files. Unpublished report. 68 pp.

Mansfield, T. M. 1994. Politics of managing large predators: Mountain lions in California. Proceedings of the Western Association of Fish and Wildlife Agencies 74:53–57.

Mansfield, T. M., and K. G. Charlton. 1998. Trends in mountain lion depredation and public safety incidents in California. Proceedings of the Vertebrate Pest Conference 18:118–21.

Marburger, R. G., R. M. Robinson, J. W. Thomas, M. J. Andregg, and K. A. Clark. 1972. Antler malformation produced by leg injury in white-tailed deer. Journal of Wildlife Diseases 8:311–14.

Marchington, R. L., and K. V. Miller. 1994. The rut. Pp. 109–21 *in* D. Gerlach, S. Atwater, and J. Schnell, eds., Deer. Stackpole Books, Mechanicsburg, Pennsylvania. 384 pp.

Marshal, J. P., V. C. Bleich, N. G. Andrew, and P. R. Krausman. 2004. Seasonal forage use by desert mule deer in southeastern California. Southwestern Naturalist 49:501–505.

Marshal, J. P., P. R. Krausman, V. C. Bleich, W. B. Ballard, and J. S. McKeever. 2002. Rainfall, El Niño, and dynamics of mule deer in the Sonoran Desert, California. Journal of Wildlife Management 66:1283–89.

Martin, P. S., and R. G. Klein. 1984. Quaternary extinctions. University Arizona Press, Tucson. 892 pp.

Martinez, A., V. Molina, F. Gonzales, J. S. Marroquín, and J. Navar. 1997. Observations of white-tailed deer and cattle diets in Mexico. Journal of Range Management 50:253–57.

Matschke, G. H., and R. D. Roughton. 1977. Delayed antler development and sexual maturity among yearling male white-tailed deer. Proceedings of the Annual Conference of Southeastern Fish and Wildlife Agencies 31:51–56.

Matson, G. 1988. Matson's Laboratory Progress Report No. 10. Winter 1988. 8 pp.

McCabe, R. A., and A. S. Leopold. 1951. Breeding season of the Sonora white-tailed deer. Journal of Wildlife Management 15:433–34.

McCaffery, K. R. 1985. On "crippling" semantics. Wildlife Society Bulletin 13:360–61.

McClymont, R. A., M. Fenton, and J. R. Thompson. 1982. Identification of cervid tissues and hybridization by serum albumin. Journal of Wildlife Management 46:540–44.

McCulloch, C. Y. 1955. Field observations of deer in the Three Bar vicinity. Arizona Chaparral Deer Study, Federal Aid Project W-71-R-2, WP3, Job 1. Arizona Game and Fish Department, Phoenix. 24 pp.

McCulloch, C. Y. 1961. The influence on carrying capacity of experimental water conservation measures. Federal Aid Project W-78-R-5, WP5, Job 7. Arizona Game and Fish Department, Phoenix. 16 pp.

McCulloch, C. Y. 1968. Transplanted mule deer in Arizona. Journal of the Arizona Academy of Sciences 5:43–44.

McCulloch, C. Y. 1972. Deer foods and brush control in southeastern Arizona. Journal of the Arizona Academy of Sciences 7:113–19.

McCulloch, C. Y. 1973. Seasonal diets of mule and white-tailed deer. Pp. 1–37 *in* Deer nutrition in Arizona chaparral and desert habitats. Special Report No. 3. Arizona Game and Fish Department, Phoenix. 68 pp.

McCulloch, C. Y. 1978. Statewide deer food preferences. Federal Aid Project W-78-R-5, WP4, Job 15. Arizona Game and Fish Department, Phoenix. 29 pp.

McCulloch, C. Y. 1986. History of predator control and deer productivity in northern Arizona. Southwestern Naturalist 31:215–20.

McCulloch, C. Y., and R. H. Smith. 1982. Evaluation of summer deer habitat on the Kaibab Plateau. Federal Aid Project W-78-R, WP4, Job 12. Arizona Game and Fish Department, Phoenix. 20 pp.

McCulloch, C. Y., and R. H. Smith. 1987. Relationship of weather and other variables to the condition of the Kaibab deer herd. Federal Aid Project W-78-R, WP 2, Job 12. Arizona Game and Fish Department, Phoenix. 34 pp.

McCulloch, C. Y., and R. H. Smith. 1991. Relationship of weather and other variables to the condition of the Kaibab deer herd. Technical Report No. 11. Arizona Game and Fish Department, Phoenix. 98 pp.

McCullough, D. R. 1987. The theory and management of *Odocoileus* populations. Pp. 535–49 *in* C. M. Wemmer, ed., Biology and management of Cervidae. Smithsonian Institution Press, Washington, D.C. 577 pp.

McGinnes, B. S., and R. L. Downing. 1977. Factors affecting the peak of white-tailed deer fawning in Virginia. Journal of Wildlife Management 41:715–19.

McGregor, J. C. 1965. Southwestern archaeology. University of Illinois Press, Urbana. 501 pp.

McIntosh, B. J., and P. R. Krausman. 1982. Elk and mule deer distributions after a cattle introduction in northern Arizona. Pp. 545–52 *in* J. M. Peek and P. D. Dalke, eds., Wildlife livestock relationships symposium. University of Idaho, Moscow.

McKinney, T. 2003. Precipitation, weather, and mule deer. Pp. 219–37 *in* J. C. deVos Jr., M. R. Conover, and N. E. Headrick, eds., Mule deer conservation: Issues and management strategies. Berryman Institute Press, Utah State University, Logan. 240 pp.

McMichael, T. J. 1967. Investigation of factors influencing deer populations. Federal Aid Project W-78-R-11, WP 4, Job 5. Arizona Game and Fish Department, Phoenix. 11 pp.

McMichael, T. J. 1972. Effect of chemical brush control on deer distribution. Federal Aid Project W-78-R-16, WP 4, Job 8. Arizona Game and Fish Department, Phoenix. 14 pp.

Mearns, E. A. 1897. Preliminary diagnosis of new mammals of the genera *Mephitis, Dorcelaphus,* and *Dicotyles* from the Mexican border of the United States. Proceedings of the U.S. National Museum Bulletin 20:467–71.

Mearns, E. A. 1907. Mammals of the Mexican boundary of the United States. U.S. National Museum Bulletin 56:1–530.

Mech, L. D., and M. E. Nelson. 2000. Do wolves affect white-tailed buck harvest in northeastern Minnesota? Journal of Wildlife Management 64:129–36.

Mendez, E. 1981. Mexico and Central America. Pp. 513–24 *in* L. K. Halls, ed., White-tailed deer ecology and management. Stackpole Books, Harrisburg, Pennsylvania. 870 pp.

Merrell, C. L., and D. N. Wright. 1978. A serological survey of mule deer and elk in Utah. Journal of Wildlife Diseases 14:471–78.

Merriam, C. H. 1901. Seven new mammals from Mexico, including a new genus of rodents. Proceedings of the Washington Academy of Sciences 3:559–63.

Merriam, C. H. 1918. Review of the grizzly and big brown bears of North America. North American Fauna No. 41. U.S. Department of Agriculture, Bureau of Biological Survey. Washington, D.C.

Michael, E. D. 1964. Birth of white-tailed deer fawns. Journal of Wildlife Management 28:171–73.

Mierau, G. W. 1972. Studies on the biology of an antlered female mule deer. Journal of Mammalogy 53:403–404.

Mierau, G. W., and J. L. Schmidt, eds. 1981. The mule deer of the Mesa Verde National Park. Mesa Verde Research Series, Paper No. 2. Mesa Verde Museum Association, Mesa Verde National Park, Colorado. 67 pp.

Miller, K. V., K. E. Kammermeyer, R. L. Marchington, and E. B. Moser. 1987. Population and habitat influences on antler rubbing by white-tailed deer. Journal of Wildlife Management 51:62–66.

Miller, K. V., R. L. Marchington, and W. M. Knox. 1991. White-tailed deer signposts and their role as a source of priming pheromones: A hypothesis. Pp. 455–58 *in* B. Bobek, K. Perzanowski, and W. L. Regelin, eds., Global trends in wildlife management. Eighteenth International Union of Game Biologists Congress, Vol. 1. Jagiellonian University, Kraków, Poland. 659 pp.

Miller, M. W., E. S. Williams, C. W. McCarty, T. R. Spraker, T. J. Kreeger, C. T. Larsen, and E. T. Thorne. 2000. Epizootiology of chronic wasting disease in free-ranging cervids in Colorado and Wyoming. Journal of Wildlife Diseases 36:676–90.

Moen, A. N. 1973. Wildlife ecology: An analytical approach. W. H. Freeman, San Francisco. 458 pp.

Morlan, R. E. 1986. The Pleistocene archaeology of Beringia. Pp. 267–307 *in* M. H. Nitecki and D. V. Nitecki, eds., The evolution of human hunting. Plenum Press, New York. 464 pp.

Morris, E. A. 1980. Basketmaker caves in the Prayer Rock district, northeastern Arizona. University of Arizona Press, Tucson. 158 pp.

Müller-Schwarze, D. 1972. Social significance of forehead rubbing in black-tailed deer (*Odocoileus hemionus columbianus*). Animal Behaviour 20:788–97.

Müller-Schwarze, D., R. Alterieri, and N. Porter. 1984. Alert odor from skin gland in deer. Journal of Chemical Ecology 10:1707–29.

Neff, D. J. 1968. The pellet-group technique for big game trend, census, and distribution: A review. Journal of Wildlife Management 32:597–614.

Neff, D. J. 1974. Forage preferences of trained mule deer on the Beaver Creek watershed. Special Report No. 4. Arizona Game and Fish Department, Phoenix. 61 pp.

Neff, D. J., C. Y. McCulloch, D. E. Brown, C. H. Lowe, and J. F. Barstad. 1979. Forest, range, and watershed management for enhancement of wildlife habitat in Arizona. Special Report No. 7. Arizona Game and Fish Department, Phoenix. 109 pp.

Nettles, V. F., F. A. Hayes, and W. M. Martin. 1976. Observation on injuries in white-tailed deer. Proceedings of the Conference of Southeastern Fish and Wildlife Agencies 13:474–80.

New Mexico Department of Game and Fish. 1996. New Mexico mule deer: Today and tomorrow. Special report. New Mexico Department of Game and Fish, Santa Fe. 5 pp.

New Mexico Department of Game and Fish. 1997. Long-range plan for mule deer management in New Mexico, 1997–2002. New Mexico Department of Game and Fish, Santa Fe. 17 pp.

Nichol, A. A. 1938. Experimental feeding of deer. University of Arizona Technical Bulletin 75:1–39.

Nicholson, M. C., R. T. Bowyer, and J. G. Kie. 1997. Habitat selection and survival of mule deer: Tradeoffs associated with migration. Journal of Mammalogy 78:483–504.

Noon, T. H., S. L. Wesche, J. R. Heffelfinger, A. F. Fuller, G. A. Bradley, C. Reggiardo. Hemorrhagic disease in Arizona deer. Journal of Wildlife Diseases 38:177–81.

Nowak, R. M. 1999. Walker's mammals of the world. 6th ed. Vols. 1 and 2. Johns Hopkins University Press, Baltimore, Maryland. 1936 pp.

Obergfell, F. A. 1957. Vergleichende Untersuchungen an den Dentitionen und Dentale altburdigaler Cerviden von Wintershof-West in Bayern und rezenter Cerviden. (Eine phylogenetische Studie.) Paleontographica (Stuttgart) 106 Abt. A 3/6:71–166.

O'Brien, G. P. 1984. Getting the most out of white-tailed deer surveys in Arizona. Pp. 87–91 *in* P. R. Krausman and N. S. Smith, eds., Deer in the Southwest: A workshop. Arizona Cooperative Wildlife Research Unit and University of Arizona, Tucson. 131 pp.

O'Brien, S. J., and E. Mayr. 1991. Bureaucratic mischief: Recognizing endangered species and subspecies. Science 251:1187–88.

Ockenfels, R. A., and D. E. Brooks. 1994. Summer diurnal bed sites of Coues white-tailed deer. Journal of Wildlife Management 58:70–75.

Ockenfels, R. A., and D. E. Brooks. 1997. Coues white-tailed deer dietary overlap with cattle in southern Arizona. Pp. 89–96 *in* J. C. deVos, ed., Proceedings of the 1997 Deer/Elk Workshop, Rio Rico, Arizona. 224 pp.

Ockenfels, R. A., D. E. Brooks, and C. H. Lewis. 1991. General ecology of Coues white-tailed deer in the Santa Rita Mountains. Technical Report No. 6. Arizona Game and Fish Department, Phoenix. 73 pp.

O'Conner, J. 1939. Game in the desert. Derrydale Press, New York. 167 pp.

Odend'hal, S., K. V. Miller, and D. M. Hoffmann. 1992. Prepupital glands in the white-tailed deer (*Odocoileus virginianus*). Journal of Mammalogy 73:299–302.

Olsen, J. W. 1990. Vertebrate faunal remains from Grasshopper Pueblo, Arizona. University of Michigan Press, Ann Arbor. 192 pp.

Olson, C. A., E. W. Cupp, S. Luckhart, J. M. C. Ribeiro, and Craig Levy. 1992. Occurrence of *Ixodes pacificus* (Parasitiformes: Ixodidae) in Arizona. Journal of Medical Entomology 29:1060–62.

Ordway, L. L., and P. R. Krausman. 1986. Habitat use by desert mule deer. Journal of Wildlife Management 50:677–83.

Ott, J. R., J. T. Baccus, S. W. Roberts, D. E. Harmel, E. Fuchs, and W. E. Armstrong. 1998. The comparative performance of spike- and fork-antlered yearling white-tailed deer: The basis for selection. Pp. 22–32 *in* D. Rollins, ed., The role of genetics in white-tailed deer management. Texas A&M University System and Texas Chapter of Wildlife Society, College Station. 102 pp.

Owen, R. 1853. Descriptive catalogue of the osteological series contained in the museum of the Royal College of Surgeons of England. Mammalia: Placentalia. Taylor and Francis, London. 914 pp.

Ozoga, J. J. 2000. Deer research: Lunar forces don't control whitetail breeding. Deer and Deer Hunting Magazine. October.

Ozoga, J. J., and L. J. Verme. 1985. Comparative breeding behavior and performance of yearling vs. prime-age white-tailed bucks. Journal of Wildlife Management 49:364–72.

Ozoga, J. J., L. J. Verme, and S. C. Bienz. 1982. Parturition behavior and territoriality in white-tailed deer: Impact on neonatal mortality. Journal of Wildlife Management 46:1–11.

Pederson, J. C. 1983. Presence of maxillary canine teeth in mule deer in Utah. Great Basin Naturalist 43:445–46.

Pederson, J. C., L. A. Jensen, and F. L. Anderson. 1985. Prevalence and distribution of *Elaeophora schneideri* Wehr and Dikmans, 1935 in mule deer in Utah. Journal of Wildlife Diseases 21:66–67.

Pence, D. B., and G. G. Gray. 1981. Elaeophorosis in barbary sheep and mule deer from the Texas panhandle. Journal of Wildlife Diseases 17:49–56.

Perez-Gil Salcido, R. 1981. A preliminary study of the deer from Cedros Island, Baja California, Mexico. Master's thesis, University of Michigan, Ann Arbor.

Phillips, J. L., and C. W. Hanselka. 1975. The annual behavioral cycle of desert mule deer (*Odocoileus hemionus crooki*) in relation to vegetative use. RAS Research Series, spring. Sul Ross University, Alpine, Texas. 16 pp.

Pierce, B. M., V. C. Bleich, and R. T. Bowyer. 2000. Social organization of mountain lions: Does a land-tenure system regulate population size? Ecology 81:1533–49.

Pierson, R. E., J. Storz, A. E. McChesney, and D. Thake. 1974. Experimental transmission of malignant catarrhal fever. American Journal of Veterinary Research 35:523–25.

Pious, M. 1989. Forage composition and physical condition of southern mule deer, San Diego County, California. Master's thesis, Humboldt State University, California. 61 pp.

Pittman, M. T., and T. Bone. 1987. Mule deer reproduction. Final report, Federal Aid Project W-109-R-10, Job 51. Texas Parks and Wildlife Department, Austin. 13 pp.

Pollock, K. H., and W. L. Kendall. 1987. Visibility in aerial surveys: A review of estimation procedures. Journal of Wildlife Management 51:502–10.

Pratt, J. J. 1966. White flags of Apacheland. Vantage Press, New York. 114 pp.

Pregler, C. E. 1974. Kaibab mule deer productivity estimates based on ovarian examination. Master's thesis, University of Arizona, Tucson. 44 pp.

Prestwood, A. K., T. P. Kistner, F. E. Kellogg, and F. A. Hayes. 1974. The 1971 outbreak of hemorrhagic disease among white-tailed deer of the southeastern United States. Journal of Wildlife Diseases 10:217–24.

Pursley, D. 1977. Illegal big game harvest during closed season. Proceedings of the Western State Game and Fish Commissions 57:67–71.

Quay, W. B. 1959. Microscopic structure and variation in the cutaneous glands of the deer, *Odocoileus virginianus*. Journal of Mammalogy 40:114–28.

Quay, W. B., and D. Müller-Schwarze. 1970. Functional histology of integumentary glandular regions in black-tailed deer (*Odocoileus hemionus columbianus*). Journal of Mammalogy 51:675–94.

Quay, W. B., and D. Müller-Schwarze. 1971. Relations of age and sex to integumentary glandular regions in Rocky Mountain mule deer (*Odocoileus hemionus hemionus*). Journal of Mammalogy 52:670–85.

Ragotzkie, K. E. 1988. Desert mule deer ecology and habitat use on cattle-grazed grass-shrub range. Master's thesis, Colorado State University, Fort Collins. 51 pp.

Ragotzkie, K. E., and J. A. Bailey. 1991. Desert mule deer use of grazed and ungrazed habitats. Journal of Range Management 44:487–90.

Ratliff, D. D. 1980. Seasonal food habits of desert mule deer on the Stockton Plateau, Texas. Master's thesis, Sul Ross State University, Alpine, Texas. 40 pp.

Raught, R. W. 1960. Deer population trends. Federal Aid in Wildlife Restoration Project W-93-R-1, Work Plan 2, Job 1. New Mexico Department of Game and Fish, Santa Fe. 19 pp.

Raught, R. W. 1967. White-tailed deer. Pp. 52–60 *in* New Mexico wildlife management. New Mexico Department of Game and Fish, Albuquerque. 250 pp.

Raught, R. W. 1969. Deer population trends, habitat condition evaluation and harvest information. Federal Aid in Wildlife Restoration Project W-93-R-11, Work Plan 2, Job 6. New Mexico Department of Game and Fish, Santa Fe. 205 pp.

Raught, R. W., and L. G. Frary. 1961. Deer population trends. Federal Aid in Wildlife Restoration Project W-93-R-2, Work Plan 2, Job 1. New Mexico Department of Game and Fish, Santa Fe. 11 pp.

Rautenstrauch, K. R. 1987. Ecology of desert mule deer in southwest Arizona. Ph.D. dissertation, University of Arizona, Tucson. 37 pp.

Rautenstrauch, K. R., and P. R. Krausman. 1986. Preventing desert mule deer drownings in the Mohawk Canal, Arizona. Final report, Contract 9–07–30-x0069. U.S. Bureau of Reclamation. 96 pp.

Rautenstrauch, K. R., and P. R. Krausman. 1989. Influence of water availability and rainfall on movements of desert mule deer. Journal of Mammalogy 70:197–201.

Raymond, G. J., A. Bossers, L. D. Raymond, K. I. O'Rourke, L. E. McHolland, P. K. Bryant III, M. W. Miller, E. S. Williams, M. Smits, and B. Caughey. 2000. Evidence of a molecular barrier limiting susceptibility of humans, cattle, and sheep to chronic wasting disease. EMBO Journal 19:4425–30.

Reat, E. P., O. E. Rhodes Jr., J. R. Heffelfinger, and J. C. deVos. 1999. Regional genetic differentiation in Arizona pronghorn. Proceedings of the Biennial Pronghorn Antelope Workshop, Arizona Game and Fish Department, 18:25–31.

Reed, J. E. 2004. Diets of free-ranging Mexican wolves in Arizona and New Mexico. Master's thesis, Texas Tech University, Lubbock.

Reeve, A. F., and S. H. Anderson. 1993. Ineffectiveness of swareflex reflectors at reducing deer-vehicle collisions. Wildlife Society Bulletin 21:127–32.

Reid, J. J., and S. Whittlesey. 1997. The archaeology of ancient Arizona. University of Arizona Press, Tucson. 297 pp.

Reiter, D. K., M. W. Brunson, and R. H. Schmidt. 1999. Public attitudes toward wildlife damage management and policy. Wildlife Society Bulletin 27:746–58.

Relyea, R. A., and S. Demarais. 1994. Activity of desert mule deer during the breeding season. Journal of Mammalogy 75:940–49.

Relyea, R. A., R. K. Lawrence, and S. Demarais. 2000. Home range of desert mule deer: Testing the body-size and habitat-productivity hypothesis. Journal of Wildlife Management 64:146–53.

Rhodes, O. E., Jr., M. H. Smith, and R. K. Chesser. 1992. Prenatal reproductive losses in white-tailed deer. Pp. 390–97 in R. D. Brown, ed., The biology of deer. Springer-Verlag, New York. 596 pp.

Rhyan, J., K. Aune, B. Hood, R. Clarke, J. Payeur, J. Jarnagin, and L. Stackhouse. 1995. Bovine tuberculosis in a free-ranging mule deer (Odocoileus hemionus) from Montana. Journal of Wildlife Diseases 31:432–35.

Richardson, L. W., H. A. Jacobson, R. J. Muncy, and C. J. Perkins. 1983. Acoustics of white-tailed deer (Odocoileus virginianus). Journal of Mammalogy 64:245–52.

Riley, C. L. 1997. Introduction. Pp. 1–28 in R. Flint and S. C. Flint, eds., The Coronado expedition to Tierra Nueva. University of Colorado Press, Niwot. 442 pp.

Robinette, W. L. 1956. Productivity—the annual crop of mule deer. Pp. 415–29 in W. P. Taylor, ed., The deer of North America. Stackpole, Harrisburg, Pennsylvania. 668 pp.

Robinette, W. L., C. H. Baer, R. E. Pillmore, and C. E. Knittle. 1973. Effects of nutritional change on captive mule deer. Journal of Wildlife Management 37:312–26.

Robinette, W. L., N. V. Hancock, and D. A. Jones. 1977. The Oak Creek mule deer herd in Utah. Publication No. 77–15. Utah Division of Wildlife Resources, Salt Lake City. 148 pp.

Robinette, W. L., and D. A. Jones. 1959. Antler anomalies of mule deer. Journal of Mammalogy 40:96–108.

Robinette, W. L., D. A. Jones, G. Rogers, and J. S. Gashwiler. 1957. Notes on tooth development and wear for Rocky Mountain mule deer. Journal of Wildlife Management 21:134–53.

Rodgers, K. J. 1977. Seasonal movement of mule deer on the Santa Rita Experimental Range. Master's thesis, University of Arizona, Tucson. 63 pp.

Rodgers, K. J., P. F. Ffolliott, and D. R. Patton. 1978. Home range and movement of five mule deer in a semidesert grass-shrub community. Research Note RM-355. Rocky Mountain Forest and Range Experiment Station, Fort Collins, Colorado. 6 pp.

Rodgers, L. L. 1987. Seasonal changes in defecation rates of free-ranging white-tailed deer. Journal of Wildlife Management 51:330–33.

Romer, A. S. 1966. Vertebrate paleontology. University of Chicago Press, Chicago, Illinois. 408 pp.

Romer, A. S. 1968. Notes and comments on vertebrate paleontology. University of Chicago Press, Chicago, Illinois. 304 pp.

Romin, L. A., and L. B. Dalton. 1992. Lack of response by mule deer to wildlife warning whistles. Wildlife Society Bulletin 20:382–84.

Rosenstock, S. S., W. B. Ballard, and J. C. deVos, Jr. 1999. Viewpoint: Benefits and impacts of wildlife water developments. Journal of Range Management 52:302–11.

Rosenstock, S. S., M. J. Rabe, C. S. O'Brien, and R. B. Waddell. 2004. Studies of wildlife water developments in southwestern Arizona: Wildlife use, water quality, wildlife diseases, wildlife mortalities, and influences on native pollinators. Technical Guidance Bulletin No. 8. Arizona Game and Fish Department, Research Branch, Phoenix. 15 pp.

Rosier, W. S. 1987. Trends in percent cover and botanical composition of mule deer habitat at Fort Stanton, New Mexico. Master's thesis, New Mexico State University, Las Cruces. 56 pp.

Russ, W. B. 1984. Muleys and whitetails side by side. Texas Parks and Wildlife 42(11):22–28.

Russ, W. B. 1993. Mule deer age determination. Federal Aid Project Number W-127-R-2, Job 53. Texas Parks and Wildlife Department, Austin. 9 pp.

Russo, J. P. 1964. The Kaibab North deer herd: Its history, problems, and management. Wildlife Bulletin No. 7. Arizona Game and Fish Department, Phoenix. 159 pp.

Ryel, L. A. 1963. The occurrence of certain anomalies in Michigan white-tailed deer. Journal of Mammalogy 44:79–98.

Salwasser, H. J. 1974. North Kings deer herd fawn production and survival study. Federal Aid Project W-51-R. California Department of Fish and Game, Sacramento. 24 pp.

Salwasser, H. J., S. A. Holl, and G. A. Ashcraft. 1978. Fawn production and survival in the North Kings River deer herd. California Fish and Game 64:38–52.

Sánchez-Rojas, G., and S. Gallina. 2000. Mule deer (*Odocoileus hemionus*) density in a landscape element of the Chihuahuan desert, Mexico. Journal of Arid Environments 44:357–68.

Sasse, D. B. 2003. Job-related mortality of wildlife workers in the United States, 1937–2000. Wildlife Society Bulletin 31:1015–20.

Schaefer, R. J. 1999. Biological characteristics of mule deer in California's San Jacinto Mountains. California Fish and Game 85:1–10.

Schaefer, R. J., S. G. Torres, and V. C. Bleich. 2000. Survivorship and cause-specific mortality in sympatric populations of mountain sheep and mule deer. California Fish and Game 86:127–35.

Schafer, J. A., and S. T. Penland. 1985. Effectiveness of swareflex reflectors in reducing deer-vehicle accidents. Journal of Wildlife Management 49:774–76.

Schildwachter, G. T. M., R. L. Marchington, and C. Hall. 1989. Deer whistles: Do they work? Annual Meeting of the Southeastern Deer Study Group 12:15.

Schmitt, S. M. 2001. Bovine tuberculosis in Michigan's deer. United States Animal Health Association Newsletter 28:4–11.
Scott, K. M., and C. M. Janis. 1987. Phylogenetic relationships of the Cervidae, and the case for a superfamily "Cervoidea." Pp. 3–20 *in* C. M. Wemmer, ed., Biology and management of Cervidae. Smithsonian Institution Press, Washington, D.C. 577 pp.
Scott, M. D. 1981. Fluorescent orange discrimination by wapiti. Wildlife Society Bulletin 9:256–60.
Scott, W. B. 1937. A history of land mammals in the western hemisphere. Macmillian, New York. 786 pp.
Scribner, K. T., M. H. Smith, and P. E. Johns. 1989. Environmental and genetic components of antler growth in white-tailed deer. Journal of Mammalogy 70:284–91.
Severinghaus, C. W. 1949. Tooth development and wear as criteria of age in white-tailed deer. Journal of Wildlife Management 13:195–216.
Severinghaus, C. W., and E. L. Cheatum. 1961. Life and times of the white-tailed deer. Pp. 57–186 *in* W. P Taylor, ed., The deer of North America. Stackpole, Harrisburg, Pennsylvania. 668 pp.
Severson, K. E., and P. J. Urness. 1994. Livestock grazing: A tool to improve wildlife habitat. Pp. 232–49 *in* M. Vavra, W. A. Laycock, and R. D. Pieper, eds., Ecological implications of livestock herbivory in the West. Society of Range Management, Denver, Colorado. 297 pp.
Shaw, H. G. 1977. Impact of mountain lion on mule deer and cattle in northwestern Arizona. Pp. 17–32 *in* R. L. Phillips and C. Jonkel, eds., Proceedings of the 1975 Predator Symposium. Montana Forestry and Conservation Experiment Station, University of Montana, Missoula.
Shaw, H. G. 1980. Ecology of the mountain lion in Arizona. Federal Aid in Wildlife Restoration Project W-78-R, WP2, Job 13. Arizona Game and Fish Department, Phoenix. 14 pp.
Short, H. L. 1964. Post-natal stomach development of white-tailed deer. Journal of Wildlife Management 28:445–58.
Short, H. L. 1977. Food habits of mule deer in a semidesert grass-shrub habitat. Journal of Range Management 30:206–209.
Short, H. L. 1979. Deer in Arizona and New Mexico: Their ecology and a theory explaining recent population decreases. General Technical Report RM-70. Rocky Mountain Forest and Range Experiment Station, Fort Collins, Colorado. 25 pp.
Short, H. L. 1981. Nutrition and metabolism. Pp. 99–127 *in* O. C. Wallmo, ed., Mule and black-tailed deer of North America. University of Nebraska Press, Lincoln. 605 pp.
Short, H. L., W. Evans, and E. L. Boeker. 1977. The use of natural and modified pinyon pine-juniper woodlands by deer and elk. Journal of Wildlife Management 41:543–59.
Smith, B. L., D. J. Skotko, W. Owen, and R. J. McDaniel. 1989. Color vision in white-tailed deer. Psychological Record 39:195–202.
Smith, J. G. 1952. Food habits of mule deer in Utah. Journal of Wildlife Management 16:148–55.
Smith, N. S. 1984. Reproduction in Coues white-tailed deer relative to drought and cattle stocking rates. Pp. 13–20 *in* P. R. Krausman, and N. S. Smith, eds., Deer in the Southwest: A workshop. Arizona Cooperative Wildlife Research Unit and University of Arizona, Tucson. 131 pp.
Smith, R. H. 1976. Hematology of deer. Federal Aid Project W-78-R, WP3, Job 4. Arizona Game and Fish Department, Phoenix. 9 pp.
Smith, R. H., and A. LeCount. 1979. Some factors affecting survival of desert mule deer fawns. Journal of Wildlife Management 43:657–65.
Smith, R. H., T. J. McMichael, and H. G. Shaw. 1969. Decline of a desert deer population. Wildlife Digest, Abstract 3. Arizona Game and Fish Department, Phoenix. 8 pp.
Sowell, B. F. 1981. Nutritional quality of mule deer diets in the Texas Panhandle. Master's thesis, Texas Tech University, Lubbock. 62 pp.

Sowell, B. F., B. H. Koerth, and F. C. Bryant. 1985. Seasonal nutrient estimates of mule deer diets in the Texas panhandle. Journal of Range Management 38:163–67.

Spier, L. 1933. Yuman tribes of the Gila River. University of Chicago Press, Chicago, Illinois. 433 pp.

Staknis, M. A., and D. M. Simmons. 1990. Ultrastructure of the eastern whitetail deer retina for color perception. Pennsylvania Academy of Science 64:8–10.

Steiner, G. 1982. Wisconsin's magic madstones. Wisconsin Natural Resources 6:32–35.

Stephen, C. L., J. C. deVos Jr., J. R. Heffelfinger, and O. E. Rhodes Jr. 2001. Genetic distinction of Sonoran pronghorn. Proceedings of the Biennial Pronghorn Antelope Workshop 20:72–83.

Stewart, R. 1957. Deer food habits study. Federal Aid Project W-75-R-4, WP13, Job 5. New Mexico Department of Game and Fish, Santa Fe. 7 pp.

Stirton, R. A. 1944. Comments on the relationships of the cervoid family Palaeolomerycidae. American Journal of Science 242:633–55.

Stone, W. 1905. On a collection of birds and mammals from the lower Colorado River delta, Lower California. Proceedings of the Academy of Natural Sciences of Philadelphia 107:676–90.

Stonehouse, B. 1968. Thermoregulatory function of growing antlers. Nature 218:870–72.

Stubblefield, S. S., D. B. Pence, and R. J. Warren. 1987. Visceral helminth communities of sympatric mule and white-tailed deer from the Davis Mountains of Texas. Journal of Wildlife Diseases 23:113–20.

Stubblefield, S. S., R. J. Warren, and B. R. Murphy. 1986. Hybridization of free-ranging white-tailed and mule deer in Texas. Journal of Wildlife Management 50:688–90.

Sullivan, T. L., A. F. Williams, T. A. Messmer, L. A. Hellinga, and S. Y. Kyrychenko. 2004. Effectiveness of temporary warning signs in reducing deer-vehicle collisions during mule deer migrations. Wildlife Society Bulletin 32:907–15.

Swank, W. G. 1956. Analysis of data from various Arizona deer herds. Federal Aid Project W-71-R-3, WP6, Job 2. Arizona Game and Fish Department, Phoenix. 34 pp.

Swank, W. G. 1958. The mule deer in the Arizona Chaparral. Wildlife Bulletin No. 3, Arizona Game and Fish Department, Phoenix. 109 pp.

Szuter, C. R. 1989. Hunting by prehistoric horticulturalists in the American Southwest. Ph.D. dissertation, University of Arizona, Tucson. 497 pp.

Tafoya, J. J., V. W. Howard Jr., and J. C. Boren. 2001. Diets of elk, mule deer and Coues white-tailed deer on Fort Bayard in southwestern New Mexico. Report No. 52. Range Improvement Task Force, Agricultural Experiment Station, Cooperative Extension Service, New Mexico State University, Las Cruces. 54 pp.

Tallon, J. 1965. Any-deer fear. Wildlife Views Magazine. Arizona Game and Fish Department, Phoenix. November–December.

Taylor, J. R. 1961. Deer food habits study. Federal Aid Project W-75-R-8, WP18, Job 8. New Mexico Department of Game and Fish, Santa Fe. 10 pp.

Taylor, J. R. 1962. Deer food habits study. Federal Aid Project W-75-R-9, WP20, Job 12. New Mexico Department of Game and Fish, Santa Fe. 16 pp.

Taylor, J. R. 1963. Deer food habits study. Federal Aid Project W-96-R-4, WP20, Job 4. New Mexico Department of Game and Fish, Santa Fe. 13 pp.

Teer, J. G., J. W. Thomas, and E. A. Walker. 1965. Ecology and management of white-tailed deer in the Llano Basin of Texas. Journal of Wildlife Management Monograph 15. 62 pp.

Texas Parks and Wildlife Department. 2003. Big game harvest estimates, 2003–2004. Texas Parks and Wildlife Department, Austin. 702 pp.

Thompson, J. R., and V. C. Bleich. 1993. A comparison of mule deer survey techniques in the Sonoran desert of California. California Fish and Game 79:70–75.

Trainer, D. O., and M. M. Jochim. 1969. Serologic evidence of bluetongue in wild ruminants of North America. American Journal of Veterinary Research 30:2007–11.

Trefethen, J. B. 1967. The terrible lesson of the Kaibab. National Wildlife. June–July.

Truett, J. C. 1971. Ecology of the desert mule deer, *Odocoileus hemionus crooki* Mearns, in southeastern Arizona. Ph.D. dissertation, University of Arizona, Tucson. 64 pp.

Tull, J. C., P. R. Krausman, and R. J. Steidl. 2001. Bed-site selection by desert mule deer in southern Arizona. Southwestern Naturalist 46:354–57.

Ullery, D. E. 1983. Nutrition and antler development in white-tailed deer. Pp. 49–60 *in* R. D. Brown, ed., Antler development in Cervidae. Caesar Kleberg Wildlife Research Institute, Kingsville, Texas.

United States Department of Agriculture. 1998. Screwworm. U.S. Department of Agriculture—Animal, Plant, and Health Inspection Service Technical Bulletin. 5 pp.

Unsworth, J. W., D. F. Pac, G. C. White, and R. M. Bartmann. 1999. Mule deer survival in Colorado, Idaho, and Montana. Journal of Wildlife Management 63:315–26.

Urness, P. J. 1969. Nutritional analysis and in vitro digestibility of mistletoes browsed by deer in Arizona. Journal of Wildlife Management 33:499–505.

Urness, P. J., W. Green, and R. K. Watkins. 1971. Nutrient intake of deer in Arizona chaparral and desert habitats. Journal of Wildlife Management 35:469–75.

Utah Division of Wildlife Resources. 2001. Statewide wildlife management, investigations and surveys. Federal Aid in Wildlife Restoration Report W-65-R-48. Utah Division of Wildlife Resources, Salt Lake City. 177 pp.

Uzzell, P. B. 1958. Trans-Pecos game management survey—deer food habits study. Final report, Federal Aid W-57-R-5, Job 7. Texas Parks and Wildlife Department, Austin. 21 pp.

Van Campen, H., J. Ridpath, E. Williams, J. Cavender, J. Edwards, S. Smith, and H. Sawyer. 2001. Isolation of bovine viral diarrhea virus from a free-ranging mule deer in Wyoming. Journal of Wildlife Diseases 37:306–11.

Verme, L. J. 1962. Mortality of white-tailed deer fawns in relation to nutrition. Proceedings of the National White-tailed Deer Disease Symposium 1:15:38.

Verme, L. J. 1965. Reproduction studies on penned white-tailed deer. Journal of Wildlife Management 29:74–79.

Verme, L. J., J. J. Ozoga, and J. T. Nellist. 1987. Induced early estrus in penned white-tailed deer does. Journal of Wildlife Management 51:54–56.

Villa, B. 1954. Contribución al conocimiento de las epocas de caida y nacimiento de la cornamenta y de su terciopelo en venados cola blanco (*Odocoileus virginianus*) de San Cayetano, Estado de México, México. Anales del Instituto de Biologia México 25:451–61.

Volkman, N. J. 1981. Some aspects of olfactory communication of black-tailed and white-tailed deer: Responses to forehead, antorbital, and metatarsal secretions. Master's thesis, Syracuse University, New York. 58 pp.

Waldron, D. F. 1998. A critical evaluation of studies on the genetics of antler traits of white-tailed deer. Pp. 72–79 *in* D. Rollins, ed., The role of genetics in white-tailed deer management. Texas A&M University System and Texas Chapter of Wildlife Society, College Station. 102 pp.

Waldron, D. F., C. A. Morris, R. L. Baker, and D. L. Johnson. 1993. Maternal effects for growth traits in beef cattle. Livestock Production Science 34:57–70.

Wallace, M. C., and P. R. Krausman. 1987. Elk, mule deer, and cattle habitats in central Arizona. Journal of Range Management 40:80–83.

Wallmo, O. C. 1960. Trans-Pecos game management survey—Big Bend ecological survey. Federal Aid Project W-57-R-8. Texas Parks and Wildlife Department, Austin. 46 pp.

Wallmo, O. C. 1961. Trans-Pecos game management survey—Big Bend ecological survey. Federal Aid Project W-57-R-9. Texas Parks and Wildlife Department, Austin. 29 pp.

Wallmo, O. C. 1978. Mule and black-tailed deer. Pp. 31–41 *in* J. L. Schimdt and D. L. Gilbert, eds., Big game of North America: Ecology and management. Stackpole Books, Harrisburg, Pennsylvania. 494 pp.

Wallmo, O. C. 1981. Distribution and habits. Pp. 1–26 *in* O. C. Wallmo, ed., Mule and black-tailed deer of North America. Wildlife Management Institute and University of Nebraska Press, Lincoln. 605 pp.

Wampler, G. E. 1981. Seasonal changes in home range and habitat preferences of desert mule deer in Pecos County, Texas. Master's thesis, Sul Ross State University, Alpine, Texas. 60 pp.

Warren, L. S. 1997. The hunter's game: Poachers and conservationists in twentieth-century America. Yale University Press, New Haven, Connecticut. 227 pp.

Watkins, R. K., and P. J. Urness. 1972. Maxillary canine and supernumerary incisors in Arizona Coues white-tailed deer. Southwestern Naturalist 17:211–13.

Way, P. R. 1960. Overland via "jackass mail" in 1858: The diary of Phocian Way. Ed. William A. Duffen. Arizona and the West 2:35–53, 147–64, 279–92, 353–70. University of Arizona Press, Tucson.

Webb, P. A., R. G. McLean, G. C. Smith, J. H. Ellenberger, D. B. Francy, T. E. Walton, and T. P. Monath. 1987. Epizootic vesicular stomatitis in Colorado, 1982: Some observations on the possible role of wildlife populations in an enzootic maintenance. Journal of Wildlife Diseases 23:192–98.

Webb, P. M. 1971. Deer management information. Federal Aid Project W-53–21, WP2, Job 3. Arizona Game and Fish Department, Phoenix. 45 pp.

Webb, S. D. 2000. Evolutionary history of new world Cervidae. Pp. 38–64 *in* E. S. Vrba and G. B. Schaller, eds., Antelopes, deer, and relatives: Fossil record, behavioral ecology, systematics, and conservation. Yale University Press, New Haven, Connecticut. 341 pp.

Weber, M., P. Rosas-Becerril, A. Morales-Garcia, C. Galindo-Leal. 1995. Biologia reproductiva del venado cola blanca en Durango, México. Pp. 111–27 *in* C. Vaughan and M. Rodriguez, eds., Ecologia y manejo del Venado cola blanca en México y Costa Rica. Editorial de la Universidad Nacional, Mexico City.

Welch, J. M. 1960. A study of seasonal movements of white-tailed deer (*Odocoileus virginianus couesi*) in the Cave Creek basin of the Chiricahua Mountains. Master's thesis, University of Arizona, Tucson. 79 pp.

Welles, P. 1959. The Coues deer at Coronado National Memorial. Unpublished manuscript. Coronado National Memorial, National Park Service. 29 pp.

White, G. C. 1984. Ideas on estimating parameters for small isolated populations. Pp. 124–27 *in* P. R. Krausman and N. S. Smith, eds., Deer in the Southwest: A workshop. Arizona Cooperative Wildlife Research Unit and University of Arizona, Tucson. 131 pp.

White, G. C. 2001. Effect of adult sex ratio on mule deer and elk productivity in Colorado. Journal of Wildlife Management 65:543–51.

White, G. C., and R. M. Bartmann. 1997. Mule deer management—what should be monitored? Pp. 104–18 *in* J. C. deVos Jr., ed., Proceedings of the 1997 Deer/Elk Workshop, Rio Rico, Arizona. Arizona Game and Fish Department, Phoenix. 224 pp.

White, R. W. 1957. An evaluation of white-tailed deer (*Odocoileus virginanus couesi*) habitats and foods in southern Arizona. Master's thesis, University of Arizona, Tucson. 60 pp.

Whitney, M. D., D. L. Forester, K. V. Miller, and R. L. Marchington. 1992. Sexual attraction in white-tailed deer. Pp. 327–33 *in* R. D. Brown, ed., The biology of deer. Springer-Verlag, New York. 596 pp.

Williams, E. S., J. K. Kirkwood, and M. W. Miller. 2001. Transmissible spongiform encephalopathies. Pp. 292–301 *in* E. S. Williams and I. K. Barker, eds., Infectious diseases of wild mammals. 3d ed. Iowa State University Press, Ames. 776 pp.

Williams, J. D., W. F. Krueger, and D. H. Harmel. 1994. Heritabilities for antler characteristics and body weight in yearling white-tailed deer. Heredity 73:78–83.

Wilson, E. O., and W. L. Brown, Jr. 1953. The subspecies concept and its taxonomic application. Systematic Zoology 2:97–111.

Wislocki, G. B. 1954. Antlers in female deer, with a report of three cases in *Odocoileus*. Journal of Mammalogy 35:486–95.

Witzel, D. A., M. D. Springer, and H. H. Mollenhauer. 1978. Cone and rod photoreceptors in the white-tailed deer, *Odocoileus virginianus*. American Journal of Veterinary Research 39:699–701.

Wood, J. E., T. S. Bickle, W. Evans, J. C. Germany, and V. W. Howard Jr. 1970. The Fort Stanton mule deer herd. New Mexico State University Agricultural Experiment Station Bulletin 567. Las Cruces, New Mexico. 33 pp.

Wood, W. F., T. B. Shaffer, and A. Kubo. 1995. (E)-3-Tridecen-2-one, an antibiotic from the interdigital glands of black-tailed deer, *Odocoileus hemionus columbianus*. Experientia 51:368–69.

Woods, L. W., R. S. Hanley, P. H. W. Chiu, M. Burd, R. W. Nordhausen, M. H. Stillian, and P. K. Swift. 1997. Experimental adenovirus hemorrhagic disease in yearling black-tailed deer. Journal of Wildlife Diseases 33:801–11.

Woods, L. W., P. K. Swift, B. C. Barr, M. C. Horzinek, R. W. Nordhausen, M. H. Stillian, J. F. Patton, M. N. Oliver, K. R. Jones, and N. J. MacLachlan. 1996. Systematic adenovirus infection associated with high mortality in mule deer (*Odocoileus hemionus*) in California. Veterinary Pathology 33:125–32.

Woodson, D. L., E. T. Reed, R. L. Downing, and B. S. McGinnes. 1980. Effect of fall orphaning on white-tailed deer fawns and yearlings. Journal of Wildlife Management 44:249–52.

Wright, V. L. 1980. Use of randomized response technique to estimate deer poaching. Wildlife Society Bulletin 8:342–44.

Youatt, W. G., L. J. Verme, and D. E. Ullrey. 1965. Composition of milk and blood in nursing white-tailed does and blood composition of their fawns. Journal of Wildlife Management 29:79–84.

Young, C. C. 2002. In the absence of predators. University of Nebraska Press, Lincoln, 269 pp.

Zwank, P. J. 1976. Mule deer productivity—past and present. Pp. 79–86 *in* G. W. Workman and J. B. Low, eds., Mule deer decline in the west: A symposium. Utah State University, Logan. 134 pp.

Index

Page numbers in *italic* typeface refer to illustrations.

Abies concolor. See fir, white
Abutilon sp. *See* Indian mallow
acacia, 103–105, 107, 109, 115; catclaw 103, 105, 107, 109; white thorn, 103
Acacia beriandier. See guajillo
Acacia constricta. See acacia, white thorn
Acacia greggii. See acacia, catclaw
Acalypha aspera. See copperleaf, California
Acalypha pringlei. See copperleaf
Acer grandidentatum. See bigtooth maple
Adenostoma fasciculatum. See chamise
Agastache spp. *See* horsemint
agave, 103, 112, 116
Agave lechuguilla. See lechuguilla
Agoseris glauca. See mountain dandelion
Agropyron spp. *See* wheatgrass
Agrostis exarata. See spikebent
albinos, 64–65
Amaranthus almery. See pigweed
Amaranthus palmeri. See careless weed
Amelanchier oreophila. See serviceberry
Amorpha californica. See false indigo
Anasazi culture, 34–35
anemone, 106
antelope bitterbrush, 108, 110
Antonio de Espejo, Don, 38
Apache, 37–38
apache plume, 105–106, 108–10, 114
Arbutus arizonicus. See madrone
Arbutus glandulosa. See madrone
archaic period, 33, 36, 38
Arcostaphylos bicolour. See manzanitia, mission
Arctostaphylos pungens. See manzanita
Arctostaphylos uva-ursi. See bearberry
Argemone spp. *See* prickly poppy
Arizona Game and Fish Department, 43–53, 58, 117, 218, 239

Arizona Game Protective Association, 48
arrowweed, 103
Artemisia californica. See sagebrush, coastal
Artemisia filifolia. See sagebrush, sand
Artemisia ludoviciana. See sagebrush, white
Artemisia tridentate. See sagebrush, big
Artimisia frigida. See fringed sagewort
aspen, trembling, 7–8, 102, 106–108, 110–11, 114
aster, 108
Astragalus recurvus. See milkvetch, slender
Astragalus spp. *See* milkvetch
Athabascan-speaking tribes, 34, 37
Atriplex canescens. See four-wing saltbush
Atriplex spp. *See* saltbush
ayenia, 101, 103
Ayenia filiformis. See ayenia

baccharis, 102
Baccharis spp. *See* baccharis
bachelor groups, 143
bahia, 105
Bahia pedata. See bahia
bearberry, 107
beaver trappers, 38
bed sites, 139, 141
behavior: alarm, 69–70; breeding, 18, 70, 87–88, 143–48; maternal, 69
Berberis repens. See Oregon grape
Betula occidentalis. See birch, water
bezoar, 63
bigtooth maple, 108
birch, water, 108
birdsbill dayflower, 106–107, 109
births. *See* partition
bitterbrush, 108
Black Canyon Game Refuge, 52, 238
black willow, 103
bladderpods, 104–105, 108
Blastomeryx, 26
bluegrass, 108

Boone and Crockett Score, 84
borage, 103
boundary survey. *See* United States-Mexico boundary survey
Bouteloua spp. *See* grama grass
Brickellia californica. *See* brickellia
brickellia, 106, 114
brittlebush, 103
Bromus rubens. *See* red brome
Bromus spp. *See* grass, brome
broom snakeweed, 105
browse monitoring, 161, 222, 230, 232
buck-doe ratios: effects on reproduction, 162–164; use in management, 208–209, 211, 217, 219–20, 222, 224, 228, 231, 233, 235
buckwheat, 101–104, 106–11, 113–16
buffalo gourd, 106
buffel grass, 162
bura deer. *See* burro deer
burro deer, 9, 10
burroweed, 103

cactus: barrel, 103, 109, 113, 116; fruits of, 102–103, 105, 109, 116; prickly pear, 103, 105–106, 109, 115–16
cactus bucks, 92–93
Calliandra eriophylla. *See* fairy duster
Calylophus serrulatus. *See* half-shrub sundrop
canines, maxillary, 61, 62
careless weed, 102
Carex spp. *See* sedge
Carmen Mountain white-tailed deer, 14, 15, 17, 124
Castilleja spp. *See* Indian paint brush
ceanothus, 101–106, 111, 113–16; desert, 101, 103–105, 113–14; fendler, 102, 106, 113
Ceanothus fendleri. *See* ceanothus, fendler
Ceanothus greggii. *See* ceanothus, desert
Ceanothus spp. *See* ceanothus
Cedros Island, 10, 58
Celtis pallida. *See* hackberry, desert
Celtis reticulata. *See* hackberry, mountain
cenizo, 105, 115
Cercedium spp. *See* palo verde
Cercocarpus betuloides. *See* mountain mahogany, birchleaf
Cercocarpus spp. *See* mountain mahogany
cervids, 5
chamise, 102
Chenopodium fremontii. *See* lamb's quarters
chokecherry, 107–108, 110
cholla, 103, 105, 109, 116; jumping, 103
Chrysothamnus spp. *See* rabbitbrush
cliffrose, 107–108, 110–11

clover, 107–108, 110
Clovis culture, 32
Cochise culture, 33–34, 36
Commelina dianthifolia. *See* birdsbill dayflower
computer models, 168–69, 217, 222, 224, 234
condalia, 102
Condalia ericoides. *See* javelina bush
Condalia hookeri. *See* condalia
conservation movement, 42–53, 237–38
Conservation Reserve Program, 9
copperleaf, 102–103, 111, 113; California, 102
Coronado, Francisco Vásquez de, 38
corpora: albicantia, 155, *156–59*; lutea, 155
cottonwood, 107
Coues, Elliot, 11, *13*, 40
courtship. *See* behavior, breeding
Cowania mexicana. *See* cliffrose
Cranioceras, 26, *27*
Crook's blacktail, 7, 9
Croton pottsii. *See* leatherweed croton
crown-beard, 114
Cucurbita foetidissima. *See* buffalo gourd
Cucurbita spp. *See* gourd

Dactylis glomerata. *See* grass, orchard
daisy, 107, 110
dalea, 102–106, 109, 113, 115–16
Dasylirion leiophyllum. *See* sotol
deer weed, 101, 104
Delphinium andesicola. *See* tall larkspur
density, deer, 8, 126–31, 208, 212–13, 226, 228, 233–34
Descurainia spp. *See* tansymustard
desert hibiscus, 105
desert vine, 103, 111, 113
desert yaupon, 115
Desmanthus cooleyi. *See* James bundleflower
Diacodexis, 25
Dichelostemma pulchellum. *See* grass nuts
Dicrocerus, 28, *29*
disease: adenovirus hemorrhagic, 184; bacterial, 188–90; bovine brucellosis, 190; bovine tuberculosis, 188; bovine viral diarrhea, 187; brain abscess, 189; Chronic Wasting Disease, 190–92; dermatitis, 188; foot-and-mouth, 185–86; hemorrhagic, 181–84; infectious bovine rhinotracheitis, 187; leptospirosis, 189–90; lumpy jaw, 189; malignant catarrhal fever, 187; parainfluenza virus 3, 187; skin fibromas, 184–85, *185;* vesicular stomatitis, 186; viral, 181–88
Disturbance: by cattle, 140–42, 162; by humans, 140–42, 164
Ditaxis neomexicana. *See* ditaxis

ditaxis, 105
doe harvest, 161, 237–40
dogweed, common, 102, 105, 109
Dorantes, Estevanico, 38
Dremotherium, 26–27, 28
drought, 41, 122, 168
Dyssodia pentachaeta. See dogweed, common

Ebenopsis ebano. See ebony, Texas
ebony, Texas, 102
elderberry, 107–108
elephant tree, 102
Elymus elymoides. See grass, squirreltail
Emory, William H., 39
Eocoileus, gentryorum, 29, 30
ephedra, 102, 105
Ephedra aspera. See ephedra
Eragrostis lehmanniana. See Lehmann's lovegrass
eriastrum, 103
Erigeron spp. See daisy, fleabane
Eriogonum spp. See buckwheat
Erodium cicutarium. See filaree
Eschscholtzia spp. See poppy
estrus, 144, 146–49, 159
Eumeryx, 25, 26
Euphorbia spp. See spurge
evergreen sumac, 105
Eysenhardtia polystachya. See kidney wood

fairy duster, 101–104, 109, 111, 113
Fallugia paradoxa. See apache plume
false indigo, 111
fawn-doe ratios, 160, 208–209, 211, 219–20, 222, 224, 228, 233
fawning. See partrition
fawns: birth weight, 154; disappearance of spots, 154
federal agencies, 205–207
Federal Aid in Wildlife Restoration Act, 53, 98
Federal Game Law of 1951, 54
fendlera, 102, 111, 114, 115
Fendlera rupicola. See fendlera
fendlerbush. See fendlera
Ferocactus spp. See cactus, barrel
fertility, 155–59
fescue, 108
Festuca spp. See fescue
filaree, 101–104
fir: white, 106–108, 110; Douglas, 107
flax, 105
fleabane, 108, 115
flehman, 68–69, 147
Folsom hunters, 33

Fouquieria splendens. See ocotillo
four-wing saltbush, 105
fringed sagewort, 108

galleta, 103
Game Protective Fund, 48
Garrya wrightii. See silktassel
genetic: analysis, 7, 22, 31–32, 216; effect on antler abnormalities, 88–89; effect on antler size, 81–84; effect on rut timing, 148
gestation period, 153
giant hyssop. See horsemint
globemallow, 101, 103, 105–106, 109, 113, 115–16
goldeneye, 104–106, 109, 114, 116; heart-leaf, 106; skeletonleaf, 104–105, 109
gourd, 106, 115
grama grass, 115
grass: brome, 108; orchard, 107; squirreltail, 107
grass nuts, 104, 113
green sprangletop, 114
ground cherry: fendler, 106; yellow, 115
guajillo, 115
guayacan, 105, 115
gumhead, 104, 109
Gutierrezia sarothrae. See broom snakeweed
Gymnosperma glutinosum. See gumhead

hackberry: desert, 102, 114; mountain, 102–103, 111
half-shrub sundrop, 104, 109
Hariot, Thomas, 5
hearing, frequency range of, 67
Hedyotis acerosa. See needleleaf bluets
Hibiscus coulteri. See desert hibiscus
Hohokam culture, 34, 36
holly-leaf buckthorn, 104, 110–11, 113
home range, 129, 130–34, 140, 143
honeysuckle, Utah, 106
hormones: effects on antler abnormalities, 92–93; regulation of hair molt, 63
horsemint, 107
huisache, Mexican, 115–16
Hunt, George, 50
hunter questionnaires, 214, 218, 221, 223–24, 226, 228, 230, 232, 234
hybridization, 7, 18, 31–32, 216; in captivity, 21; in the wild, 15, 19, 21–22, 24, 60, 152
hybrids: behavior of, 21; characteristics of, 19, 21, 24; genetic analysis of, 22–23; sterility of, 21; survival of, 21, 24, 32

Indian mallow, 111, 113
Indian paint brush, 107

Ipomea spp. *See* morning glory
ironwood, 101–103

james bundleflower, 106–107
Janusia gracilis. See desert vine
jojoba, 101–104, 111, 113, 116
juniper, 102–10, 113–15; alligator, 103, 105, 107, 113, 115; one-seed, 114; Utah, 107–108
Juniperus deppeana. See juniper, alligator
Juniperus monosperma. See juniper, one-seed
Juniperus osteosperma. See juniper, Utah

Kaibab Plateau: antler development, 81; deer density, 127, 129–30; deer diet, 100, 107–108; deer disease, 183; deer weights, 58–59; habitat use, 139; history of, 48–52, 215, 238; home range, 132; importance of water, 122; migration, 134–35; predator control, 177–78; rut and parturition, 150, 156, 158, 161
Kearny, Stephen, 39
Kennerly, C. B. R., 11, 39
kidney wood, 101, 103, 113–16
Kino, Eusebio Francisco, 38
Krameria erecta. See ratany

lamb's quarters, 102
laurel sumac, 102
leatherstem, 115
leatherweed croton, 105
lechuguilla, 105, 109, 115–16
lehmann's lovegrass, 162
lemonade berry, 102
lentisco, 102
Leopold, Aldo, 118
Leptochloa dubia. See green sprangletop
Leptomeryx, 25
LeRaye, Charles, 5
Lesquerella spp. *See* bladderpods
Leucophyllum frutescens. See cenizo
lewisia, 107
Lewisia spp. *See* lewisia
lichen, 106
Linnaeus, Carl von, 3–4
Linum sp. *See* flax
locoweed. *See* milkvetch
longevity, deer, 167
longstalk greenthread, 104, 109
Lotus spp. *See* vetch, deer
lupine, 101, 103, 108, 113, 115–16
Lupinus spp. *See* lupine
Luzula parviflora. See woodrush, small-flowered

madrone, 102, 114–16
mallow, 113

Malosma laurina. See laurel sumac
Malvacae family, 113
manzanita, 102, 104, 115–16; mission, 102
Marcos de Niza, Fray, 38
meadow rue, 114
Mearns, Edgar A., 7, 41
Megaloceros, 88
Melilotus albus. See white sweetclover
menodora, 105
mesquite, 103–104, 113, 115–16
Metastelma arizonicum. See metastelma
metastelma, 101, 103
metatarsal glands: differences between species, 10, 18, *20;* function of, 72
migration, 129, 134–35, 138
milkvetch, 107, 108, 110
milkvetch, slender, 107
milkwort, 104–105, 109
Mimosa dysocarpa. See mimosa, velvet-pod
mimosa, 115; velvet-pod, 111, 113
Mimulus spp. *See* monkey flower
mistletoe, 98, 102, 111, 115–16
Mogollon culture, 34–36
molt, hair, 63, *64,* 148
monkey flower, 102
moon phase: effects on movements, 139–41; effects on rut timing, 149
morning glory, 106, 113, 115–16
mortality rates, 165–70, 209, 224
mountain dandelion, 107, 110
mountain mahogany, 101, 103–11, 114–16; birchleaf, 104–107, 113
movements, 36–39
mutton grass, 110

Navajo, 35, 37
needleleaf bluets, 104, 109
New Mexico Department of Game and Fish, 44–48
Nichol, A. A., 98, 116
nightshade, 113
nontypical antlers, 88
nutrition, 11, 57, 62, 97, 118, 137, 140, 216–17; effects on antler development, 78, 81, 83–84, 87–90, 216; effects on reproduction, 145, 149, 153, 155, 158–62; effects on survival, 168–71

oak, 102, 103–11, 113–16, 138; Arizona white, 114; Emory, 105, 109, 111, 114; Gambel, 106, 107, 108, 110, 138; gray, 106; mohr shrub, 104–105, 109; turbinella, 103–104, 107, 110, 113; wavyleaf, 105–106, 109
ocotillo, 103, 113, 116
Odocoileus hemionus, 7, 9, 10,

Odocoileus virginianus, 11, 14
Olneya tesota. See ironwood
Opuntia bigelovii. See cholla, jumping
Opuntia engelmannii. See cactus, prickly pear
Oregon grape, 107–108, 110
Oshara culture, 33
over-grazing, 41, 49, 118, 127, 140–42, 161, 172

Pachycormus discolor. See elephant tree
Painteria leptophyllum. See huisache, Mexican
palo verde, 8, 101–103
paperflower, 103
parasites: abdominal worm, 198–99; elaeophorosis, 197–98; fleas and lice, 196; foot or leg worm, 198; gastrointestinal nematodes, 198; louse flies, 195–96; lungworm, 199; mites, 194; nasal bots, 193; screwworms, 196–97; tapeworm, 199–200, *200;* ticks, 194–95, *195*
parturition, 117, 153–54; changes to home range, 133, 154; need for water, 117, 123
Patayan culture, 34, 37
Pattie, James Ohio, 39
Paunsaugunt, 135, 169
peninsula mule deer, 10
Pennisetum ciliare. See buffel grass
penstemon, 101, 107–108, 110
Phoradendron spp. *See* mistletoe
photoperiod: effect on antler development, 80; effect on molt, 63–64; effect on rut timing, 148
Physalis fendleri. See ground cherry, fendler
Physalis viscosa. See ground cherry, yellow
piebald, 65
pigweed, 106
pine: pinyon, 106–108, 110, 114; ponderosa, 7, 107, 110
Pinus edulis. See pine, pinyon
Pinus ponderosa. See pine, ponderosa
Pittman-Robertson Act. *See* Federal Aid in Wildlife Restoration Act
Plantago insularis. See plantain
plantain, 103–104
Pleistocene, 10, 29–32
Pleuraphis rigida. See galleta
poaching, 201–202
Poa fendleriana. See mutton grass
Poa spp. *See* bluegrass
Polygala spp. *See* milkwort
poppy, 102
Populus spp. *See* cottonwood
Populus tremuloides. See aspen, trembling
Porlieria angustifolia. See guayacan
Porophyllum gracile. See deer weed
Portulaca oleracea. See purslane

predation, 166, 169, 172–80
predators: control of, 176–80; impact of, 173–76
preorbital glands: differences between, 18, 73, *74;* function of, 73–75
prickly poppy, 113
Procervulus, 27, 28
Prosopis glandulosa. See mesquite
Prunus virginiana. See chokecherry
Psilotrophe spp. *See* paperflower
Purshia mexicana. See cliffrose
Purshia spp. *See* bitterbrush
Purshia tridentata. See antelope bitterbrush
purslane, 102, 107

Quercus arizonica. See oak, Arizona white
Quercus emoryi. See oak, Emory
Quercus gambelii. See oak, Gambel
Quercus grisea. See oak, gray
Quercus mohriana. See oak, mohr shrub
Quercus turbinella. See oak, turbinella
Quercus undulata. See oak, wavyleaf

rabbitbrush, 108, 110
rainfall, importance of, 160–61
Ramoceras, 26, 27
ratany, 101, 103–104, 111, 113–14
redberry, 102
red brome, 162
reproductive rate, 154–62
Rhamnus crocea. See holly-leaf buckthorn
Rhamnus sp. *See* redberry
Rhus choriophylla. See sumac, Mearns
Rhus integrifolia. See lemonade berry
Rhus lentii. See lentisco
Rhus microphylla. See sumac, littleleaf
Rhus ovata. See sumac, sugar
Rhus trilobata. See skunkbush. *See also* sumac, threeleaf
Rhus virens. See evergreen sumac
Rosa spp. *See* rose
rose, 107, 115
Rothrock, Joseph, 11
rub-urination, 71, 145
rut timing, 148–53, 162–64

sage: black, 102; white, 102–105, 109, 113
sagebrush, 102–104, 108–11; big, 108; coastal, 102; sand, 104, 109; white, 103
Salix nigra. See black willow
saltbush, 103
Salvia apiana. See sage, white
Salvia mellifera. See sage, black
Sambucus spp. *See* elderberry
San Dieguito-Pinto culture, 33

San Jose Island, 10, 58
Schaefferia cuneifolia. See desert yaupon
scrapes, 146
sedge, 102, 104, 108, 110
SEMARNAP. *See* SEMARNAT
SEMARNAT, 54, 55, 207
serviceberry, 107–108, 110
Sierra Seri, 9
silktassel, 105–107, 109, 111, 113–15
Simmondsia chinensis. See jojoba
Sinagua culture, 35
Sioux, 5
Sitgreaves, Lorenzo, 40
skunkbush, 104–107, 109, 113–14
snowberry, 108, 110, 115–16; mountain, 108, 110
Solanum sp. *See* nightshade
Sonchus oleraceus. See sowthisle
Sonoran Fantail, 7, 13
sotol, 102, 105
southern mule deer, 10
sowthisle, 102
Spanish, contact with Native Americans, 37–38
Sphaeralcea spp. *See* globemallow
spikebent, 113
spurge, 101, 103–106, 109–110, 113, 115–16
Stephanocemas, 28, 29
Stephanomeria pauciflora. See wire lettuce
stomach stone. *See* bezoar
stotting, 17, 21
subspecies, 3–7, 10, 15, 31, 216
sumac: evergreen, 115; littleleaf, 104–105, 109, 115–16; Mearns, 114; sugar, 113; threeleaf (*See* skunkbush)
surveys, deer, 126–29, 208–14; by fecal pellet counts, 127–28, 141, 208, 220, 223; by fixed-wing aircraft, 213, 218–19, 225–26; by foot and horseback, 210, 218–20, 231; by helicopter, 126–29, *212*, 219–20, 224, 226, 228, 230, 233; by spotlight count, 219–20, 226; by vehicle, 210–11, 219–20, 224–26, 231
Symphoricarpos oreophilus. See snowberry, mountain
Symphoricarpos spp. *See* snowberry
Synthetoceras, 26, *27*

talinum, 102
tall larkspur, 113, 116
tansymustard, 106–107, 109
taxonomy, 3–4, 7–8
testosterone, 78–79, 143–44
Thalictrum fendleri. See meadow rue
Thelesperma longipes. See longstalk greenthread
Three-leaf sumac. *See* skunkbush
Thymophylla pentachaeta. See dogweed, common
Tiburón Island, 10, 58
tidestromia, 103
Tidestromia spp. *See* tidestromia
Trifolium spp. *See* clover

United States–Mexico boundary survey, 39–41

vehicle collisions, 202–03
venado: bura, 7; cola blanca, 7
Verbena bipinnatifida. See vervain
Verbesina rothrockii. See crown-beard
vervain, 106
vetch deer, 101–103, 106–108, 110, 114–16
Viguiera cordifolia. See goldeneye, heart-leaf
Viguiera spp. *See* goldeneye
Viguiera stenoloba. See goldeneye, skeletonleaf
vision, 66
Vitis arizonica. See grape

Wallace, Henry C., 49
water developments, 117–19, *120–21*, 122–24, 243
Way, Phocian, 40
wheatgrass, 110
Whipple, Amiel W., 40
white sweetclover, 106
Willard, G., M., 50
wire lettuce, 103, 114
wolves, 173
woodrush, small-flowered, 107

yucca, 105

Zuñi, 38